Reviewing Mathematics

James Tate
Administrator
Albany City School District
Albany, New York

John Schoonbeck
Mathematics Chair
LaSalle School
Albany, New York

Marilyn Davis
Contributing Editor

 AMSCO

Amsco School Publications, Inc.
315 Hudson Street, New York, N.Y. 10013

Dedication

This book is dedicated to Cindy Tate, who served as a typist and assistant throughout the developmental period of the project. Her time and commitment have been greatly appreciated.

James Tate has always been intrigued by mathematics. He believes that all students have the ability and the need to learn mathematical concepts and apply them to problem solving. He started his career teaching at the elementary level and then spent many years teaching middle school/junior high school students in the Albany City School District. In addition to being an administrator in the Albany City School District, he is on the board of directors of the New York State Association of Mathematics Supervisors and on the executive committee of the Hudson Mohawk Valley Area Mathematics Association. He has also organized workshops for upstate New York teachers with the Capital Area School Development Association at the University of New York at Albany.

John Schoonbeck received his B.S. in Secondary Education/Mathematics and Science from the State University College at Plattsburgh, New York. He is currently the Mathematics Department Chair at LaSalle School in Albany, New York. During the past twenty-five years, he has been teaching high school mathematics and science, as well as being a college administrator.

Reviewers

Catherine C. Deline
    Math Specialist (K-12),
    Lyme Central School
    Chaumont, NY

Susan M. Edwards
    Math Educator
    Greece Central School District, NY

Carol Fatta
    Math/Computer Instructor,
    Chester Jr./Sr. High School
    Chester, NY

Richard P. Flinn
    Secondary Math Teacher
    Sachem Central School District, NY

Terri Husted
    Math Department Chairperson,
    DeWitt Middle School
    Ithaca, NY

Gretchen Jewiss
    Math Consultant
    Marathon Central School District, NY

Rita Kabasakalian, Ed.D.
    Staff Developer,
    C.I.S. 303
    Bronx, NY

Nancy H. Wagner
    Math Teacher
    Uniondale Schools
    Uniondale, NY

Judith Wood
    Teacher, 8th Grade Mathematics
    Northeastern Clinton Central School
    Champlain, NY

Cover and text design by One Dot Inc.
Composition and line art by Nesbitt Graphics Inc.

Please visit our Web site at: **www.amscopub.com**

When ordering this book, please specify:
either **R 764 W** or REVIEWING MATHEMATICS.

ISBN 1-56765-557-2/NYC Item 56765-557-1

Printed in the United States of America

1 2 3 4 5 6 7 8 9   07 06 05 04 03

Contents

Getting Started   vii

About the Book   vii

Diagnostic Test   ix

Diagnostic Test Answer Key with Analysis Table   xv

CHAPTER 1

Problem-Solving Strategies and Mathematical Reasoning   1

1.1	Problem Solving	1
1.2	Identifying Information	4
1.3	Guessing and Checking	7
1.4	Working Backward	9
1.5	Using Pictures and Diagrams	12
1.6	Using Tables and Lists	15
1.7	Recognizing Patterns	18
1.8	Solving a Simpler Related Problem	20
1.9	Using Logical Reasoning	23
1.10	Using Estimation	27
	Chapter Review	29

CHAPTER 2

Whole Numbers and Decimals   31

2.1	Place Value	31
2.2	Powers of 10 and Scientific Notation	35
2.3	Comparing and Ordering	39
2.4	Rounding	42

2.5	Properties	45
2.6	Estimation and Addition	47
2.7	Estimation and Subtraction	50
2.8	Estimation and Multiplication	53
2.9	Division of Whole Numbers	57
2.10	Division of Decimals	61
	Chapter Review	65
	Cumulative Review	67

CHAPTER 3 **Fractions** **70**

3.1	Factors, Primes, and Composites	70
3.2	Divisibility and Greatest Common Factor	73
3.3	Equivalent Fractions	76
3.4	Mixed Numbers and Improper Fractions	79
3.5	Fractions and Decimals	82
3.6	Least Common Multiple and Least Common Denominator	86
3.7	Comparing and Ordering Fractions	89
3.8	Adding Fractions and Mixed Numbers	93
3.9	Subtracting Fractions and Mixed Numbers	96
3.10	Multiplying Fractions and Mixed Numbers	100
3.11	Dividing Fractions and Mixed Numbers	103
	Chapter Review	107
	Cumulative Review	109

CHAPTER 4 **Rational Numbers** **111**

4.1	Using Signed Numbers	111
4.2	Comparing and Ordering Signed Numbers	115
4.3	Adding Signed Numbers	118
4.4	Subtracting Signed Numbers	121
4.5	Multiplying Signed Numbers	123
4.6	Dividing Signed Numbers	126
4.7	Squares and Square Roots	129
4.8	Sets of Numbers	132
	Chapter Review	135
	Cumulative Review	137

CHAPTER 5 **Writing and Solving Equations** **139**

5.1	Numerical Expressions and Order of Operations	139
5.2	Writing Algebraic Expressions	142
5.3	Evaluating Algebraic Expressions	145
5.4	Writing Equations and Inequalities	147

5.5	Solving Equations with One Operation	151
5.6	Solving Equations with Two Operations	154
5.7	Graphing and Solving Inequalities	158
5.8	Working with Formulas	161
	Chapter Review	165
	Cumulative Review	166

CHAPTER 6 Ratio, Proportion, and Percent — 169

6.1	Ratio	169
6.2	Proportion	172
6.3	Scale Drawing	176
6.4	Percent	179
6.5	Solving Percent Problems	183
6.6	Percent of Increase and Decrease	188
6.7	Applications of Percent	190
	Chapter Review	194
	Cumulative Review	196

CHAPTER 7 Measurement — 198

7.1	Customary Units of Measure	198
7.2	Computing with Customary Units	202
7.3	Metric Units of Measure	206
7.4	Temperature	210
7.5	Time and Time Zones	213
7.6	Precision in Measurement	218
	Chapter Review	223
	Cumulative Review	225

CHAPTER 8 Geometry — 227

8.1	Lines and Angles	228
8.2	Triangles and Angles	233
8.3	Polygons and Special Quadrilaterals	237
8.4	Congruent Figures	242
8.5	Similar Figures	245
8.6	Perimeter and Circumference	249
8.7	Area	255
8.8	Surface Area	260
8.9	Volume	264
8.10	The Pythagorean Theorem	268
8.11	Trigonometry of the Right Triangle	272
	Chapter Review	278
	Cumulative Review	282

CHAPTER 9 | **Coordinate Graphing and Geometric Constructions** | **284**
9.1	Using Coordinates to Graph Points	284
9.2	Distance Between Two Points on a Coordinate Plane	289
9.3	Translations	293
9.4	Reflections and Symmetry	298
9.5	Rotations	304
9.6	Constructing and Bisecting Angles	308
9.7	Constructing Perpendicular Lines	310
	Chapter Review	312
	Cumulative Review	315

CHAPTER 10 | **Data Analysis and Probability** | **319**
10.1	Displaying Data with Graphs	319
10.2	Frequency Tables and Histograms	324
10.3	Finding the Range, Mean, Median, and Mode	327
10.4	Stem-and-Leaf Plots and Box-and-Whisker Plots	333
10.5	The Meaning of Probability	337
10.6	Counting Outcomes and Compound Events	342
10.7	Permutations and Combinations	348
10.8	Experimental Probability	352
	Chapter Review	357
	Cumulative Review	361

CHAPTER 11 | **Patterns and Functions** | **365**
11.1	Relations and Functions	365
11.2	Graphing Linear Equations	371
11.3	Graphing Inequalities	375
11.4	Slope	379
11.5	Exploring Nonlinear Functions	383
	Chapter Review	387
	Cumulative Review	390

Mathematics Reference Sheet | **393**

Practice Tests | **394**
Practice Test 1	394
Practice Test 2	401
Practice Test 3	408

Using a Scientific Calculator | **416**

Index | **419**

Getting Started

About the Book

Reviewing Mathematics is designed to give students the mathematical foundation they need to succeed in secondary mathematics courses. The book covers the following topics:

1. Mathematical Reasoning
2. Numbers and Numeration
3. Operations
4. Modeling/Multiple Representation
5. Measurement
6. Uncertainty
7. Patterns and Functions

This review book features:

- An introductory chapter on Problem-Solving Strategies and Mathematical Reasoning
- A Diagnostic Test with Answer Key and Analysis Table to pinpoint areas of weakness and direct students to the pages they need to review
- Vocabulary lists for each chapter
- Model Problems with detailed solutions
- Practice sets for each topic
- Chapter Reviews to reinforce skills and concepts covered in each chapter
- Cumulative Reviews to reinforce and integrate skills and concepts covered from the beginning of the book through the chapter

- Three Practice Tests with multiple-choice, short-response, and extended-response questions
- Highlighted uses of the scientific calculator, ruler, and protractor in Model Problems and Practice sets
- *Using a Scientific Calculator*, a demonstration of the calculator features students are likely to use
- Test Taking Strategies to offer students advice on answering both multiple-choice and open-response questions

The Practice sets, Chapter Reviews, and Cumulative Reviews throughout the book include multiple-choice, short-response, and extended-response questions. Multiple-choice questions challenge students to demonstrate and apply mathematical knowledge and skills. Students must show their ability to use mathematics by applying facts, definitions, and arithmetic skills to problem-solving situations. Short-response and extended-response questions ask students to communicate their answers by showing the steps they use and by writing explanations of the concepts involved and of their problem-solving strategies.

The *Mathematical Reference Sheet* on page 393 is designed for use with the Practice Tests. It gives more advanced formulas (such as the area of a cylinder) and includes a trigonometric table.

Test Taking Strategies

General

- Read all directions and questions carefully. Although this is a mathematics test, you need to use your reading skills to understand what you are being asked to do.
- If there are any directions that you don't understand, raise your hand and ask your teacher to explain them.
- Plan your time so that you can answer as many questions as possible. Take a quick look at the whole test to get an idea of what is ahead of you. Don't spend all your time on difficult questions. Do your best and move on. Go back to the difficult questions if you have time at the end.
- Use a protractor and ruler to help you solve problems whenever you think it is appropriate.

Multiple-Choice Questions

- Always look at the answer choices to see what kind of answer is required. Should the answer be a fraction or a decimal? Is it in inches or feet?
- Be careful about making wild guesses. If a question is too difficult, move on and return to it later. Save your wild guesses for the

last few minutes of the test, when you fill in answers for any questions you have left blank.

- Keep track of your place on the answer sheet. If you skip a question, make sure you leave the answer line blank for that question. For example, if you skip question 8, don't mistakenly put the answer for question 9 on that line.
- Don't assume that your answer is correct just because it appears among the choices. The wrong choices usually represent common student errors (like finding the circumference of a circle instead of the area, or multiplying by 2 instead of squaring). After you find the answer, reread the question to make sure you have answered the question that was asked, not just the question you have in your mind.
- Do the math. This is the ultimate strategy! Don't waste your time searching for tricks and gimmicks. For most questions, the best way to get the right answer is to do the math and solve the problem.

Short–Response and Extended-Response Questions

- After you understand the problem and have planned a strategy, use your calculator if it will help you arrive at the solution.
- Answer all parts of the question.
- Be sure to show all your work when asked. You may receive partial credit for work you show even if your final answer is incorrect.
- Showing work can include writing down calculations performed on the calculator, making diagrams, writing sentences, or a combination of all three.
- Make sure that the work you show is written down in a way that other people can understand. Label your graphs and diagrams, write neatly, and number the steps you take to solve the problems.

Diagnostic Test

Multiple-Choice Questions

1. Which of the numbers given below will be a term of this pattern?

   1, 3, 7, 15, 31, . . .

   A. 65   B. 128
   C. 255   D. 516

2. Renee has the same number of pennies as dimes. Which of the following sums could she have?

   F. $0.90   G. $1.15
   H. $1.54   J. $1.69

3. Which expression does not have the same value as the others?

   A. $5 \times 5 \times 5 \times 5 \times 5 \times 5$
   B. 1.5625×10^5
   C. 5^6
   D. 25×625

4. Which of the following has a product between 500 and 1,000?

   F. 3.7×86.94
   G. 48.2×60.7
   H. 205×1.39
   J. $3,692.8 \times 0.23$

5. What is the prime factorization of 100?

   A. $2^2 \times 5^2$ B. $2 \times 5 \times 10$
   C. 4×25 D. 10^2

6. How much greater is $6\frac{1}{8}$ than $2\frac{5}{6}$?

   F. $3\frac{3}{12}$ G. $3\frac{7}{24}$

   H. $4\frac{1}{12}$ J. $4\frac{17}{24}$

7. $|6| - |-14| =$

   A. -20 B. -8
   C. 8 D. 20

8. The product of two integers is -36. The quotient is -4. What are the integers?

   F. $-9, 4$ G. $6, -6$
   H. $-12, 3$ J. $24, -6$

9. Simplify $50 - 20 \div 5 + 5 \times 3$.

   A. 21 B. 23
   C. 61 D. 144

10. What value of x will make the statement below true?
$$4(x - 3) = 24$$

   F. 36 G. 9
   H. 6 J. 5.25

11. In the inequality below, which of the given numbers could replace the variable x?

$$\frac{1}{8} < x < 0.75$$

   A. $\frac{1}{10}$ B. 0.125

   C. $\frac{1}{7}$ D. 0.76

12. Dave and Len bought a large anchovy pizza and cut it into 12 equal slices. Dave ate 25% of the pizza and Len ate 5 slices. What part of the pizza was left?

   F. $\frac{1}{6}$ G. 25%

   H. $\frac{1}{3}$ J. 40%

13. At a 20%-off sale, Allegra paid $30.36 for a sweater. What was the original price of the sweater?

   A. $35.50   B. $35.98
   C. $36.45   D. $37.95

14. Triangle ABC is similar to $\triangle DEF$. $\overline{AB}$ corresponds to $\overline{DE}$. Find the perimeter of $\triangle ABC$.

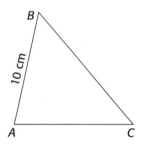 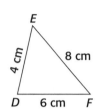

   F. 27 cm
   G. 33.5 cm
   H. 36 cm
   J. 45 cm

15. Complete: $8.4 \text{ m} \div 6 = $ _____ cm.

   A. 1.4 cm
   B. 14 cm
   C. 140 cm
   D. 1,400 cm

16. The most reasonable temperature for a bowl of hot soup is

   F. 50°F
   G. 95°C
   H. 98°F
   J. 210°C

17. Each square on the grid represents 9 ft². What is the perimeter of the figure shown on the grid?

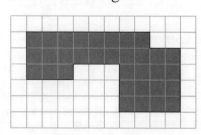

   A. 33 ft
   B. 96 ft
   C. 99 ft
   D. 297 ft

18. What is the volume of the cylinder? Use $\pi \approx 3.14$.

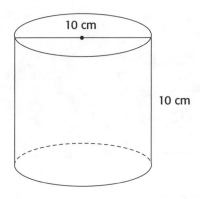

10 cm

10 cm

F. 785 cm²   G. 1177 cm²
H. 1570 cm²   J. 3140 cm²

19. Which set of measurements could be the lengths of the sides of a right triangle?

A. 8 m, 15 m, 20 m
B. 8 m, 11 m, 14 m
C. 12 m, 18 m, 24 m
D. 12 m, 16 m, 20 m

20. Which set of ordered pairs could be the endpoints of a line segment parallel to the y-axis?

F. $P(0, 5), Q(5, 0)$
G. $R(-1, 6), S(4, 6)$
H. $T(-2, -7), V(-2, 3)$
J. $M(3, -4), N(6, -8)$

21. Triangle *PQR* is translated 3 units left and 4 units up. The new coordinates of the vertices are

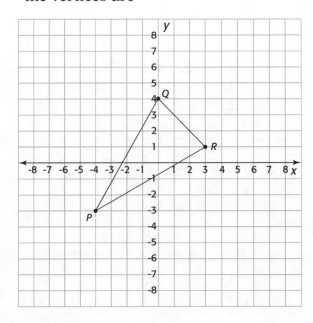

A. $P'(-1, 1) \; Q'(3, 8) \; R'(6, 5)$
B. $P'(-7, 1) \; Q'(-3, 8) \; R'(0, 5)$
C. $P'(-7, 0) \; Q'(3, -8) \; R'(0, 3)$
D. $P'(0, 0) \; Q'(4, 7) \; R'(7, 4)$

22. During a gymnastics competition, the judges awarded the following scores:
     7.7   7.0   9.6   10.0   8.8
     9.2   9.8   8.7   8.8   9.4
Find the mean and median of the scores.

F. mean = 7.9, median = 8.9
G. mean = 8.2, median = 9.0
H. mean = 8.9, median = 8.8
J. mean = 8.9, median = 9.0

23. The weekday lunch special at Soup Stop includes soup, bread, and a piece of fruit. How many different lunch combinations can customers order at Soup Stop?

Soup Stop Menu		
Soup	**Bread**	**Fruit**
Chicken Noodle	Roll	Apple
Split Pea	Toast	Banana
Vegetable	Bagel	Pear
Beef Barley		Orange
Clam Chowder		
Matzo Ball		
Tomato Rice		

A. 14
B. 36
C. 84
D. 210

24. If the spinner shown is spun twice, what is the probability of getting 4, then 4 again?

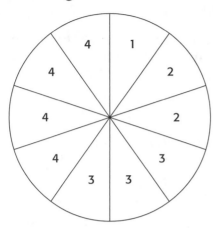

F. $\dfrac{1}{100}$

G. $\dfrac{4}{25}$

H. $\dfrac{2}{5}$

J. $\dfrac{4}{5}$

25. The table below shows the results of an experiment in which a number cube was tossed 20 times.

Outcome of Toss	1	2	3	4	5	6
Number of Times Outcome Occurred	4	1	5	5	2	3

Based on these results, if the same number cube was tossed 240 times, about how many times could the outcome 5 be expected?

A. 24   B. 40
C. 48   D. 100

26. Which table satisfies the equation $y = 2x - 2$?

F.

x	y
1	−1
0	−2
−1	0

G.

x	y
−2	−6
3	4
5	8

H.

x	y
−3	−8
1	0
4	4

J.

x	y
−4	−10
−2	2
6	12

27. The $6 admission to a fair includes 3 free rides. Each additional ride costs $2. Which equation can be used to find the total cost T of a visit to the fair with x rides, where $x > 3$?

A. $T = 6 + 2(3 - x)$
B. $T = 6 + 2x$
C. $T = 6 + 2(x - 3)$
D. $T = 6 + 3(x - 2)$

Short-Response Questions

 You may use a calculator to solve problems on this part of the test.

28. Six towns form the vertices of a hexagon. There is a road connecting each town with every other town. How many roads are there?

29. Monarch Motors expects a delivery of 10 new silver Adventas. The cars must be distributed among Monarch's three branches: North, South, and Central. Each branch must get at least two of the new cars. How many different ways are there to distribute the cars?

30. During No Tax Week, Margaret bought 3 turtlenecks for $14.44 each and a jacket for $39.99. She gave the cashier $100. How much change should Margaret receive?

31. Rent-A-Ride charges $25 per day plus $0.25 for each mile over 125 miles per day. On Monday, Mr. Shia rented a car and drove 212 miles. He kept the car overnight and drove another 183 miles on Tuesday. How much did Mr. Shia pay when he returned the car?

32. Write a fraction and a decimal to name each point on the number line below.

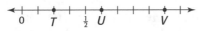

  a. T
  b. U
  c. V

33. It takes $8\frac{1}{2}$ gallons of paint to paint 3 identical offices.

  a. How many gallons of paint are needed to paint 5 more identical offices?
  b. How much paint will be used for all 8 offices?

34. The temperature at 7:00 A.M. was −11°F. By noon, the temperature had risen by 23°, and by 7:00 P.M. it had dropped 15° from noontime.

  a. What was the temperature at 7:00 P.M.?
  b. What was the difference between the 7:00 P.M. temperature and the 7:00 A.M. temperature?

35. This table shows how membership in Volunteers for Parks changed each month.

Month	Jan.	Feb.	Mar.	Apr.
Change in Membership	−3	−5	+18	+10
Month	May	June	July	Aug.
Change in Membership	+2	−8	−11	+1

  a. In which month did membership decrease the most?
  b. If there were 42 members at the end of December of the previous year, how many members were there at the end of August this year?

36. Solve and graph: $4x - 3 < 13$.

37. A formula that relates the number of times per minute (N) that a cricket chirps and the temperature in degrees Fahrenheit (F) is:

$$F = \frac{N}{4} + 32$$

  a. Find the temperature when the number of chirps is 40 per minute.
  b. Find the number of chirps per minute when the temperature is 60°F?
  c. What is the relationship between the temperature and the number of chirps per minute?

38. The scale of this drawing of Vincent's art studio is $\frac{1}{2}$ in. = 8 ft. Use a ruler to measure the drawing of the art studio.

w

ℓ

a. What is the actual length of the studio?
b. What is the actual width of the studio?
c. A supply closet in the art studio is 6 feet long and 2 feet deep. What would the closet's dimensions be on the drawing?

39. The circle graph shows Ned's budget. He earns $380 a month at a part time job.

Ned's Budget

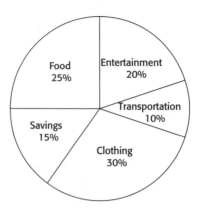

a. How much money does Ned allow for clothing?
b. How much more does Ned spend on entertainment than he saves?

Extended-Response Questions

 You may use a calculator to solve problems on this part of the test.

40. A pair of shoes that Zia liked was on sale for 30% off the regular price of $50. When she took the shoes to the register, the cashier said that since it was Bonus Hour an additional 10% would be deducted from the sale price.

a. Did Zia save 40% off the regular price of the shoes? Explain you answer.
b. What single discount is equivalent to the two discounts Zia received? Show your work.

41. A small pizza has a diameter of 8 inches and costs $5.25. A large pizza has a diameter of 12 inches and costs $10.50. Assuming that both pizzas are comparable with regards to crust, cheese, sauce, and so on, which pizza is the better buy? Explain how you decided.

42. The shaded surface of the access ramp will be covered with non-slip rubber. Find the area of non-slip rubber needed. Show your work.

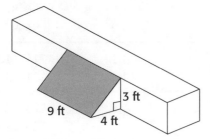

9 ft   3 ft   4 ft

43. The volume of the covered box shown is 630 cubic inches.

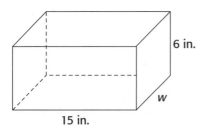

6 in.   w   15 in.

a. Find the width w of the box.
b. Find the total surface area of the box.

44. a. Graph quadrilateral *ABCD* with vertices $A(-8, 1)$, $B(-8, 5)$, $C(-2, 5)$, and $D(-2, 1)$.
b. Identify the type of quadrilateral. Give all names that apply.
c. Graph the image $A'B'C'D'$ of *ABCD* after a translation 10 units right and 3 units up.
d. Graph figure *FGHJ*, which is similar to *ABCD*. The ratio of the sides of *FGHJ* to *ABCD* should be 1:2, and vertex *F* should be at $(3, -3)$.
e. What is the ratio of the area of *FGHJ* to the area of *ABCD*?

45. Timothy's scores on the first three Communication Arts tests were 88, 91, and 83. He needs at least a 90 average to earn an A. Ms. Matthews has announced that the score on the final will be counted twice, as if it were two tests. What is the lowest score Timothy can get on the final if he wants to earn an A in Communication Arts? Show your work.

Diagnostic Test Answer Key with Analysis Table

Use the Answer Key below to score your Diagnostic Test. Record the number of any question that you left blank or answered incorrectly. Use the analysis table to determine which sections require additional work. Review the recommended sections and complete the practice problems in those sections.

Answer Key	Practice Sections	Pages
1. C	1.7	18–19
2. H	1.3	7–8
3. B	2.2	35–37
4. J	2.4	42–43
	2.8	53–55

Answer Key	Practice Sections	Pages
5. A	3.1	70–72
6. G	3.6 3.9	86–87 96–97
7. B	4.1 4.4	111–113 121–122
8. H	4.5 4.6	123–124 126–127
9. C	5.1	139–140
10. G	5.3	145–146
11. C	3.5 5.7	82–84 158–159
12. H	3.3 6.4	76–77 179–181
13. D	6.5 6.6	183–185 188
14. J	6.2 8.5	172–174 245–246
15. C	7.3	206–208
16. G	7.4	210–211
17. B	4.7 8.6	129–130 249–252
18. F	2.8 8.9	53–55 264–265
19. D	4.7 8.10	129–130 268–270
20. H	9.1	284–287
21. B	4.3 4.4 9.3	118–119 121–122 293–296
22. J	2.6 2.9 10.3	47–49 57–60 327–329
23. C	10.6	342–344
24. G	10.6	342–344

Answer Key	Practice Sections	Pages
25. A	6.2 10.8	172–174 352–353
26. G	5.3 11.1	145–146 365–368
27. C	5.4 11.1	147–149 365–368
28. 15 roads	8.3 10.7	237–239 348–349
29. 15 ways	10.7	348–349
30. $16.69	1.1	1–3
31. $86.25	1.1 1.2	1–3 4–5
32. a. $\frac{1}{4}$, 0.25 b. $\frac{5}{8}$, 0.625 c. $\frac{9}{8}$ or $1\frac{1}{8}$, 1.125	2.3 3.5	39–40 82–84
33. a. $14\frac{1}{6}$ gal b. $22\frac{2}{3}$ gal	3.8 6.2	93–94 172–174
34. a. $-3°$F b. $+8°$ or 8 degrees warmer	4.3 4.4 7.4	118–119 121–122 210–211
35. a. July b. 46 members	4.2 4.3 4.4	115–116 118–119 121–122
36. $x < 4$	5.7	158–159
37. a. 42°F b. 112 chirps per minute c. As temperature increases, chirps increase. The relation is a function and the chirps vary directly with the temperature.	5.8 11.1	161–163 365–368

Answer Key	Practice Sections	Pages
38. a. 76 ft b. 32 ft c. $\frac{3}{8}$ in. by $\frac{1}{8}$ in.	6.2 6.3	172–174 176–177
39. a. $114 b. $19	6.2 6.4 6.5	172–174 179–181 183–185
40. a. No. The second discount of 10% was off 70% of the regular price and amounted to 7% of the regular price. b. 37% off	6.2 6.5 6.6	172–174 183–185 188
41. Use area to determine unit cost. Small area ≈ 50.24 in.^{2} Small cost ≈ 10.5¢ per in.^{2} Large area ≈ 113.04 in.^{2} Large cost ≈ 9.3¢ per in.^{2} Large is a better buy.	2.10 8.7	61–63 255–257
42. Use the Pythagorean Theorem: $3^2 + 4^2 = w^2$ $9 + 16 = w^2$ $5 = w$ To find area: $9 \times 5 = 45$ sq ft	8.7 8.10	255–257 268–270
43. a. 7 in. b. $(2 \times 15 \times 7) + (2 \times 7 \times 6) + (2 \times 15 \times 6) = 474$ sq in.	8.8 8.9	260–261 264–265
44. a, c, d. b. parallelogram, rectangle e. 1:4	6.1 8.3 8.5 8.7 9.1 9.3	169–170 237–239 245–246 255–257 284–287 293–296

Answer Key	Practice Sections	Pages
45. $\dfrac{88 + 91 + 83 + x + x}{5} = 90$	5.2	142–143
$262 + 2x = 450$	5.6	154–156
$2x = 188$	10.3	327–329
$x = 94$		
The lowest score he can get is 94.		

CHAPTER 1

Problem-Solving Strategies and Mathematical Reasoning

Chapter Vocabulary

problem solving
pattern
estimate

inverse operation
logic

diagram
Venn diagram

1.1 Problem Solving

People often say, "I have a problem," when they are faced with an unfamiliar or difficult situation and there is no obvious way to find an answer. **Problem solving** is the process by which this new situation is analyzed and resolved. It begins with an understanding of all the aspects of the problem and ends when a satisfactory answer has been found. Problem solving is not just an exercise carried out in the classroom, but a skill that is used continually in business and daily life.

This chapter will help improve your skill at solving problems. Often, there is more than one way to approach a problem, and not every problem will be solved on the first try. Persistence is one of the qualities that is needed to become a good problem solver. The four basic steps that follow can help you to examine a problem situation. If you practice applying these steps to easier problems, you will develop the thinking skills needed for solving more difficult problems.

Step 1   Understand the problem.

Read the problem carefully. Be sure you understand the situation and the question being asked. What information are you given? Are any facts missing or extra? Is there any special vocabulary? Look for important clues that may lead to a solution.

Step 2   Formulate a plan for a solution.

Organize the given data by using notes, lists, tables, and diagrams. Identify the mathematical operations that you will need to perform. If the plan involves many steps, list them in order. Estimate an answer if possible.

Step 3   Solve the problem.

Carry out the steps in your plan. Record the work for each step because it may be necessary to review and revise these steps at a later time.

Step 4   Check the solution.

Check your calculations. Did you make any errors? Does your answer make sense? Has the question been answered properly? If not, ask yourself what is wrong and how it can be fixed. If your check shows that you have successfully solved the problem, clearly identify the answer.

> **These four steps can be remembered as**
>
> Understand, Plan, Solve, Check.

Model Problem

At lunch, Linda bought a taco for $3.50, juice for $1.49, and fruit salad for $1.75. If she gave the cashier a $10 bill, how much change should she receive?

Solution   Use the problem-solving steps.

Understand   Linda bought three items for lunch and paid with a $10 bill. The problem asks how much change she should get back. The information given is

- Taco cost $3.50
- Juice cost $1.49
- Fruit salad cost $1.75
- Gave cashier $10.00

Plan   How can the answer be found? Add the costs of the three items. Then subtract the total from $10.00. Estimate: $4 + $1 + $2 = $7 and $10 − $7 = $3.

Solve   Carry out the plan.

Add.   Subtract.

$3.50 $10.00
 1.49   − 6.74
+1.75 $3.26
$6.74

Check   Has the question been answered? Is the answer accurate and reasonable?

Check the subtraction by adding. $3.26
                                   + 6.74
                                   $10.00

The estimated answer was $3. The actual answer after following the plan is $3.26. This is a reasonable answer since it is close to the estimate.

Answer   Linda should receive $3.26 change.

Practice

Solve   Show how you use the problem–solving steps.

1. For a recycling drive, Nina's ninth grade class filled 20 bags with aluminum cans. 65 cans fit in one bag. The class also filled 30 bags with plastic bottles. Only 45 bottles fit in a bag. Did the class collect more aluminum cans or plastic bottles? How many more are there?

2. Mr. Penny gives each of his 24 employees presents at the end of the year. Last year he gave each of them a $20 cassette player, a $5 box of cookies, and a $50 gift certificate to a local department store. How much did Mr. Penny spend on the gifts in all?

3. Shavel bought a suit, a blouse, and a belt at Celebrity Fashions. The price of the suit was $89 and the blouse cost $27. The total cost of all three items was $125. How much did the belt cost?

4. Lana buys oil by the case for her garage. Each case costs her $11.00 and contains 24 cans of oil. If she sells each can for $1.50, how much profit does she earn on a case?

5. Colleen has a newspaper route. The table shows the number of papers she delivers each day. Colleen earns $0.10 for each weekday and Saturday paper and $0.50 for each Sunday paper. How much does Colleen earn in a week?

Colleen's Newspaper Deliveries	
Monday-Friday	84
Weekends	116

6. The temperature rose 2° each hour that Joel was at the beach. When he arrived at 10:00 A.M., the temperature was 68°F. What was the temperature when he left at 4:00 P.M.?

7. A machine can cap 52 jars of pasta sauce per minute. How many jars of sauce can 8 machines cap in 15 minutes?

8. A car's odometer read 36,392 miles at the start of a trip and 36,686 miles at the end of the trip. If the trip took 6 hours, what was the driver's average speed in miles per hour?

9. Before the day of a dog show, 153 advance-sales tickets were ordered at $9 each. How much more money must be taken in to bring the total sales to $3,000?

10. The Natural Way sells dried apricots for $4.95 a pound, dried apple rings for $3.49 a pound, and dried pineapple slices for $4.25 a pound. Ani bought 2 pounds of each fruit and gave the cashier two $20 bills. How much change did she receive?

1.2 Identifying Information

To solve some problems, you must select the information that is needed from a source that may also contain extra information. Being able to recognize the unnecessary information will make you a more efficient problem-solver. On the other hand, if a problem does not provide sufficient information, you will not be able to solve it unless you can find the missing facts. Sometimes, information may not be stated outright, but you are expected to know it. For example, you might be expected to know that there are:

- 12 items in a dozen
- 4 aces in a standard deck of cards
- 365 days in 1 year
- 24 hours in a day

 **Model Problems**

1. Mr. Nolan bought one half-dozen cordless phones on sale for $22.00 each. The original price was $29.00. He gave the cashier $140.00. How much change did he receive?

Solution

Understand You are expected to know that one half-dozen is 6. He bought 6 phones for $22.00 each. The question is "How much change did he receive?" It is asking for a difference between the total cost and the money given to the cashier. The original price of the phone is not needed to find the answer.

Plan   Multiply $22.00 by 6, and then subtract the total from $140.

Solve $22.00 × 6 = $132
             $140 − $132 = $8

Check   The original question has been answered. He received $8.00 change.

Answer $8.00

2. Alyssa Chavez earns $2,300 a month. How much does she earn in a year?

Solution

Understand   You are expected to know that a year has 12 months. You must figure out how much she earns in one year.

Plan   Multiply $2,300 × 12. Estimate: $2,000 × 12 = $24,000.

Solve $2,300
            × 12
            4,600
       +23,000
       $27,600

Check   The answer is close to the estimate, so it is reasonable.

Answer   Alyssa earns $27,600 in one year.

3. There are 238 students going on a trip to Milky White's Dairy Farm. The buses will depart from the school at 9:00 A.M., and return by 4:00 P.M. How many buses will be needed to transport all the students?

Solution

Understand   The problem asks you to figure out the number of buses needed. The times are extra information. You do not need to know them to solve the problem. You do, however, need to know how many students each bus holds. This important piece of information is missing.

Plan   If this were a real–life problem, you could call the bus company to find the missing information. Otherwise, you can estimate the number of students who can ride on one bus and use that number to estimate how many buses will be needed.

Estimate that each bus holds 40 students. Divide to find out how many groups of 40 students there are.

238 ÷ 40 is close to 240 ÷ 40 = 6.

It is reasonable to estimate that about 6 buses will be needed.

Answer 6 buses

If there is enough information, solve the problem. Identify any extra information or unstated information that you are expected to know. If there is not enough information, state what is missing.

1. Small boxes of fudge contain 16 pieces. Large boxes contain 30 pieces. Small boxes sell for $5 each and large boxes are $9 each. On Saturday, the store sold 10 small boxes and 15 large boxes. How much money did the store take in?

2. Inez is 3 inches shorter than Peter and 4 inches taller than Lynn. Lynn is 2 inches taller than Barry. How tall is Peter?

3. A taxi company bought new tires for each of its 18 cars. Each tire cost $80. How much did the company spend for tires?

4. Mrs. Simmons drove 392 miles and used 14 gallons of gasoline. She paid $1.72 per gallon at a station 2 miles from her home. How much did she spend for gasoline?

5. On an assembly line, it takes 5 minutes to produce and package a popular board game. There are 62 workers at the factory and there are two 8-hour shifts each day. How many games can be produced and packaged in one day?

6. Marilyn paid $1.40 for the 2-mile bus ride to the train station. She then paid $6.25 for a 19-mile train ride to Baltimore. She paid the same amounts on her return trip home. How much did Marilyn spend for transportation?

7. The Brooklyn-Battery Tunnel in New York City was completed in 1950. It is located under the East River and is 9,117 feet long. The Holland Tunnel was completed 23 years before the Brooklyn-Battery Tunnel and runs beneath the Hudson River for 8,557 feet. How much longer is the Brooklyn-Battery Tunnel than the Holland Tunnel?

8. Notebooks come in packages of 3 and cost $4 per package. Pens come in packages of 4 and cost $2 per package. How much does it cost to buy a dozen notebooks and a dozen pencils?

9. The coach ordered new uniforms for the cheerleading squad. Each uniform cost $59 and each pair of pompoms cost $7.50. The school had budgeted $800 for the new items. Did the order exceed the budget or was money left over? If money was left over, how much?

10. Each side of a square flower garden measures 10 feet. Fencing costs $7 per foot and a No Trespassing sign costs $3. What is the total cost of enclosing the garden and putting up the sign?

1.3 Guessing and Checking

One way to begin working on a problem is to make a reasonable guess at the answer. You must then test the guess to see if it satisfies the conditions of the problem. If it does not, observing whether the guess is too high or too low helps you to make a better second guess. Continue guessing and checking until you find the correct answer.

Model Problems

1. Hector has 55 nickels, dimes, and quarters in his piggy bank. There are three times as many nickels as dimes, and half as many quarters as nickels. What is the value of Hector's coins?

Solution

Understand   The total number of coins must equal 55.
The number of nickels = 3 × number of dimes.

The number of quarters = $\frac{1}{2}$ × number of nickels.

Find how many of each coin, then find the total value.

Plan   Guess the number of dimes. Use the conditions of the problem to figure out the number of nickels and quarters. Check if the total number of coins is equal to 55.

Solve

First guess: 8 dimes

The number of nickels = 3 × 8 = 24.

The number of quarters = $\frac{1}{2}$ × 24 = 12.

Total number of coins = 8 + 24 + 12 = 44.   Too low!

Second guess: 12 dimes

The number of nickels = 3 × 12 = 36.

The number of quarters = $\frac{1}{2}$ × 36 = 18.

Total number of coins = 12 + 36 + 18 = 66.   Too high!

Third guess: 10 dimes

The number of nickels = 3 × 10 = 30.

The number of quarters = $\frac{1}{2}$ × 30 = 15.

Total number of coins = 10 + 30 + 15 = 55.   Correct!

Next, find the value of the coins.

$30 \times \$0.05 = \$1.50$

$10 \times \$0.10 = \$1.00$

$15 \times \$0.25 = \$3.75$

Add.

$$\begin{array}{r} \$1.50 \\ 1.00 \\ +\ 3.75 \\ \hline \$6.25 \end{array}$$

Check The conditions of the problem are satisfied and the question was answered by finding the value of the coins.

Answer Hector has $6.25.

2. Two different numbers have the same two digits. The sum of the digits is 5 and the difference between the numbers is 27. Find the numbers.

Solution

Plan Identify combinations of digits that have a sum of 5. Form the numbers and check if the difference is 27.

Solve First guess: $3 + 2 = 5$. Try 32 and 23.

$$\begin{array}{r} 32 \\ -\ 23 \\ \hline 9 \end{array}\quad \text{No!}$$

Second guess: $4 + 1 = 5$. Try 41 and 14.

$$\begin{array}{r} 41 \\ -\ 14 \\ \hline 27 \end{array}\quad \text{Yes!}$$

Check The numbers 41 and 14 meet the conditions of the problem.

Answer 41 and 14

Practice

1. Mr. Schick has 42 shirts that are either white, blue, or striped. He has twice as many striped shirts as white shirts. He has 6 more blue shirts than white shirts. How many of each type of shirt does he have?

2. Cynthia has 80 pennies, nickels, dimes, and quarters in her bank. There are four times as many dimes as nickels and one third as many quarters as dimes. There are 8 more nickels than pennies. How much money does Cynthia have in her bank?

3. A dictionary is opened at random. To which pages is it opened if the product of the facing page numbers is 2,162?

4. Together, Val and Ray have $50. Val has $7 more than Ray. How much money does each person have?

5. There were 100 tickets sold to the drama club's performance of *Phantom of the Opera*. Adult tickets cost $5. Student tickets cost $2. In all, $410 was collected. How many of each kind of ticket were sold?

6. Monique's mother is 23 years older than Monique. The product of their ages is 518. How old are Monique and her mother?

7. Fruitful Farms gave away 50 gift baskets as prizes. The baskets contained either 3 pears or 4 apples. How many of each type of basket were given away if 185 pieces of fruit were used?

8. Each student on the trip to Cooperstown paid the same amount. The total collected from the students was $589. Pilar paid for her ticket using exactly three bills.
   a. How much did each ticket cost.
   b. How many students went on the trip?

9. The candidates in a union election were Lang, Breen, and Kovic. There were 500 votes cast. Breen received 39 votes more than Lang. Kovic lost the election by 25 votes. How many votes did each candidate receive?

10. Place the numbers 1 through 9 in the circles so that the sum along each of the five lines is 17.

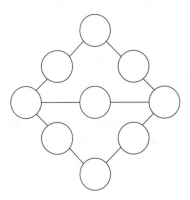

1.4 Working Backward

Some problems tell you a final result and the steps it took to get there. These problems ask you to figure out what the starting point must have been. Working backward is a strategy that works well for this type of problem. For example, a problem might tell you that after going to the store and buying milk for $2, Tony had $3 in his wallet. The problem asks you how much money Tony must have had *before* he bought the milk. Working backward, you use the opposite or inverse of the operation that was performed when Tony bought the milk. Since spending $2 is subtraction, you do the **inverse operation** and add $2 back to the $3. You have figured out that Tony must have started with $5.

An inverse operation *undoes* whatever the original operation did. For example, *unzipping* your jacket is the inverse operation of *zipping* your jacket.

Addition and subtraction are inverse operations.

Multiplication and division are inverse operations.

 # Model Problems

1. During the month of March, Ruth deposited $450 in her checking account, wrote checks for $118 and $246, and was charged $12 for new checks. If her balance at the end of March was $973, what was her balance at the end of February?

Solution

Understand During the month, Ruth put money into her account and took money out. You have to figure out how much money she had at the beginning of the month.

Plan Start with the ending balance and work backward, using inverse operations.

Solve

Start with ending balance	$973
Add back $12 charge	+ 12
	985
Add back $246 check	+ 246
	1,231
Add back $118 check	+ 118
	1,349
Subtract $450 deposit	− 450
	$899 ← Balance at end of February

Check Begin with $899 and use the original operations.

$899 + $450 = $1,349

$1,349 − $118 = $1,231

$1,231 − $246 = $985

$985 − $12 = $973

Answer Ruth's beginning balance was $899.

2. Students at Seneca High School are dismissed at 2:55 P.M. During the day there is a 15-minute homeroom period, 7 class periods of 40 minutes each, and a 35-minute lunch period. If 5 minutes are allowed for passing in the halls between each period, what time does school start?

Solution

Understand   Given all the information about the school schedule, you have to figure out what time school starts.

Plan   Calculate the time for class periods and time between changes. Add homeroom and lunch periods. Use the total time to work backward from dismissal time.

Solve

Class periods 7×40 min $= 280$ min

Between periods $8 \times 5 = 40$ min ← Remember to include time between homeroom and class.

Lunch   35 min

Homeroom   15 min

Add. $280 + 40 + 35 + 15 = 370$ min

Think! 1 hour $= 60$ minutes, so 370 min $= 6$ hours and 10 minutes.

Answer   6 hours 10 minutes earlier than 2:55 P.M. is 8:45 A.M.

Practice

1. Mr. Damon opened a box of cookies and used half for his children's lunches. Later, Mr. Damon and a neighbor each ate two cookies. When Mrs. Damon arrived home, she ate the last three cookies. How many cookies had been in the box?

2. During a four-day sale, a store sold twice as many televisions on the second day as on the first day. On the third day, 21 televisions were sold and 13 were sold on the last day. If 109 televisions were sold during the sale, how many were sold on the first day?

3. The cable guy stood on the middle rung of a big ladder. He moved up three rungs to fasten a wire, then backed down five rungs. A few minutes later, he climbed up seven rungs to check a connection. Then it was time for lunch, so he climbed the six remaining rungs and went onto the roof of the building. How many rungs does the ladder have?

4. Jerry subtracted 6 from his age, multiplied the result by 3, added 8, and then divided by 2. The result was 19. How old is Jerry?

5. Olivia purchased an $800 gold necklace using money she had earned working at the library. Olivia gets paid $6.00 an hour. She also had saved birthday gifts of $30 from her aunt and $50 from her grandfather to use for the necklace. How many hours did Olivia work to earn enough for her purchase?

6. During August, the price of a grapefruit tripled. In September, the price dropped by $0.40. When the new crop started to arrive in October, the price was halved. In November, the price rose $0.14 to $0.69 per grapefruit. Did the grapefruit cost more or less at the end of November than the beginning of August? How much more or less did it cost?

7. Mr. Zeeman received a tax refund. He spent $40 on take-out food for a family dinner and used one-third of the remaining money to buy a new vacuum cleaner. After he deposited half of what was left in his checking account, he still had $80. How much was his tax refund?

8. A shuttle bus makes a loop around the downtown business area. When the 10:00 A.M. bus left the Transit Center, there were already several passengers aboard. At the Galaxy Mall, five passengers got off the bus and eight boarded. At the Sportsdome, ten passengers got off and two boarded. At the Triumph Tower, nine people got on and seven got off. The remaining eight passengers left the bus when it returned to the Transit Center. How many passengers were on the bus when it started at 10:00 A.M.?

9. Paula exercises for 20 minutes each morning, then spends an hour getting ready for work. She walks 5 minutes to a café where she spends 15 minutes eating breakfast. It takes her 2 minutes to walk to the ferry where, after a 3-minute wait, she boards an express for the 34-minute ride into the city. Then she walks the distance to her office in 6 minutes and is at her desk at 9:00 A.M. What is the latest Paula can wake up to get to her job on time?

10. On Monday, the price of a share of Amtex stock fell 3 points. On Tuesday, the price soared up 5 points, but it fell 2 points on Wednesday and again on Thursday. On Friday, the stock gained 1 point to close the week at 32. What was the price of a share of Amtex at the opening on Monday?

1.5 Using Pictures and Diagrams

When problems involve physical situations, the strategy of drawing a picture or **diagram** can help you visualize the relationships in the problem. Once you have a better understanding of the problem, you can begin to plan a solution.

1. Diana began making a tablecloth by sewing together 6 squares of printed cotton. She made 2 rows with 3 squares in a row. She now wants to attach a border all around using solid cotton squares the same size as the others. How many solid squares will she need? How many rows will there be? How many squares will be in each row?

Solution

Plan   Draw a diagram of the printed squares and then complete the diagram by adding solid squares all around.

Solve   Draw the arrangement of 6 printed squares.

Complete the drawing by putting border squares all around.

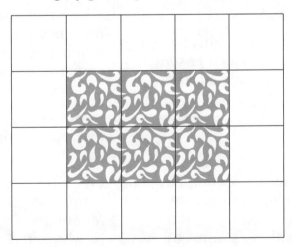

Count the solid squares.

Answer   Diana needs 14 solid squares. The tablecloth will have 4 rows with 5 squares in each row.

2. A hotel installed lamps along one side of the driveway leading to its entrance. If the distance from the first to the last lamp is 400 feet and the lamps are 40 feet apart, how many lamps were used?

Solution

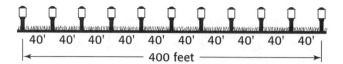

40'	40'	40'	40'	40'	40'	40'	40'	40'	40'

← 400 feet →

Answer The hotel used 11 lamps. Note that without the diagram, you might mistakenly think the answer was simply the quotient of 400 and 40.

Practice

1. Diana uses 2 rows of 5 printed squares for the center of another tablecloth. How many solid squares will she need for the border? How many rows will there be and how many squares in each row?

2. The highway commission decided to fence one side of a 60-foot road using posts that will be placed 5 feet apart. How many posts will they need?

3. Gino is building a fence around his 14-foot square patio. He has placed posts 2 feet apart along each side of the patio.
   a. How many posts are along each side?
   b. How many posts are there in all?

4. Barney asks the clerk at the post office for four stamps. The clerk will get the stamps from a large sheet of 100 stamps. Barney wants the 4 stamps to be attached, not loose, so they will not get lost. How many *different* ways can the clerk give him the stamps? For example, the two arrangements below would be considered the same. Draw the other ways.

5. A case of iced tea holds 24 bottles in 4 rows of 6 each. Show how there can be 18 bottles of tea in the case so that each row and each column holds an even number of bottles.

6. Mrs. Norris wants to display four post-cards on her refrigerator door using magnets. If she puts them up individually, she will need 16 magnets. If she allows the postcards to overlap, she will need fewer magnets. What is the least number of magnets she will need if the corner of every postcard is secured by a magnet? Draw the arrangement.

7. If Mrs. Norris uses 16 magnets, what is the *greatest* number of postcards she can display?

8. Five cities are connected by five roads: Lynn and Mountaindale, Newly and Mountaindale, Ogden and Preston, Newly and Preston, and Ogden and Mountaindale. Draw a diagram and describe all the routes from Preston to Lynn.

9. The supermarket is 5 blocks east of Veronica's house. Amina lives 3 blocks west of the movie theater, which is 4 blocks south and 2 blocks west of the supermarket. How many blocks and in what direction is Amina's house from Veronica's house?

10. The Kents are giving a dinner party for 24 guests including themselves. They will use square card tables that seat one guest per side. The tables will be pushed together to form one long row. What is the least number of tables the Kents can use?

1.6 Using Tables and Lists

Tables and lists are useful for organizing information that relates to a problem. The entries displayed on a table can be examined and compared with the conditions of a problem. The solution to a problem may be one, some, or all of the entries on the table or list. Making a table or list may be combined with another strategy, such as guessing and checking or drawing a diagram.

Model Problems

1. Emily has the same number of nickels and dimes. The coins total $2.55. How many of each coin does she have?

Solution

Plan   Make a table showing how many of each coin and the total value.

Solve   Think about the total to find a starting point for the number of coins. A reasonable choice is to begin with 10 of each coin.

Nickels	Value	Dimes	Value	Total Value	
10	$0.50	10	$1.00	$1.50	Much, much too little.
12	$0.60	12	$1.20	$1.80	Much too little.
14	$0.70	14	$1.40	$2.10	Still too little.
16	$0.80	16	$1.60	$2.40	Getting close.
17	$0.85	17	$1.70	$2.55	Yes!

Answer   Emily has 17 nickels and 17 dimes.

2. Josh throws three darts and all hit the target shown. What scores are possible?

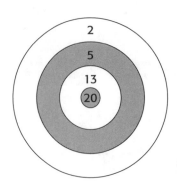

Solution

Plan   Make an organized list of all the possible combinations of the three point-values. Find each sum. Note that since the question asks only for possible scores, it is not necessary to list both 2 + 5 + 13 and 2 + 13 + 5 since the sum is the same.

Solve   List the possible combinations in an organized way.

2 + 2 + 2 = 6	2 + 13 + 13 = 28	5 + 13 + 13 = 31
2 + 2 + 5 = 9	2 + 13 + 20 = 35	5 + 13 + 20 = 38
2 + 2 + 13 = 17	2 + 20 + 20 = 42	5 + 20 + 20 = 45
2 + 2 + 20 = 24		

2 + 5 + 5 = 12	5 + 5 + 5 = 15	13 + 13 + 13 = 39
2 + 5 + 13 = 20	5 + 5 + 13 = 23	13 + 13 + 20 = 46
2 + 5 + 20 = 27	5 + 5 + 20 = 30	13 + 20 + 20 = 53
		20 + 20 + 20 = 60

Check   Check that all the combinations of three point scores are listed and that no combination is listed twice.

Answer   There are 20 possible scores: 6, 9, 12, 15, 17, 20, 23, 24, 27, 28, 30, 31, 35, 38, 39, 42, 45, 46, 53, and 60.

1. Claire has 30 coins, all nickels and dimes, that total $2.40. How many of each coin does she have?

2. Manuel is buying a granola bar from a vending machine. The granola bar costs 50¢, and the machine takes nickels, dimes, and quarters.
   a. In how many different ways can Manuel pay the correct amount?
   b. How can he pay using 7 coins?

3. A customer asks a bank teller to change a $20 bill into smaller bills. Using $1, $5, and $10 bills, in how many ways can the teller make the change?

4. How many whole numbers between 0 and 100 are there with a sum of digits equal to 7?

5. Tanya has a 10-foot piece of ribbon that she wants to cut into three pieces. The length of each piece must be a whole number of feet, no piece can be longer than 6 feet, and no ribbon should be left over. What are the different combinations of lengths she can make?

6. a. How many different four-letter codes can be made using the letters A, B, C, and D? No letter may be used more than once in any code.
   b. How many of the codes have C as the third letter?

7. Joan throws three darts and all hit the target shown.

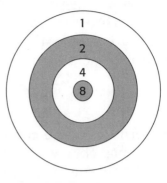

   a. What different scores are possible?
   b. Which scores can be made more than one way?

8. It takes 930 digits to number the pages of a book. How many pages does the book contain?

9. Shirley has some gumdrops. If she divides the gumdrops into groups of 3, she has 1 left over. If she divides the gumdrops into groups of 4, she has 3 left over. If she divides the gumdrops into groups of 5, she has 4 left over. What is the least number of gumdrops she could have?

10. Jumbo the Elephant was offered a part in a movie. He was given a choice of the following pay plans, A and B, plus all the peanuts he can eat:
   (A) $0.50 the first day, $1.00 the second day, $2.00 the third day, and so on, doubling each day, or
   (B) $50 a day.
   If the job lasts for 12 days, with which plan will Jumbo earn more? How much more will he earn?

1.7 Recognizing Patterns

When you are trying to identify a **pattern**, the way to begin is by examining the first few numbers, figures, or other parts of the problem. Once you recognize the pattern, you can predict other outcomes.

Some patterns are very simple, such as a series of numbers or figures that repeats. For example, the pattern below is triangle, square, circle, then repeat. So the next figure would be a square.

Other patterns are more complicated. A good way to begin is by figuring out what has changed from one term to the next. Ask yourself if the pattern is increasing or decreasing and by how much.

 **Model Problems**

1. What is the next number in the following sequence?

1, 3, 6, 10, 15, . . .

Solution

Since the numbers are *increasing* whole numbers, look for a pattern rule that involves addition or multiplication. The pattern for this problem appears to use addition rather than multiplication. Examine the differences from one number to the next.

The differences are increasing by 1. If this pattern is continued, the next number is found by adding 6 to 15 to obtain 21.

```
1    3    6    10   15   21
  +2   +3   +4   +5   +6
```

Answer   The next number in the sequence is 21.

2. What is the next number in the following sequence?

256, 128, 64, 32, 16, . . .

Solution

Since the numbers are *decreasing* whole numbers, look for a pattern rule that involves subtraction or division. The second number is half the first number, the third number is half the second number, and so on.

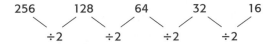

Divide by 2 to get the next number: $16 \div 2 = 8$.

Answer The next number in the sequence is 8.

3. A row of tiles along a restaurant wall uses a repeating pattern of yellow, blue, and green. What color is the 100th tile in the row?

Solution

The pattern repeats after 3 tiles, so there will be complete patterns for 3 tiles, 6 tiles, 9 tiles, and so on. The pattern ends on numbers that are multiples of 3.

The number 99 is 3×33, so the pattern will be complete on the 99th tile. The next tile will start the pattern again, so the 100th tile will be yellow.

Answer Yellow

Practice

1. Find the next number in the sequence: 2, 4, 8, 14, 22, 32, …

2. Find the next number in the sequence: 486, 162, 54, 18, …

3. Fiona plans to put $2 in her savings account this week, $5 next week, $8 the following week, $11 the week after that, and so on. If the pattern continues
   a. How much money will she put into her account on the twelfth week?
   b. How much money will she have in all for the entire twelve weeks?

4. A radio station has an hourly contest and the prize money increases each hour that there is no winner. There has been no winner since 6 A.M. and the prizes are shown below.

Time	6:00	7:00	8:00	9:00	10:00
Prize	$10	$30	$70	$150	$310

If there is no winner, what prize will be offered during the noon hour?

5. Find the next two numbers in the sequence: 5, 10, 8, 13, 11, 16, 14, . . .

6. If $4 \ominus 3 = 6$, and $8 \ominus 6 = 24$, and $9 \ominus 4 = 18$, what is the value of $5 \ominus 10$?

7. Visitors to a flower show receive a flower as they enter. The flowers are given out in this order: daisy, carnation, rose, daffodil. Which flower will each visitor receive?
a. the 27th visitor
b. the 41st visitor
c. the 62nd visitor

8. A digital clock now reads 9:00. What will it read 51 hours from now?

9. Look at the following:
$12 \times 11 = 132$
$13 \times 11 = 143$
$14 \times 11 = 154$
Predict the value of 18×11.

10. Draw the next figure in the pattern.

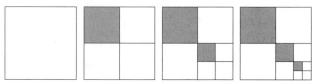

1.8 Solving a Simpler Related Problem

Sometimes you can find the answer to a problem by solving another problem that has simpler numbers or fewer cases. Using smaller or easier numbers, like whole numbers instead of decimals or fractions, can help you focus on the operations that are needed to plan a solution. Or, by starting with fewer cases and working up, you may be able to identify a pattern that will lead to a solution.

Model Problems

1. There will be 40 representatives from different nations at an international conference. The name of the countries will be alphabetized and then each representative will be assigned a number starting with 1. The representatives will be equally spaced and seated in order around a large, round table. What is the number of the representative who will be sitting directly opposite representative 14?

Solution

Start with fewer people and look for patterns. Make a diagram.

4 guests   2 pairs (4,2)   Differ by 2
                   (1,3) $4 \div 2 = 2$

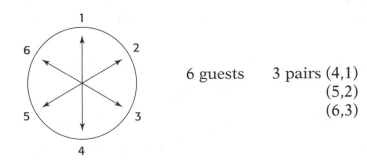

6 guests   3 pairs (4,1)   Differ by 3
                   (5,2) $6 \div 2 = 3$
                   (6,3)

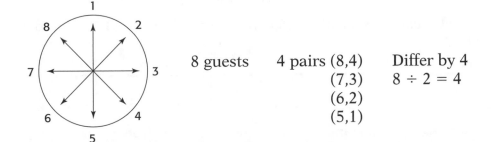

8 guests   4 pairs (8,4)   Differ by 4
                   (7,3) $8 \div 2 = 4$
                   (6,2)
                   (5,1)

The patterns are now clear. The number of pairs is half the number of people. The difference in the numbers of opposites is also half the number of people. So, for 40 people, the difference between opposite numbers will be

$40 \div 2 = 20$.

Use the pattern to solve the more difficult problem.

Answer   The number of the representative who will be sitting directly opposite number 14 is $14 + 20 = 34$.

2. What is the remainder when $6 \times 6 \times 6 \times 6 \times 6 \times 6 \times 6 \times 6 \times 6$ is divided by 5?

Solution

Solve some simpler problems.

$6 \div 5 = 1 \text{ r } 1$

$6 \times 6 = 36$ and $36 \div 5 = 7 \text{ r } 1$

$6 \times 6 \times 6 = 216$ and $216 \div 5 = 43 \text{ r } 1$

Answer There is a pattern. The remainder is always 1. So, if the given product is divided by 5, the remainder will also be 1. (Check using a calculator.)

3. Cashews sell for $5.79 per pound and almonds sell for $4.85 per pound. Anita bought 3.4 pounds of cashews and 2.8 pounds of almonds. Show the steps you would use to find the total amount Anita spent.

 Solution

Think about the problem using whole numbers. Suppose the price of cashews were $5 per pound, the price of almonds were $4 per pound, and Anita bought 3 pounds of cashews and 2 pounds of almonds. How would you solve the simpler problem? Use the same operations for the actual problem.

Simpler Problem	**Actual Problem**

Step 1 Multiply 3 × $5 (total for cashews) → Multiply 3.4 × $5.79

Step 2 Multiply 2 × $4 (total of almonds) → Multiply 2.8 × $4.85

Step 3 Add the products above.   Add the products above.

You may wish to solve the actual problem using a calculator.

Answer The cashews cost $19.69, the almonds cost $13.58, and the total is $33.27.

Practice

1. Refer to the situation in Model Problem 1. What is the number of the representative who will sit opposite representative 27?

2. Suppose that at a larger conference there will be 54 representatives. The numbering and seating arrangement will be similar to before. What is the number of the representative who will sit opposite
   a. representative 13?
   b. representative 48?

3. What is the remainder when
   a. 4 is divided by 3?
   b. 4 × 4 is divided by 3?
   c. 4 × 4 × 4 is divided by 3?
   d. 4 × 4 × 4 × 4 × 4 × 4 × 4 × 4 × 4 × 4 is divided by 3?

4. What is the remainder when
   a. 30 is divided by 9?
   b. 300 is divided by 9?
   c. 3,000 is divided by 9?
   d. 300,000,000 is divided by 9?

5. Find each sum.
   a. 1 + 3
   b. 1 + 3 + 5
   c. 1 + 3 + 5 + 7
   d. 1 + 3 + 5 + 7 + 9
   e. Predict the sum: 1 + 3 + 5 + 7 + 9 + 11 + 13 + 15 + 17 + 19 + 21 + 23 + 25 + 27 + 29 + 31.

6. In how many different ways could 11 identical buttons be distributed among 3 boxes, if each box must have at least 2 buttons? For example, there could be 4 buttons in 2 of the boxes and 3 buttons

in the other, or there could be 2, 3, and 6 buttons in the boxes. (Start with 6 buttons, then 7, and look for a pattern.)

7. What is the ones digit in $9 \times 9 \times 9 \times 9 \times 9 \times 9 \times 9 \times 9 \times 9 \times 9 \times 9 \times 9 \times 9$? Solve the problem by examining simpler cases such as 9×9, $9 \times 9 \times 9$, $9 \times 9 \times 9 \times 9$, and so on. Explain the pattern you find and give the answer to the question.

8. Edwin has 100 baseball cards. He wants to place the cards in three stacks so that the second stack has twice as many cards as the first, and the third stack has twice as many cards as the second. How many cards will be in each stack? How many will be left over?
   a. Try a simpler problem. How many cards are needed to make the smallest possible stacks that satisfy the given conditions?
   b. Multiply each of the small stacks in part a by the same number. Try different numbers until you find the greatest stacks possible.

9. A classroom has 6 desks and 6 chairs. Ms. Lewis wants to arrange the desks and chairs so that each desk is paired with a chair. How many such arrangements are there?

10. The cafeteria manager looks in the refrigerator and finds $2\frac{1}{2}$ packages of cheese. Each package of cheese weighs $3\frac{3}{4}$ pounds. The cook said that $9\frac{1}{2}$ pounds of cheese were needed for tomorrow's lunch. Is there enough cheese? If not, how much more is needed?
    a. What whole numbers would you substitute into this problem to make it a simpler problem?
    b. Show the steps you would use to solve the simpler problem.
    c. Describe the steps you would use to solve the actual problem.

1.9 Using Logical Reasoning

Logic is the study of reasoning. All problem solving, whether in mathematics or daily life, involves reasoning, but there are some problems for which special approaches are very effective. These problems generally present several facts that can be used together to reach a conclusion. Combining reasoning skills with other strategies, such as drawing a diagram or making a table, will help you find the desired solution.

Model Problems

1. Of 200 travelers surveyed at an airport, 156 carried a cell phone, 72 carried a laptop computer, and 49 had both items.
   a. How many travelers had only a cell phone?
   b. How many travelers had neither item?

Solution

The information can be pictured using a **Venn diagram.**

Draw a rectangle to represent all of the travelers surveyed. Inside the rectangle, draw one circle to represent travelers with phones and another circle to represent travelers with computers. The circles should overlap since the 49 people who had both items must be included in both circles. The region outside both circles represents travelers who carried neither item.

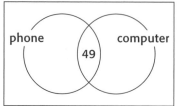 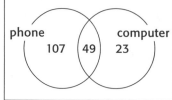 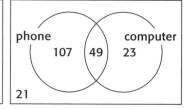

Find out how many travelers had only phones and only computers by subtracting the number of travelers who had both.

156 – 49 = 107 had only phones

72 – 49 = 23 had only computers

To find out how many had neither item, add the number of people with only phones, only computers, and both.

107 + 23 + 49 = 179

Since 200 travelers were surveyed, subtract the total above.

200 – 170 = 21 had neither item

Answer   107 travelers had only phones and 21 had neither item.

2. Three musicians appeared at a concert. Their last names were Benton, Lanier, and Rosario. Each plays only one of the following instruments: guitar, piano, or saxophone.

   • Benton and the guitar player arrived at the concert together.
   • The saxophone player performed before Benton.
   • Rosario wished the guitar player good luck.

Who played each instrument?

Solution

Make a table listing each name and each instrument. Use the given information to find what is true about each player. You may use the information in any order. If one fact is more definite than the others, start with that one. In this case, no single fact allows you to reach a final conclusion.

Since Benton came with the guitar player, this means Benton cannot play guitar. Use an **x** to show this.

	Guitar	Piano	Saxophone
Benton	x		
Lanier			
Rosario			

The second fact shows that Benton also cannot be the saxophone player. Place an ✗ in the box for saxophone. This leaves only piano for Benton. Use a ✓ to show this. Moreover, use ✗'s to show that nobody else plays piano.

	Guitar	Piano	Saxophone
Benton	✗	✓	✗
Lanier		✗	
Rosario		✗	

Using the third fact, you may conclude that Rosario is not the guitar player. This leaves only the saxophone for Rosario.

	Guitar	Piano	Saxophone
Benton	✗	✓	✗
Lanier		✗	
Rosario	✗	✗	✓

Finally, since the other instruments are assigned, Lanier must play guitar.

	Guitar	Piano	Saxophone
Benton	✗	✓	✗
Lanier	✓	✗	✗
Rosario	✗	✗	✓

To check, compare each original fact with the completed chart.

Answer Benton plays piano, Lanier plays guitar, and Rosario plays the saxophone.

Practice

1. Mike has 300 stamps in his collection. Of these, 158 are from European countries, 143 have pictures of flags, and 87 are from European countries and have pictures of flags.

   a. How many stamps are from European countries, but do not have pictures of flags?
   b. How many are neither European nor have a flag picture?

2. A soft drink company conducted a taste test of two new products, Zoom and Magna. Of the 461 people who were surveyed, 238 liked Zoom, 127 liked both Zoom and Magna, and 48 did not like either drink. How many people liked Magna?

3. Use the Venn diagram shown to answer the questions below.

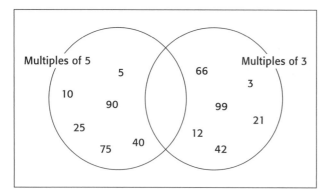

    a. Write two numbers that could be placed in the overlapping part. Explain why they belong there.

    b. Write two numbers that could be placed outside both circles. Explain why they belong there.

4. The diagram below shows subjects studied by a group of students at the Greenvale School.

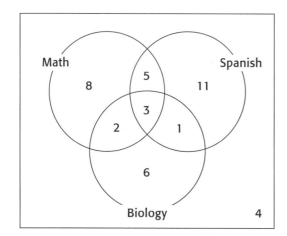

How many students take
a. math and Spanish?
b. only biology?
c. Spanish and biology, but not math?
d. all three subjects?
e. How many students are represented in the diagram?

5. Use a Venn diagram with three circles to solve this problem. The 100 guests at a luncheon chose from a buffet that included soup, salad, and sandwiches. Of the guests, 50 had salad, 66 had sandwiches, 48 had soup, 30 had soup and sandwiches, 28 had soup and salad, 22 had salad and sandwiches, and 12 had all three foods. How many guests did not eat any of the three foods? How many had only soup?

6. Three friends, Adrienne, Brenda, and Crystal each have one brother whose name is either Tommy, Danny, or Harry. Brenda said that Tommy and her brother jogged together. Crystal went to the movies with her brother and Harry. Danny asked Adrienne and Crystal to help him buy a gift for his sister. Identify each brother and sister.

7. Five friends tried solving a puzzle and each took a different amount of time. The friends are: Denise, Sam, Keisha, Max, and Julio. The times were: 3 minutes, 6 minutes, 9 minutes, 10 minutes, and 12 minutes.

- Keisha took more time than Julio.
- Denise took 3 minutes longer than Sam.
- Only one person took longer than Julio.
- Sam did not have the best time.

How long did each person take?

8. Dwayne, Reneé, and Fred each have a different hobby and a different pet. One has a dog, one a cat, and one a bird.

- The person who likes to cook does not have a bird.
- Dwayne builds model rockets and does not have a dog.
- The person who likes to garden has a cat.
- Fred never cooks.

Find the hobby and pet for each person.

9. The heights of six students are 51 in., 56 in., 59 in., 64 in., 66 in., and 68 in. Only one student is taller than Scott. Annie is taller than Gary. Cindy is the shortest. Trisha is 5 inches taller than Cindy. Molly is 5 inches shorter than Gary. How tall is each student?

10. The deepest lake in the United States is in Oregon. Lake Mendota is in Wisconsin. What conclusion can you draw from these two facts?

1.10 Using Estimation

An **estimate** tells you about how much. In later sections, you will learn and apply several estimation strategies. The idea behind all estimation strategies is to use numbers close to the numbers in the problem so that you can work with them mentally. Knowing when to estimate is an important skill in itself.

You can estimate when

an exact answer is not needed.

the problem can be solved just by knowing "about how much."

paper and pencil or a calculator are not available.

you want to check if an answer is reasonable. A reasonable answer is one that makes sense for the given situation.

Once you decide to estimate, you may also need to think about whether an overestimate or underestimate would be more useful.

An overestimate is an estimate you know is greater than the exact answer. For example, to be sure that you take enough money, you may wish to overestimate the total cost of groceries.

An underestimate is an estimate you know is less than the exact answer. For example, you may want to underestimate the total amount of paint remaining in some started cans to be sure there is enough to finish a new job.

In some situations it does not matter if the estimate is high or low, such as when you are estimating the number of oranges on a tree.

Model Problems

For problems 1 and 2, indicate if an exact answer is needed or if an estimate is enough.

1. The numbers of tickets sold for three showings of a movie were 642, 573, and 518. Were more than 1,500 tickets sold in all?

Answer Since you only want to know if the sum is more than 1,500, you can estimate. The sum $600 + 500 + 500 = 1,600$ is more than 1,500. Since each number in the estimated sum is less than the actual number, the actual sum is certainly greater than 1,500.

2. A customer buys a sandwich for $3.75, a soft drink for $0.85, and a piece of fruit for $0.69 from a lunch cart. How much should the cart operator ask the customer to pay?

Answer An exact sum must be found. The customer does not want to pay too much and the cart operator does not want to charge too little.

For problems 3 and 4, indicate whether an underestimate or overestimate should be used.

3. You want to know how many pounds of shrimp to buy for a party. You will estimate the amount for one person, then multiply by the number of expected guests.

Answer If you underestimate, you may not have enough food. It is better to overestimate the amount per guest.

4. You are estimating how many miles you can travel on the gas left in your car's tank. There is a filling station nearby, and then there isn't one for a long distance.

Answer It is better to underestimate. If you overestimate, you may run out of gas at an inconvenient location.

Practice

For problems 1–5, indicate if an exact answer is needed or if an estimate is enough.

1. A pet store owner wants to know how much food to order for the animals for a month.

2. Mr. Cobb wants to know how much money he earned last year so he can determine his federal income tax.

3. Headbands cost $3.89 each. Sheila wants to know how many she can buy for $10.00.

4. Dr. John wants to determine the dosage of a medicine based on a child's weight.

5. Vince wants to know how much money he needs for a day at the beach.

For problems 6–9, indicate which estimate should be used and explain.

6. Ms. Shine has three errands to do before a doctor's appointment at 11:00 A.M. To figure out what time to leave her house, should she underestimate or overestimate the total time it will take to complete the errands?

7. Jeanette is replacing the fabric on the seats of the dining room chairs. Should she underestimate or overestimate the amount of fabric needed per chair?

8. Ahmed wants to know how long it will take him to earn the money to pay for a $700 ocean cruise. He will schedule the cruise based on this calculation. Should he underestimate or overestimate the amount of tips he will earn weekly from his restaurant job?

9. Isabel needs to order programs for a play her theater group is putting on. There will be 6 performances. Should she underestimate or overestimate the number of people attending each performance?

10. Describe a problem situation that could be solved by estimation. Indicate whether you would underestimate or overestimate.

Chapter 1 Review

Solve each problem using any strategy. Show your work.

1. There are 8 contestants on a television quiz show. Before the show began, each contestant shook hands with every other contestant. How many handshakes were there?

2. On Friday, a store had 24 tennis rackets in stock. On Thursday, a delivery had doubled the number of rackets. On Wednesday, the store had sold 10 rackets. On Tuesday, 5 rackets had been returned. On Monday, the store had sold half its stock of rackets. How many rackets had the store started with on Monday?

3. A company has 96 employees. Six more employees work on the second floor than on the first. Nine more employees work on the third floor than on the second. How many employees work on each floor?

4. Of the 200 students surveyed, 127 owned a dog, 40 owned a dog and a cat, and 31 had neither animal. How many students owned cats?

5. On what day of the week must a non-leap year start in order for May 27 to fall on a Thursday?

6. There are 12 editors at a publishing company. The 5 senior editors have each been with the company for 7 years or longer. Each senior editor is paid $50,000 per year. The remaining assistant editors are between 25 and 30 years old and work 40 hours per week. Each assistant editor is paid $40,000 per year. The company has been in business since 1994. How much does the company spend for editors' salaries in one year?

7. Vivian has $4.05, all in quarters and dimes. She has twice as many dimes as quarters. How many of each coin does she have?

8. Mario, Julia, Eve, and Wen live on the same street. Their ages are 10, 11, 13, and 14.

   - Julia and the 11-year old live across from each other.
   - The 10-year old watches Mario and Wen play basketball.
   - Mario and Julia went to the movies with the 14-year old.
   - Julia is older than Mario.

   How old is each person?

9. How many different squares are there in the picture? Squares can be any size. (*Hint:* Make a list. Use letters to keep track of the parts of each square.)

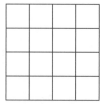

10. Bus routes connect six towns. Bromly is connected to Clinton. There are routes connecting Clinton to Dennis and to Eady. There are routes connecting Franklin to Glendon and to Eady. A route connects Glendon and Eady. List all the ways to travel from Glendon to Clinton. (No route should pass through any town more than once.)

CHAPTER 2

Whole Numbers and Decimals

Chapter Vocabulary

expanded notation	period	exponent
base	power	cubed
scientific notation	standard form	rounded number
estimate	difference	factors
product	partial product	quotient
divisor	dividend	

2.1 Place Value

If you have two bills in your wallet, it makes a big difference whether they are ones or tens or hundreds. The same is true for numbers. The value of each digit in a number depends upon the place that the digit occupies. A place-value chart can help you read and write whole numbers and decimals. When moving to the left of the decimal point, each place is 10 times the value of the place to its right. When moving to the right of the decimal point, each place is one tenth of the value of the place to its left. Since the value of each place is a power of 10, any number can be written as the sum of the values of its digits. This is called **expanded notation**.

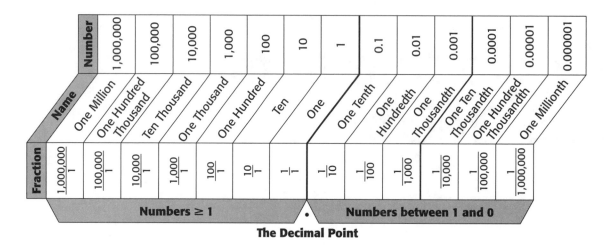

| | | | | | | Numbers ≥ 1 | | • | | | Numbers between 1 and 0 | | | | |

The Decimal Point

Reading whole numbers that are written as numerals

The places are grouped into sets of three, separated by commas as shown below. Each group is called a **period**, and each period has a name such as ones, thousands, and millions.

Billions			Millions			Thousands			Ones		
hundred billions	ten billions	billions	hundred millions	ten millions	millions	hundred thousands	ten thousands	thousands	hundreds	tens	ones

To read the number 64,206,011:

1. Begin with the largest period, which is farthest to the left. Read the number in this period followed by the word *million*, that is: *sixty-four million*.
2. Read the number in the thousands period followed by the word *thousand*, that is: *two hundred and six thousand*.
3. Read the number in the ones period, without the period name: *eleven*.

All together, you read: *sixty-four million, two hundred and six thousand, eleven*.

Examples

1. 2,342 is read *two thousand, three hundred forty-two*.

   Expanded notation: 2,342 = (2 × 1,000) + (3 × 100) + (4 × 10) + (2 × 1)

2. 4,236,764.325 is read *four million, two hundred thirty-six thousand, seven hundred sixty-four and three hundred twenty-five thousandths*.

Expanded notation: $4{,}236{,}764.325 = (4 \times 1{,}000{,}000) + (2 \times 100{,}000) + (3 \times 10{,}000) + (6 \times 1{,}000) + (7 \times 100) + (6 \times 10) + (4 \times 1) + \left(3 \times \frac{1}{10}\right) + \left(2 \times \frac{1}{100}\right) + \left(5 \times \frac{1}{1{,}000}\right)$

3. 34.57 is read *thirty-four and fifty-seven hundredths*.

Expanded notation: $34.57 = (3 \times 10) + (4 \times 1) + \left(5 \times \frac{1}{10}\right) + \left(7 \times \frac{1}{100}\right)$

4. 673.15 is read *six hundred seventy-three and fifteen hundredths*.

Expanded notation: $673.15 = (6 \times 100) + (7 \times 10) + (3 \times 1) + \left(1 \times \frac{1}{10}\right) + \left(5 \times \frac{1}{100}\right)$

Practice

Multiple-Choice Questions

1. The number 1,436.06 is read as

   A. one thousand, four hundred thirty-six
   B. one thousand, four hundred thirty-six and six tenths
   C. one thousand, four hundred thirty-six and six hundredths
   D. fourteen hundred thirty-six and six tenths

2. 1,436.06 written in expanded notation is

   F. $(1 \times 100) + (4 \times 10) + (3 \times 1) + (6 \times 1) + \left(\frac{1}{100}\right)$
   G. $(1 \times 1{,}000) + (4 \times 100) + (3 \times 10) + (6 \times 1) + \left(\frac{6}{100}\right)$
   H. $(1 \times 10{,}000) + (4 \times 1{,}000) + (3 \times 100) + (6 \times 10) + (6 \times 1)$
   J. $(1 \times 10) + (4 \times 100) + (3 \times 1{,}000) + (6 \times 10{,}000) + (6 \times 100{,}000)$

3. The expanded notation of 193.004 is

   A. $(1 \times 1{,}000{,}000) + (9 \times 100{,}000) + (3 \times 10{,}000) + (4 \times 10)$
   B. $(1 \times 100) + (9 \times 10) + (3 \times 1) + (4 \times 1)$
   C. $(1 \times 1{,}000) + (9 \times 100) + (3 \times 10) + (4 \times 1)$
   D. $(1 \times 100) + (9 \times 10) + (3 \times 1) + \left(4 \times \frac{1}{1{,}000}\right)$

4. 243.708 written in expanded notation is

   F. $(2 \times 1{,}000) + (4 \times 100) + (3 \times 10) + \left(7 \times \frac{1}{10}\right) + \left(8 \times \frac{1}{100}\right)$
   G. $(2 \times 1{,}000) + (4 \times 100) + (3 \times 10) + \left(7 \times \frac{1}{1{,}000}\right) + \left(8 \times \frac{1}{100}\right)$
   H. $(2 \times 100) + (4 \times 10) + (3 \times 1) + \left(7 \times \frac{1}{10}\right) + \left(8 \times \frac{1}{1{,}000}\right)$
   J. $(2 \times 100) + (4 \times 10) + (3 \times 1) + \left(7 \times \frac{1}{10}\right) + \left(8 \times \frac{1}{100}\right)$

5. The number 1.0001 is read as

   A. one and one ten thousandth
   B. one and one thousandth
   C. one and one millionth
   D. one and one hundredth

6. In 473.169, which digit is in the tenths place?

   F. 1   G. 3
   H. 6   J. 7

7. Given the number 6,954,209.01, identify the digit in the ten-thousands place.

   A. 9   B. 5
   C. 4   D. 1

8. Identify the digit in the tens place in the number 7,426.

   F. 7   G. 4
   H. 2   J. 6

9. What number does $(6 \times 100{,}000)$ + $(9 \times 1{,}000) + (5 \times 100) + (2 \times 1)$ $+ \left(3 \times \dfrac{1}{100}\right)$ represent?

   A. 6,952.3   B. 69,052
   C. 609,502.03   D. 695,002.03

10. What is the value of the digit 8 in the number 3,098,430.651?

    F. eighty thousand
    G. eight thousand
    H. eight hundred
    J. eighty thousandths

11. In the world today there are about 6,083,000,000 people. This number would be read as

    A. 6 trillion, 83 billion
    B. 6 million, 83 hundred thousand
    C. 6 billion, 83 million
    D. 6 trillion, 83 million

12. 5,545,000,000 written in expanded notation is

    F. $(5 \times 1{,}000{,}000{,}000)$
       $+ (5 \times 100{,}000{,}000)$
       $+ (4 \times 10{,}000{,}000)$
       $+ (5 \times 1{,}000{,}000)$
    G. $(5 \times 1{,}000{,}000{,}000) + (545 \times 1{,}000)$
    H. $(5 \times 1{,}000{,}000{,}000)$
       $+ (5 \times 1{,}000{,}000) + (4 \times 100{,}000)$
       $+ (5 \times 10{,}000)$
    J. $(5 \times 1{,}000{,}000{,}000)$
       $+ (5 \times 100{,}000{,}000)$
       $+ (4 \times 1{,}000{,}000) + (5 \times 100{,}000)$

13. By how much will the number 745.096 be increased if the digit 6 is changed to an 8?

    A. two thousandths
    B. twenty hundredths
    C. two tenths
    D. twenty

14. Which number is 11 thousandths more than 2.6715?

    F. 11,002.6715   G. 2.6726
    H. 2.6825   J. 2.7815

Short-Response Questions

15. Write each number in words.

    a. 209   b. 3,494
    c. 10,006   d. 305,009
    e. 7.62   f. 42.6
    g. 0.0001   h. 3.0

16. Write each number in standard form.

    a. five hundred sixty-five
    b. three thousand, five hundred
    c. ten million, four thousand two
    d. seventy thousand, nine hundred one and one tenth
    e. two and ninety-nine hundredths

2.2 Powers of 10 and Scientific Notation

An **exponent** tells the number of times a **base** is used as a factor.

$$10^3 = 10 \times 10 \times 10 = 1{,}000$$

10 is the base.
3 is the exponent, the number of times 10 is used as a factor.
1,000 is the third **power** of 10, or 10 **cubed**.

Powers of 10

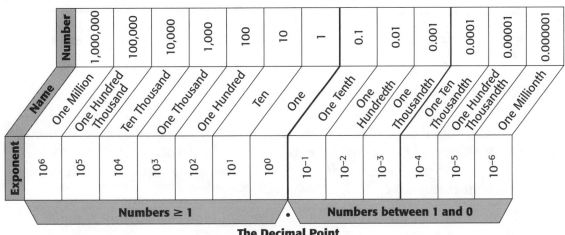

What patterns can you find in the table above?

$10^0 = 1$
$10^1 = 10$
$10^2 = 10 \times 10 = 100$
$10^3 = 10 \times 10 \times 10 = 1{,}000$
$10^4 = 10 \times 10 \times 10 \times 10 = 10{,}000$

Notice that for positive powers of 10, the number of zeros in the product is equal to the number that is the exponent.

Negative powers are expressed as fractions with numerators of 1 and denominators of the positive powers.

$$10^{-1} = \frac{1}{10^1} = \frac{1}{10} = 0.1$$

$$10^{-2} = \frac{1}{10^2} = \frac{1}{100} = 0.01$$

$$10^{-3} = \frac{1}{10^3} = \frac{1}{1{,}000} = 0.001$$

Scientific Notation

Scientists often work with very large and very small numbers like the distance from Earth to the sun or the diameter of a cell. It makes the work easier to write these numbers using **scientific notation**. A number written in scientific notation has two factors. The first factor is a number greater than or equal to 1 and less than or equal to 10, and the second factor is a power of 10.

1. A micrometer is a very small unit of length equal to 0.00003937 inch. Express this number in scientific notation.

Solution

Move the decimal point to the right so the first factor is between 1 and 10.

$$0.00003937 \quad \text{becomes} \quad 3.937 \times 10^{?}$$

Count the number of places you moved the decimal point. The number of places it was moved becomes the exponent of 10. If you moved the decimal to the left, the exponent is positive. If you moved the decimal to the right, the exponent is negative.

The decimal was moved 5 places to the right, so the exponent is -5.

Answer Therefore, $0.00003937 = 3.937 \times 10^{-5}$.

For problems 2–4, write the number in scientific notation.

Solution

2. 2,000,000 $\qquad 2{,}000{,}000 = 2.0 \times 1{,}000{,}000 = 2.0 \times 10^{6}$

3. 7,800 $\qquad 7{,}800 = 7.8 \times 1{,}000 = 7.8 \times 10^{3}$

4. 0.065 $\qquad 0.065 = 7.8 \times 0.01 = 6.5 \times 10^{-2}$

Notice how 2.0, 7.8, and 6.5 are all between 1 and 10. The first factor of a number in scientific notation is always between 1 and 10. The second factor is always a power of 10.

Scientific Notation: Is the exponent positive or negative?

- A number that is equal to or greater than 1 has 10 with a *positive* exponent as one factor.

- A number between 0 and 1 has 10 with a *negative* exponent as one factor.

Numbers written in scientific notation can be written in decimal notation or **standard form** by expanding the power of 10 and multiplying.

Example

Write 2.08×10^3 in standard form.

$2.08 \times 10^3 = 2.08 \times 10 \times 10 \times 10 = 2.08 \times 1,000$ or $2,080$.

This is the same as moving the decimal point 3 places to the right.

Model Problems

Write each number in standard form.

Solution

1. 2.52×10^8

$2.52 \times 10^8 = 2.52 \times 100,000,000 = 252,000,000$
The decimal point moves 8 places to the right.

2. The distance from Earth to the sun is 9.3×10^7 miles.

$9.3 \times 10^7 = 9.3 \times 10,000,000 = 93,000,000$
The decimal point moves 7 places to the right.

3. A micron (μ) is equal to 1×10^{-3} of a millimeter.

$1 \times 10^{-3} = 1 \times \dfrac{1}{1,000} = 0.001$

4. 3.27×10^{-5}

$3.27 \times 10^{-5} = 3.27 \times \dfrac{1}{100,000} = 3.27 \times 0.00001 = 0.0000327$

Remember: When small numbers between zero and one are written in scientific notation, the power of 10 is written with a negative exponent.

Multiple-Choice Questions

1. $10^4 =$

    A. 4×10
    B. $10 + 10^3$
    C. $10 \times 10 \times 10 \times 10$
    D. 4×10^{-4}

2. Write 100 as a power of 10.

    F. 1.0×10
    G. 10×10^2
    H. $10 + 10 + 10$
    J. 10^2

3. Write 1,000 as a power of 10.

    A. 1.0×10
    B. $10 + 10 + 10$
    C. 10^3
    D. 0.1×10

4. Write 40,000 using a power of 10.

    F. 4.0×10^4
    G. 10^4
    H. $10 \times 10 \times 10 \times 10$
    J. 4.0×10^5

5. There are approximately 950 million acres of farmland in the United States. Which of these is 950 million written in scientific notation?

    A. 9.5×10^6
    B. 9.5×10^7
    C. 9.5×10^8
    D. 9.5×10^9

6. 2,000 written in scientific notation is

    F. $0.2 \times 1,000$
    G. $1,000^3$
    H. 20×10^3
    J. 2.0×10^3

7. 0.300 written in scientific notation is

    A. 3.0×1^{-4}
    B. 30×10^{-2}
    C. 3.0×10^{-1}
    D. 30^{-2}

8. 0.089 written in scientific notation is

    F. 0.08×9^{-3}
    G. 0.089×10^{-3}
    H. 8.9×10^{-2}
    J. 8.9×100^{-2}

9. 0.8296 written in scientific notation is

    A. 8.2×96
    B. 8.296×10^{-1}
    C. 8296×10^{-4}
    D. 8×296

10. 1,000,000 written in scientific notation is

    F. $1,000,000 \times 10^1$
    G. 10×10^6
    H. 1.0×10^6
    J. 6.0×10^6

For problems 11–15, find the value of n that makes the statement true.

11. $130 = 1.3 \times 10^n$

    A. -2 B. 1
    C. 2 D. 3

12. $8,300 = 8.3 \times 10^n$

    F. -3 G. -2
    H. 2 J. 3

13. $2,589,000 = 2.589 \times 10^n$

    A. -6 B. -3
    C. -1 D. 6

14. $860 = 8.6 \times 10^n$

   F. 1

   G. −2

   H. 2

   J. 3

15. $425,000 = 4.25 \times 10^n$

   A. −1

   B. 1

   C. 5

   D. 6

Short-Response Questions

16. A satellite leaves Earth and travels 5.7×10^7 miles. Express this distance in standard form.

17. To express 185,000 in scientific notation, Joey wrote 18.5×10^4. Was Joey's answer correct? Explain. If you think that Joey's answer was not correct, give the correct notation and explain the error that you think he made.

Extended-Response Question

18. Dr. Zelda designed a spaceship that has 7,321,800 cubic feet of cargo space.

  a. Express the amount of cargo space in scientific notation. Explain in words how you changed the number of cubic feet into scientific notation.

  b. If Dr. Zelda plans to transport 3.5 million energy prisms that each take up 2 cubic feet of cargo space, how much cargo space will remain empty? Show your work.

2.3 Comparing and Ordering

Whole numbers and decimals can be represented by points on a number line.

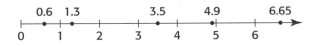

On a number line, the numbers increase from left to right. The symbol > means *is greater than*. The symbol < means *is less than*. The symbols > and < always point to the smaller number.

Examples

1. 4.9 is to the right of 3.5 so 4.9 > 3.5.

2. 0.6 is to the left of 1.3 so 0.6 < 1.3.

To compare two or more whole numbers or decimals without using a number line, align the places and work from left to right. Find the first place where the numbers differ.

Examples

1. 7.286   In the tens place, $8 > 5$ so $7.286 > 7.259$.
7.259

2. 47.3   47.30   When the numbers of decimal
47.38   47.38   places in two decimals are different,
write zeros as needed to compare.
In the hundredths place, $0 < 8$ so
$47.30 < 47.38$. Then $47.3 < 47.38$.

Model Problem

The average speeds (mph) of three cars in a race were 147.350, 150.686, and 144.317. Arrange the speeds from least to greatest.

Compare 147.350 and 150.686.	$14\underline{7}.350 < 15\underline{0}.686$	$4 < 5$
Compare 150.686 and 144.317.	$1\underline{4}4.317 < 1\underline{5}0.686$	$4 < 5$
Compare 147.350 and 144.317.	$14\underline{4}.317 < 14\underline{7}.350$	$4 < 7$

In order from least to greatest, the speeds are $144.317 < 147.350 < 150.686$.

Practice

Multiple-Choice Questions

1. Which comparison is NOT true?

A. $230 < 302$
B. $7,887 > 4,787$
C. $6,103 < 6,013$
D. $5,296 > 5,278$

2. Which group of numbers is in order from least to greatest?

F. 5,648; 5,864; 5,486; 4,847
G. 4,847; 5,486; 5,648; 5,864
H. 5,864; 5,648; 5,486; 4,847
J. 4,847; 5,648; 5,486; 5,864

3. Which number is the greatest?

A. 35,740
B. 35,470
C. 34,870
D. 35,047

4. What number is most likely represented by point x?

F. 11.25
G. 14
H. 13.5
J. 14.2

5. On a number line, which number would be to the left of 7.46?

   A. 7.5
   B. 8.02
   C. 7.39
   D. 9.11

6. A decimal with the same value as 0.59 is

   F. 0.509
   G. 0.059
   H. 0.5090
   J. 0.590

7. The decimal with the greatest value is

   A. 0.303
   B. 0.0333
   C. 0.33
   D. 0.03303

8. Which of the decimals is NOT between 0.121 and 0.123?

   F. 0.122
   G. 0.1225
   H. 0.120
   J. 0.1219

9. Which of the numbers is farthest to the right on a number line?

   A. 2,001
   B. 2,102
   C. 2,012
   D. 2,120

10. Which statement says that 17 is between 15 and 24?

   F. 17 > 15 < 24
   G. 15 < 17 < 24
   H. 15 > 17 > 24
   J. 15 < 24 > 17

11. Which group of numbers is in order from least to greatest?

   A. 6.8, 0.068, 0.68, 0.608
   B. 0.068, 0.608, 0.68, 6.8
   C. 0.608, 0.068, 0.68, 6.8
   D. 0.68, 0.608, 0.068, 6.8

12. How many whole numbers are between 3,999 and 4,011 (not including 3,999 and 4,011)?

   F. an infinite number
   G. 10
   H. 11
   J. 12

Short-Response Questions

13. Each letter on the number line shown represents a number. Replace each ◉ by < or > to write a true statement.

   a. A ◉ Y
   b. D ◉ C
   c. N ◉ Y
   d. D ◉ N ◉ C
   e. C ◉ A ◉ Y
   f. Y ◉ D ◉ A

14. The table shows the average price of a gallon of self-service regular gasoline on the same day in several cities. Put the cities in order of price from greatest to least.

City	Price
Boston	$1.11
Chicago	$1.21
Denver	$1.07
Houston	$1.06
Los Angeles	$1.52
Miami	$1.14
New York	$1.29
Seattle	$1.41

2.4 Rounding

A **rounded number** is an approximation of an exact whole number or decimal. Rounded numbers are often used for making estimates.

To round a whole number or decimal to a given place

1. Underline the place to be rounded.
2. Look at the digit to the immediate right of the underlined digit.
   - If the digit to the right is 5 or more, add 1 to the underlined digit.
   - If the digit to the right is less than 5, leave the underlined digit unchanged.
3. For whole numbers, replace all the digits to the right of the underlined digit with zeros.

   For decimals, drop all digits to the right of the underlined digit.

4. All digits to the left of the underlined digit remain unchanged.

1. Round 837,654 to the nearest ten thousand.

   Solution   The digit to be rounded is in the ten thousands place. The digit to the right is 7.

   8<u>3</u>7,654

   Since 7 > 5, add 1 to the underlined digit and replace all the digits to the right with zeros.

   Answer   8<u>4</u>0,000

2. Round 3.1415927 to the nearest
   a. tenth   b. thousandth

   Solution
   a. Underline the tenths place.

   3.<u>1</u>415927

   The digit to the right is 4. Since 4 < 5, leave the tenths digit unchanged and drop all digits to the right.

   Answer   3.1

b. Underline the thousandths place.

3.14**1**5927

The digit to the right is 5. Add 1 to the thousandths place and drop the digits to the right.

Answer   3.142

Multiple-Choice Questions

1. In 2000, the Sun Foods Company had total sales of $57,468,000. Rounded to the nearest million, total sales were

   A. $60,000,000
   B. $57,000,000
   C. $58,000,000
   D. $57,500,000

2. The area of New York State is 54,471 square miles. Round the area to the nearest hundred square miles.

   F. 50,000
   G. 54,000
   H. 54,400
   J. 54,500

3. To the nearest hundred, which number does NOT round to 6,000?

   A. 6,030
   B. 5,980
   C. 5,940
   D. 6,009

4. 2,222,222 rounded to the nearest hundred thousand is

   F. 2,000,000
   G. 2,200,000
   H. 2,220,000
   J. 2,300,000

5. 39,998 rounded to the nearest ten is

   A. 40,000
   B. 39,990
   C. 39,900
   D. 39,999

6. Mrs. Chang spent $143.78 on groceries. Round her bill to the nearest ten dollars.

   F. $144.00
   G. $143.80
   H. $150.00
   J. $140.00

7. The length of a hiking trail is 26.184 kilometers. Round the length to the nearest tenth of a kilometer.

   A. 30 km
   B. 26.2 km
   C. 26.18 km
   D. 26 km

8. A calculator displayed a player's batting average as 0.2508519. Round the batting average to the nearest thousandth.

   F. 0.251
   G. 0.200
   H. 0.300
   J. 0.250

9. 0.536 rounded to the nearest one is

   A. 1.0
   B. 0
   C. 0.5
   D. 0.54

10. 376.2945 rounded to the nearest hundredth is

   F. 400
   G. 400.3
   H. 376.29
   J. 376.3

11. When rounded to the nearest tenth, which number will NOT be 0.8?

   A. 0.777
   B. 0.8319
   C. 0.808
   D. 0.7495

12. At the hardware store, Mark bought items costing $6.35, $8.95, $3.17, and $4.49. To estimate his total, Mark rounded each amount to the nearest dollar and added. What was his estimate?

   F. $421
   G. $22
   H. $23
   J. $24

Short-Response Questions

13. The U.S. population reported from the 2000 census was 281,421,906. Round the population to the nearest

   a. hundred million
   b. ten million
   c. million
   d. hundred thousand
   e. thousand

14. Use the clues to find the number.

- If you round it to the nearest one, you get 7.

- If you round it to the nearest tenth, you get 7.0.

- If you round it to the nearest hundredth, you get 7.00.

- It is the least number that fits the clues.

2.5 Properties

Addition and multiplication have these special properties for all numbers a, b, and c.

Property	Addition	Multiplication
Commutative Property	The order in which you add two numbers does not change the sum. $9 + 5 = 5 + 9$ $a + b = b + a$	The order in which you multiply two numbers does not change the product. $8 \times 6 = 6 \times 8$ $a \times b = b \times a$
Associative Property	The way you group numbers does not change the sum. $(7 + 3) + 4 = 7 + (3 + 4)$ $(a + b) + c = a + (b + c)$	The way you group numbers does not change the product. $(2 \times 3) \times 7 = 2 \times (3 \times 7)$ $(a \times b) \times c = a \times (b \times c)$
Identity Property	If you add 0 to any number, that number remains the same. $34 + 0 = 0 + 34 = 34$ $a + 0 = 0 + a = a$	If you multiply any number by 1, that number remains the same. $78 \times 1 = 1 \times 78 = 78$ $a \times 1 = 1 \times a = a$
Zero Property		If you multiply any number by 0, the product is 0. $526 \times 0 = 0 \times 526 = 0$ $a \times 0 = 0 \times a = 0$
Distributive Property of Multiplication Over Addition or Subtraction	When you *distribute* multiplication over addition or subtraction, the result is the same whether you multiply first and then add (or subtract) or if you add (or subtract) first and then multiply. $2(3 + 5) = (2 \times 3) + (2 \times 5)$ $\quad a(b + c) = (a \times b) + (a \times c)$ $9(6 - 4) = (9 \times 6) - (9 \times 4)$ $\quad a(b - c) = (a \times b) - (a \times c)$	

Model Problems

1. Find the missing number. Name the property used.
   a. $? + 37 = 37$
   b. $(4 \times ?) \times 8 = 4 \times (9 \times 8)$

Solution
   a. Use the identity property of addition. The missing number is 0 since $0 + 37 = 37$.
   b. Use the associative property of multiplication. The missing number is 9 since $(4 \times 9) \times 8 = 4 \times (9 \times 8)$

2. The seats in a theater are divided into two sections. Section A has 15 rows of seats and Section B has 13 rows. There are 10 seats in each row. How many seats are in the theater? What property is illustrated?

Solution   There are two ways to find the total.

Method I   Multiply to find the seats in each section, then add.

Section A $10 \times 15 = 150$

Section B $10 \times 13 = 130$

Total Seats $(10 \times 15) + (10 \times 13) = 150 + 130 = 280$

Method II   Add to find the total number of rows, and then multiply by the number of seats per row.

Rows in A and B $15 + 13$

Total Seats $10(15 + 13) = 10 \times 28 = 280$

The problem illustrates that multiplication is distributive over addition since $10(15 + 13) = (10 \times 15) + (10 \times 13)$.

Practice

Multiple-Choice Questions

1. Which sentence illustrates the commutative property of addition?

A. $9 + 6 = 8 + 7$
B. $13 + 0 = 13$
C. $14 + 5 = 5 + 14$
D. $10 + (20 + 30) = 30 + 30$

2. If $87 \times M = 0$, the value of M is

F. 0
G. 1
H. -87
J. $\frac{1}{87}$

3. $5 \times 29 =$

A. $5 \times 20 \times 9$
B. $(5 + 20) \times (5 + 9)$
C. $(5 \times 20) + 9$
D. $(5 \times 20) + (5 \times 9)$

4. Which number makes the sentence true?
$(13 \times 6) \times \, ? = 13 \times (6 \times 17)$

F. 1,326
G. 19
H. 17
J. 78

5. The sentence $0 \times 41 = 0$ illustrates which property?

A. Identity property of multiplication
B. Identity property of addition
C. Distributive property of multiplication over addition
D. Zero property of multiplication

Short-Response Questions

For problems 6–10, write *true* or *false*.

6. When a number is added to 0, the sum is 0.

7. Zero multiplied by a number greater than 0 is 0.

8. $5 - 3 = 3 - 5$

9. Any number divided by 0 is that number.

10. If the product of two numbers is 0, at least one of the numbers must be 0.

For problems 11–15, find the missing number that makes each sentence true. Name the property shown.

11. $32 \times 57 = ? \times 32$

12. $79 \times 0 = ?$

13. $(21 \times 8) \times 14 = ? \times (8 \times 14)$

14. $5 \times 76 = (5 \times 70) + (5 \times ?)$

15. $83 \times ? = 83$

16. Find each product by distributing multiplication.

    a. 4×63
    b. 10×98

Extended-Response Question

17. Mia is the highest paid baby-sitter in her neighborhood. In one weekend, she earned three times the total amount that her two brothers earned baby-sitting.

    a. If one brother earned $15 and the other brother earned $12, show two ways to find Mia's total. Name the property illustrated.
    b. If Mia and her brothers continue to earn the same amounts of money for three more weekends, what will be the total difference between the amount that she earns in four weeks and the amount her brothers earn in four weeks? Show your work.

2.6 Estimation and Addition

Addition is used to find the total or sum of a group of numbers. Making an **estimate** first can help you confirm the answer to a given problem. There are several estimation methods.

To add whole numbers and decimals

1. Arrange the numbers in columns by place value. For decimals, be sure the decimal points are lined up.

2. Add the numbers in each column, from right to left, regrouping if necessary.

 **Model Problems**

1. Find the sum of 2,468 + 279 + 23.

   Solution Estimate first. For this problem, a good estimate can be found by rounding each number to the nearest hundred.

   $$2,468 \rightarrow 2,500$$
   $$279 \rightarrow 300$$
   $$+ \quad 23 \rightarrow \underline{\quad 0}$$
   $$2,800 \quad \text{Estimated sum}$$

   Another way to estimate is to use the first or front digits of the numbers. $2,000 + 200 + 20 = 2,220$

   Adding the actual numbers, we get

   $$2,468$$
   $$279$$
   $$+ \quad 23$$
   $$\overline{2,770}$$

   When you compare, you see that rounding gives a better estimate than using the first digits. The exact answer is closer to the first estimate.

2. Find the sum of 2.43 + 36 + 13.8 + 0.003.

   Solution Estimate. Round each number to the nearest ten.

   $$0 + 40 + 10 + 0 = 50$$

   To get a closer estimate, round the numbers to the nearest one.

   $$2 + 36 + 14 + 0 = 52$$

   Add to find the exact answer.

Arrange the numbers in columns with all the decimal points aligned. Place zeros in the tenths, hundredths, and thousandths place for the numbers 2.43, 36, and 13.8.

```
   2.430
  36.000
  13.800
+  0.003
  52.233
```

Remember to place the decimal point in the sum. The exact answer is close to the estimate, so it is reasonable.

Practice

Multiple-Choice Questions

1. The best estimate of 3,418 + 4,576 + 1,861 + 920 is

   A. 3,000 + 4,000 + 1,000
   B. 3,000 + 5,000 + 2,000 + 1,000
   C. 3,000 + 4,000 + 2,000 + 1,000
   D. 4,000 + 5,000 + 1,000 + 1,000

2. Susan went to the store and bought items costing $3.39, $2.68, $1.29, $0.82, and $6.79. She estimated the total by rounding each price to the nearest dollar and adding. Which group of numbers did she add?

   F. $3.00 + $2.00 + 1.00 + $0 + $6.00
   G. $3.00 + $3.00 + $2.00 + $1.00 + $6.00
   H. $3.00 + $3.00 + $1.00 + $1.00 + $7.00
   J. $4.00 + $3.00 + $2.00 + $1.00 + $7.00

3. Find the sum: 319 + 7 + 610 + 2,305.

   A. 3,241
   B. 2,221
   C. 2,755
   D. 3,934

4. In May, Mr. Koch's airline mileage statement showed the following:

New York/Dallas	1,525 miles
Dallas/Los Angeles	1,401 miles
San Francisco/New York	2,886 miles

   What was his total mileage?

   F. 4,706 miles
   G. 4,922 miles
   H. 5,812 miles
   J. 6,806 miles

5. Find the sum of the numbers below. 826.95, 1,439.14, 29.2

   A. 2,295.29
   B. 2,264.01
   C. 2,558.09
   D. 2,266.382

6. Find the sum of the numbers below. 4.6, 62.34, 0.293

   F. 401.34
   G. 65.73
   H. 6573
   J. 67.233

7. Find the sum of the numbers below.
13.2, 0.132, 0.00132

   A. 13.464   B. 26.53200
   C. 13.33332   D. 39.60000

8. Vicky walked 3.1 km on Monday, 4.86 km on Tuesday, and 4.25 km on Wednesday. Find the total distance she walked.

   F. 12.12 km   G. 12.21 km
   H. 11.11 km   J. 13.01 km

9. The sum closest to 26.53 + 8.29 is

   A. 35.0   B. 34.8
   C. 34.7   D. 34.0

10. Ralph and Calvin work part-time at a department store. Last week, Ralph earned $97.85. Calvin earned $28.55 more than Ralph did. The best estimate of how much Calvin earned is

   F. $130   G. $69
   H. $70   J. $127

Short-Response Questions

Place a decimal point in each addend to make the sentences true.

11. 6027 + 1845 = 7.872

12. 3649 + 1621 + 5591 = 25.45

13. 7863 + 5409 + 1145 = 73.403

14. 2431 + 1358 + 3366 + 5904 = 65.232

Extended-Response Questions

15. Halley's comet is visible on Earth approximately once every 76 years. It was last seen in 1986. List the years for the next six appearances. Explain the process you used to make your list.

16. Rounded to the nearest whole dollar, Steven has $135 in his savings account and $48 in his wallet. Find the greatest and least possible amounts of money that Steven could have in all. Show all work.

2.7 Estimation and Subtraction

Subtraction is used to find how much greater or less one number is than another or how much is left after an amount is spent. The result of a subtraction is called the **difference**.

To subtract whole numbers or decimals

1. Arrange the numbers in columns by place value. To align the decimal points, you may want to put zeros in any empty places. Remember that the "from" number goes on top.

2. Subtract the numbers in each column, from right to left, regrouping if necessary.

1. Subtract 359 from 7,547.

Solution
Estimate first. A good estimate is 7,500 – 400 = 7,100
Line up the numbers and subtract.

7,547 → The "from" number is on top.
– 359
7,188

Subtraction can be checked by addition.

7,188
+ 359
7,547

2. Find the difference: 6.314 – 1.29.

Solution
Estimate using front digits: 6 – 1 = 5
Line up the decimal points and subtract.

6.314
– 1.290 ← Write a zero.
5.024 ← Place the decimal point in the answer.

Check

5.024
+ 1.290
6.314

3. Peter spent $16.83 at the bookstore and $7.91 at the drugstore. How much money does he have left from $50.00?

Solution
Add to find total spent.

$16.83
+   7.91
$24.74

Subtract to find how much is left.   *Check*

$50.00 $25.26
– 24.74   + 24.74
$25.26 $50.00

Practice

Multiple-Choice Questions

1. When two numbers are added their sum is 1,513. If the first number is 998, what is the second number?

   A. 2,511
   B. 998
   C. 515
   D. 1,513

2. The difference between 128,148 and 58,668 is

   F. 69,480
   G. 186,816
   H. 58,668
   J. 6,948

3. Which expression gives the closest estimate for 5,609 − 2,894?

   A. 5,600 − 2,900
   B. 6,000 − 2,000
   C. 5,600 − 2,000
   D. 5,000 − 2,800

4. The area of Cuba is 42,804 square miles. The area of Puerto Rico is 3,339 square miles. How much larger is Cuba than Puerto Rico?

   F. 9,414 square miles
   G. 38,575 square miles
   H. 39,465 square miles
   J. 46,143 square miles

5. The average annual precipitation in Singapore is 84.6 inches. The average annual precipitation in London, England, is 29.7 inches. How many more inches of precipitation does Singapore get than London?

   A. 65.8
   B. 54.9
   C. 43.9
   D. 114.3

6. Haines, Daines, and Raines ran for mayor. Haines received 8,316 votes, and Daines received 7,495 votes. If 20,000 votes were cast, how many did Raines receive?

   F. 821
   G. 10,694
   H. 5,209
   J. 4,189

7. Mr. Leung had a balance of $365.57 in his checking account. He wrote a check for $96.80. How much money was left in his account?

   A. $279.97
   B. $268.77
   C. $266.97
   D. $168.77

8. Subtract 0.987 from 2.87.

   F. 0.700
   G. 7.00
   H. 1.97
   J. 1.883

9. How much greater than 37.58 is 51.04?

   A. 23.48
   B. 13.46
   C. 14.56
   D. 88.62

10. A sack of flour weighed 22.5 pounds. After the baker used some flour, 14.85 pounds were left. How many pounds of flour did the baker use?

    F. 8.75
    G. 8.4
    H. 7.65
    J. 37.35

Short-Response Questions

For problems 11–14, find the missing addends.

11. 467 + 328 + [ ] = 1,003

12. 1,792 + [ ] + 2,815 + 694 = 7,602

13. 1.06 + 0.582 + [ ] = 4.2

14. 0.985 + 13.4 + [ ] + 5.76 = 21.02

15. Linda bought some ground beef for $3.66, rolls for $1.59, and a container of juice for $2.87. How much change did she receive from a $10 bill?

16. The table shows the populations of two cities in 1990 and 2000. Which city had the greater population increase? How much greater?

Population		
	1990	2000
Thorton	122,859	140,692
Sheffield	130,076	147,305

Extended-Response Questions

17. In 1911, the winning car at the Indianapolis 500 averaged 74.602 mph. Fifty years later, in 1961, the winning car averaged 139.130 mph.
    a. By how many miles per hour did the average speed increase?
    b. If the average speed of the winning car improves the same amount over the fifty years after 1961, how fast will it be in 2011?
    Show your work.

18. a. The U.S. Treasury Department was created by an act of Congress in 1789. The Department of Transportation was created in 1966. How many years apart were the two departments created?
    b. The U.S. Treasury Department first issued the half-cent coin in 1793 and last issued it in 1857. For how many years was the half-cent issued?
    c. How would the value of a half-cent be written as a decimal?
    Show your work.

2.8 Estimation and Multiplication

Multiplication is a shortcut for repeated addition. The numbers that are multiplied are called **factors**. The answer to a multiplication problem is called the **product**.

To multiply whole numbers

1. To multiply by a one-digit number, begin at the ones place. Multiply each digit in the top number by the one-digit number on the bottom.

2. To multiply by a number with two or more digits, begin at the ones place. Multiply each digit in the top number by each digit in the bottom number to find **partial products**. Then add the partial products.

Examples

1.   272
        $\times$ 4   Estimate to check. Round each factor. $300 \times 4 = 1{,}200$.
      1,088

2.   236
        $\times$ 62
          472 ← 236 $\times$ 2   Estimate to check. Round each factor.
      $+$ 14,160 ← 236 $\times$ 60 $200 \times 60 = 12{,}000$
        14,632

To multiply a decimal by a whole number or another decimal

1. Multiply as if both numbers were whole numbers.

2. The number of decimal places in the product is equal to the sum of the number of decimal places in the two numbers. (A whole number has 0 decimal places.) Count from right to left to place the decimal point in the product.

Examples

3.   1.73 ← 2 decimal places
        $\times$ 29
        1557
      $+$ 3460
        50.17 ← 2 decimal places

Estimate to check the placement of the decimal point. Round both numbers so you can multiply mentally.

$2 \times 30 = 60$ The product is reasonable.

4.   4.16 ← 2 decimal places
        $\times$ 2.8 ← 1 decimal place
        3328
      $+$ 8320
      11.648 ← $2 + 1 = 3$ decimal places. Estimate to check: $4 \times 3 = 12$.

To multiply a decimal by a power of 10

1. Count the number of zeros in the power of 10.

2. Move the decimal point to the right the same number of places as there are zeros in the power of 10. Write extra zeros to the right if there are not enough places to move the decimal point.

Examples

5. $4.35 \times 100 = [4.35] = 435.0$

There are 2 zeros, so you move the decimal point 2 places right.

6. $0.29 \times 10,000 = [0.2900] = 2900.0$

There are 4 zeros, so you move the decimal point 4 places right.

Multiple-Choice Questions

1. Find the product of 507 and 86.

   A. 4,902
   B. 7,098
   C. 42,142
   D. 43,602

2. A market ordered 150 cases of soda. If each case contains 24 cans, how many cans are there?

   F. 360
   G. 900
   H. 3,600
   J. 9,000

3. Which expression gives the best estimate of 47×92?

   A. 40×90
   B. 50×90
   C. 50×100
   D. 40×100

4. The best estimate of $2.869 \times 5,017$ is

   F. less than 10,000
   G. more than 10,000
   H. more than 15,000
   J. more than 20,000

5. Multiply 6.7×9.8.

   A. 11.39
   B. 63.56
   C. 65.66
   D. 68.36

6. Nina is buying a barbecued chicken that weighs 2.65 pounds. If the price of the chicken is $2.89 per pound, what should Nina pay?

   F. $7.66
   G. $9.70
   H. $7.82
   J. $5.74

7. $0.13 \times 1{,}000{,}000 =$

    A. 1,300
    B. 13,000
    C. 130,000
    D. 1,300,000

8. Find the product of 0.058 and 0.09.

    F. 0.000522
    G. 0.00522
    H. 0.0522
    J. 0.522

9. Which product is the greatest?

    A. 63×58
    B. 39×97
    C. 76×44
    D. 82×29

10. Mr. Stone's car averages 22.3 miles per gallon. How many miles can he travel on 12.5 gallons of gasoline?

    F. 178.4
    G. 253.25
    H. 267.6
    J. 278.75

Short-Response Questions

Compare the products. Replace ( ) with <, >, or =.

11. $19 \times 38 \times 42$ ( ) $1{,}265 \times 23$

12. $116 \times 37 \times 55$ ( ) $3{,}204 \times 78$

Place the decimal point in the product.

13. $4.67 \times 8.56 = 399752$

14. $12.9 \times 6.32 = 81528$

15. $0.35 \times 18.2 = 637$

16. $10.418 \times 5.73 = 5969514$

Extended-Response Questions

Solve and show your work.

17. The cost per mile for a ticket on the Metroliner is 41¢.
    a. How much does a passenger pay to travel the 226-mile distance from New York to Washington, DC? Round to the nearest dollar.
    b. How much more will the same trip cost after a 0.6¢ per mile fare hike?

18. Luis earns $10.20 per hour and time and one-half for each hour that he works overtime. Last week, Luis worked 35 hours and 8 hours overtime. How much money did he earn?

19. A certain metal alloy is made of copper, iron, and an unknown element. The alloy is 0.18 part copper and 0.35 part iron.
    a. How many kilograms of copper and how many kilograms of iron are in 840 kilograms of the alloy?
    b. What is the weight of the unknown element in 840 kilograms of the alloy?

20. On a mathematics test, each correct answer in Part A earned 1.5 points and each correct answer in Part B earned 3.5 points. Find each student's score.

Student	Correct Answers		Score
	Part A	Part B	
Kate	16	14	
Victor	12	18	

2.9 Division of Whole Numbers

Division is used to determine how many times one number is contained in another. Division can be written in the following three ways:

$$32 \div 4 \quad \text{or} \quad 4\overline{)32} \quad \text{or} \quad \frac{32}{4}.$$

In all three cases the problem is read as *32 divided by 4*. To find out how many 4's are in 32, you would divide 32 by 4 to get an answer of 8.

quotient

divisor → 4$\overline{)32}$ ← **dividend**

8

Example

21$\overline{)12,345}$

To determine how many digits will be in your answer before you begin a division, do the following:

a. Cover the dividend with your index finger.

b. Uncover one digit at a time.

c. Ask yourself: Does 21 (the divisor) go into 1? If no, uncover another digit.

d. Does 21 go into 12? No. Uncover another digit.

e. Does 21 go into 123? Yes!

f. Put a dash over 3 and over each digit that follows it. The number of dashes tells you how many digits will be in your answer. This example will have a three-digit answer.

21$\overline{)12,\overline{\overline{\overline{345}}}}$

Division is a repeated process that involves these steps:

1. divide
2. multiply
3. subtract
4. bring down

In the example above, since you made three dashes in the quotient, you will need to repeat the division process three times, going back to "divide" after each "bring down." For each time you do the four-step process, you *must* put a digit on the next blank dash.

To complete the example:

$$\begin{array}{r} 587 \\ 21{\overline{)12{,}345}} \\ -10\ 5 \\ \hline 1\ 84 \\ -1\ 68 \\ \hline 165 \\ -147 \\ \hline 18 \end{array}$$

The final answer may be written as 587 r 18 or as $587\frac{18}{21}$.

Division Strategy

Getting started on a division problem can be difficult. How many times does 21 go into 123? If you cannot think of the answer, try the guess and check strategy. Just pick any number and multiply. $2 \times 21 = 42$. That's too small. $4 \times 21 = 84$. That's still too small. Keep guessing and checking until you get close to the product you want without going over. $5 \times 21 = 105$. That's close. $6 \times 21 = 126$. That's too big. 21 goes into 123 five times.

 **Model Problems**

1. Divide 458 by 7.

   Solution
   Cover the dividend, 458, with your index finger. Uncover the first digit. Does 7 go into 4? No. Uncover another digit.
   Does 7 go into 45? Yes. Put a dash over the 5 and over the digit that follows it, 8. The two dashes tell you that the answer must have two digits.

   Step 1   Divide. How many times does 7 go into 45?
   Step 2   Multiply. $6 \times 7 = 42$.
   Step 3   Subtract. $45 - 42 = 3$.
   Step 4   Bring down the 8.

   Go back to Step 1 and repeat the process to get the second digit of your quotient.

$$\begin{array}{r} 65 \text{ r } 3 \\ 7{\overline{)458}} \\ \underline{-42}\downarrow \\ 38 \\ \underline{-35} \\ 3 \end{array}$$

Check division with multiplication.

$$\begin{array}{r} 65 \\ \times\ 7 \\ \hline 455 \\ +\ \ 3 \\ \hline 458 \end{array}$$

Answer   65 r 3 or $65\frac{3}{7}$

2. Divide 7,751 by 37.

Solution

Estimate: $8,000 \div 40 = 200$.

$$\begin{array}{r} 209 \text{ r } 18 \\ 37{\overline{)7,751}} \\ \underline{-7\ 4}\downarrow\downarrow \\ 351 \\ \underline{-333} \\ 18 \end{array}$$

How many times does 37 divide 77? **2**
Multiply, subtract, and bring down the 5.
Since 37 does not divide 35, write **0** in the quotient. Bring down the 1.
How many times does 37 divide 351? Try **9**. Multiply. Subtract.
18 is less than 37. Write the remainder.

Answer   209 r 18 or $209\frac{18}{37}$

3. Divide 2,570 by 52.

Solution

Estimate: $2,500 \div 50 = 50$

$$\begin{array}{r} 49 \text{ r } 22 \\ 52{\overline{)2,570}} \\ -2\ 08\downarrow \longleftarrow 4 \times 52 \\ 490 \\ -468 \longleftarrow 9 \times 52 \\ 22 \end{array}$$

The answer must have two digits.
How many times does 52 divide 257? Try 5. $5 \times 52 = 260$. Since $260 >$
257, the first guess is too large. Try 4. Continue the division process.

Answer   49 r 22 or $49\frac{22}{52}$

Division Involving Zero

1. When zero is divided by any nonzero number, the quotient is zero.
   $0 \div 538 = 0$

2. Dividing a number by zero is not possible!
   Since division is the inverse of multiplication, $4 \div 0 = ?$ means that $? \times 0 = 4$.

There is no number that can replace **?** since any number $\times 0 = 0$.

Practice

Multiple-Choice Questions

1. Which expression gives the best estimate of $73,625 \div 81$?

   A. $80,000 \div 80$ B. $74,000 \div 80$
   C. $72,000 \div 90$ D. $81,000 \div 90$

2. Estimate $23,914 \div 62$.

   F. less than 400
   G. more than 400
   H. more than 500
   J. more than 600

3. Which problem has a remainder of 3?

   A. $184 \div 7$ B. $487 \div 9$
   C. $291 \div 8$ D. $269 \div 6$

4. $19\overline{)4,420}$ =

   F. 21 r 5   G. 20 r 4
   H. 232 r 12   J. 252 r 10

5. Which of the following could NOT be the quotient when a number is divided by 17?

   A. 39 r 14   B. 205
   C. 670 r 11   D. 981 r 19

6. What is the quotient of 9,024 and 48?

   F. 188   G. 189 r 2
   H. 202   J. 204 r 6

7. A distributor needs to ship 620 books. Each carton holds, at most, 16 books. What is the least number of cartons that must be shipped?

   A. 38   B. 39
   C. 40   D. 42

8. $9,125 \div 200$ =

   F. 41 r 25   G. 45 r 125
   H. 408 r 20   J. 450 r 5

9. What number should replace the ? to make a true statement?
   $17 \times ? = 561$

   A. 31   B. 33
   C. 544   D. 9,537

10. For which division is it NOT possible to find a quotient?

    F. $\dfrac{0}{21}$ G. $0 \div 18$
    H. $109\overline{)0}$ J. $0\overline{)42}$

Short-Response Questions

Divide and check.

11. $721 \div 47$
12. $8\overline{)2456}$

13. $5,022 \div 33 =$

14. $16,940 \div 97$

Supply the missing numbers.

15. $4 \times ? = 572$

16. $29 \times ? = 1,363$

17. $? \times 53 = 15,794$

18. $106 \times ? = 8,586$

Extended-Response Questions

Solve. Show all work.

19. James is planning a trip from New York City to Chicago, Illinois, a distance of 792 miles. His car gets an average of 24 miles to the gallon. At $1.69 per gallon for gas, how much should James budget for gas for the round-trip? Round to the nearest dollar.

20. Complete the table. To find the output:
    • Divide the input by 23.
    • Add 19

Input	Output
253	
529	
874	
1,472	

21. A restaurant owner paid $336 for one dozen folding chairs. How much would the owner have paid for 20 chairs?

22. The North Park School is planning a trip to the planetarium. There will be 774 people going on the trip. Each bus can seat 45 people.

    a. How many buses are needed?
    b. How many empty seats will there be?
    c. Suppose the trip leaders want there to be the same number of people on each bus. How many people will be on each one?

2.10 Division of Decimals

To divide a decimal by a whole number

• Place the decimal point in the quotient directly above the decimal point of the dividend. Then divide as with whole numbers. If necessary, use zeros as placeholders in the quotient.

To divide by a decimal

• Move the decimal point in the divisor right to make a whole number. Then move the decimal point in the dividend the same number of places to the right. Write zeros as needed. Write a decimal point in the quotient and divide as with whole numbers.

1. Dan received a paycheck for $95.25 one week. If he worked 15 hours, how much did he earn per hour?

   Solution

   ```
          6.35
   15)95.25
     - 90
        5 2
      - 4 5
         75
       - 75
          0
   ```

   Place the decimal point. 15 does not divide 9.

   The quotient starts in the ones place. Divide as with whole numbers.

   Answer He earned $6.35 per hour.

2. Divide 1.08 by 12.

   Solution

   ```
         0.09
   12)1.08
    - 1 08
          0
   ```

   12 does not divide 1 or 10.

   Write zeros in the quotient. Divide.

   Answer 0.09 is the quotient.

3. Divide 1.134 by 0.18.

   Solution

   0.18)1.134   Move each decimal point 2 places right.

   Place the decimal point and divide.

   ```
          6.3
   18)113.4
    - 108
        5 4
      - 5 4
         0
   ```

   Answer The quotient is 6.3.

4. Divide 62 by 0.248.

Solution

Move each decimal point 3 places right. Insert 3 zeros in the dividend.

$$0.248\overline{)62.000}$$

Complete the division.

```
         250.
  248)62,000.
    - 49 6
      12 40
    - 12 40
          0
```

Answer The quotient is 250.

To divide a decimal by a power of 10

- Count the number of zeros in the power of 10.
- Move the decimal point to the left the same number of places as there are zeros in the power of 10. Write extra zeros to the left if necessary.

Note: The movement of the decimal point for division is in the opposite direction as for multiplication.

Examples

1. $0.29 \div 100 = 0.0029$

100 has 2 zeros, so move the decimal point 2 places left.

2. $1.365 \div 1,000 = 0.001365$

1,000 has 3 zeros, so move the decimal point 3 places left.

Practice

Multiple-Choice Questions

1. $8\overline{)48.16}$

   A. 62   B. 6.2
   C. 6.02   D. 0.62

2. $6.4 \div 25$

   F. 0.0256   G. 0.256
   H. 2.56   J. 25.6

3. Lisa and her two brothers bought their mother a tote bag that cost $34.95. If they shared the cost equally, how much did each child pay?

   A. $11.65   B. $11.31
   C. $10.59   D. $10.13

4. Rounded to the nearest tenth, $63 \div 1.3 =$

   F. 4.85   G. 5.00
   H. 48.5   J. 48.6

5. $27.1 \div 1{,}000 =$

   A. 2.71   B. 0.271
   C. 0.0271   D. 0.00271

6. Find the missing number.
$? \div 3.7 = 2.5$

   F. 1.48   G. 0.6756
   H. 1.2   J. 9.25

7. Divide 2.628 by 1.2.

   A. 21.9   B. 2.19
   C. 2.09   D. 0.0219

8. A package of paper is 1.464 inches thick. If each sheet of paper is 0.003 inch thick, how many sheets are in the package?

   F. 48   G. 4,880
   H. 488   J. 0.488

9. How many pieces measuring 0.55 meter each can be cut from a roll with 50.6 meters of wire?

   A. 9.02   B. 9200.0
   C. 902.0   D. 92.0

10. A row of 12 tiles is 115.2 inches long. How long would a row of 7 tiles be?

   F. 182.4 inches   G. 67.2 inches
   H. 54.6 inches   J. 16.45 inches

Short-Response Questions

Find each quotient.

11. $2.5\overline{)52.25}$

12. $2.43 \div 0.75$

13. $21 \div 0.15$

14. $38\overline{)275.88}$

Find each missing number.

15. $0.9 \div 0.08 = 90 \div\ ?$

16. $2.53 \div 0.11 = 2{,}530 \div\ ?$

17. $? \div 0.02 = 76.8 \div 2$

18. $47.5 \div\ ? = 0.475 \div 0.05$

Extended-Response Questions

Solve. Show all work.

19. There were 925 guests at a charity barbecue. The total cost for food was $3,496.50. How much did food cost per person?

20. Jason drove his car 530.4 miles and used 19.5 gallons of gasoline. What was his mileage per gallon?

21. One thousand pencils cost $72. At the same rate, what is the cost of 550 pencils?

22. Suppose m is a whole number. Write $<$ or $>$ for ( ). Explain how you arrived at your answer.

   a. If $m \div 0.6 = n$, then n ( ) m.
   b. If $m \times 0.06 = n$, then n ( ) m.

Multiple-Choice Questions

1. What is the decimal form of:
   $(6 \times 10{,}000) + (3 \times 100) +$
   $\left(5 \times \dfrac{1}{10}\right) + \left(7 \times \dfrac{1}{1{,}000}\right)$?

   A. 63,750
   B. 63,000.57
   C. 60,300.507
   D. 6,003.507

2. Write the number 409,200 in scientific notation.

   F. 40.92×10^4
   G. $4.092 \times 1{,}000$
   H. 4.092×10^{-5}
   J. 4.092×10^5

3. Write in standard form:
   7.15×10^{-4}

   A. 0.0000715
   B. 0.000715
   C. 71.5000
   D. 71,500

4. Which group of numbers is in order from least to greatest?

   F. 2.617, 26.17, 21.6, 261.7
   G. 26.17, 21.6, 261.7, 2.617
   H. 2.617, 21.6, 26.17, 261.7
   J. 21.6, 261.7, 26.17, 2.617

5. Round 254.3168 to the nearest hundredth.

   A. 300   B. 254.32
   C. 254.317   D. 250.4

6. Doreen bought items at the store costing $3.27, $22.65, $11.49, and $8.08. She estimated the total cost by rounding to the nearest dollar. Her estimate was

   F. $45.00   G. $44.00
   H. $47.00   J. $40.00

7. Which sentence illustrates the distributive property?

   A. $(12 \times 29) \times 34 = 12 \times (29 \times 34)$
   B. $562 + 879 = 879 + 562$
   C. $4 \times 57 = (4 \times 50) + (4 \times 7)$
   D. $14 \times 38 \times 0 = 0$

8. $14.58 \times 0.25 =$
   F. 3.645   G. 10.206
   H. 36.45   J. 29.889

9. $2.09 - 0.786 =$

   A. 2.214   B. 1.304
   C. 1.214   D. 1.023

10. A bottle of perfume containing 0.24 ounce costs $8.40. How much does the perfume cost per ounce?

    F. $21.16   G. $211.60
    H. $35.00   J. $350.00

Short-Response Questions

For problems 11 and 12, use the price list below. Find the cost (before tax) of each order.

Office Supply Prices	
Item	Price
Storage Box	$5.09
Bookends (per pair)	$4.92
Tape Dispenser	$2.63
Report Covers (10-pack)	$7.58
Desk Calendar	$1.95
Markers (4-pack)	$2.29

11. 3 storage boxes, a tape dispenser, 2 packs of report covers

12. 4 pairs of bookends, 2 desk calendars, 12 markers

13. Write each number in scientific notation.

a. 193,800
b. 7,246,000,000
c. 0.0000059
d. 0.32461
e. 70

Extended-Response Questions

14. Leo's paycheck for a week in which he worked 27 regular hours was $199.80. If Leo works overtime, he gets paid one and one-half times his hourly rate. How much will Leo earn in a week where he works 35 regular hours and 5 overtime hours?

15. The population density of a state is found by dividing the total population by the area of the state. In the year that the population density of New York was 382.7 people per square mile, the density of South Dakota was 9.37 people per square mile. How many more people per square mile did New York have?

16. The average annual precipitation of some U.S. cities is shown in the table to the right.

a. Order the cities from least to greatest precipitation.
b. How much more precipitation did Miami have than Albany?
c. If the amounts in the table were rounded to the nearest whole inch, which cities would have the same amount of precipitation? Explain your answer.

City	Annual Precipitation (in.)
Albany, NY	34.08
Buffalo, NY	33.99
Detroit, MI	34.30
Jackson, MS	35.46
Los Angeles, CA	23.28
Miami, FL	79.30
Newark, NJ	37.67
Philadelphia, PA	36.77
St. Louis, MO	34.44
Syracuse, NY	31.34

17. Find each quotient.

a. $\dfrac{6.4 \times 1.3}{8}$

b. $\dfrac{5.76}{0.33 + 0.15}$

18. Pieces of wire 0.2 m, 0.31 m, and 0.14 m long are cut from a 1.7-m length of wire. The part remaining is cut into 5 pieces of equal length. How long is each piece?

19. A travel agent sold 92 weekend package tours of Washington, DC. The total was $28,796. What was the cost of the tour?

20. Busy Bee Messenger Service charges

- $7.20 for the first pound
- $1.30 for each additional pound up to 5 pounds
- $2.00 for each additional pound over 5 pounds

Find the delivery cost for each package.

a. 2 pounds
b. 5 pounds
c. 7 pounds

Multiple-Choice Questions

1. Estimate: $61,297 + 903 + 4,085$.

   A. 65,000
   B. 60,000
   C. 66,000
   D. 71,000

2. What is the standard form of $(4 \times 10^2) + (2 \times 10^{-1}) + (6 \times 10^{-3}) + (9 \times 10^{-4})$?

   F. 426.9
   G. 4020.0609
   H. 402.6009
   J. 400.2069

3. What is the next number in the pattern?
   0.021, 0.033, 0.057, 0.105, __?__

   A. 0.096
   B. 0.153
   C. 0.201
   D. 1.065

4. The Electric Outlet is having a sale. A television that regularly sells for $529 is on sale for $398. A music system has been reduced to $244. Customers who shop between 8 A.M. and 9 P.M. receive an extra $20 off each item.

   Which of the following questions could NOT be answered from the information given?

   F. How much does a customer who buys a television at noon save off the regular price?
   G. A customer bought a television and a music system at 8:30 A.M. How much did the customer save in all off the regular prices?
   H. In the afternoon, one salesperson sold 3 televisions and 4 music systems. What was the total for these items?
   J. Between 8 A.M. and 9 A.M., 11 televisions and 7 music systems were sold. How much more would these items have cost if they were purchased after 9 A.M.?

5. Which number is the greatest?

   A. 6,045,773
   B. 6,405,377
   C. 6,054,773
   D. 6,540,737

6.   31.1
     $\times\,0.27$

   F. 5.287
   G. 8.397
   H. 24.887
   J. 839.7

7. The table shows the number of tornadoes reported for the last five decades. During what period did the greatest increase occur?

Reported Tornadoes	
1950's	4,796
1960's	6,813
1970's	8,580
1980's	8,196
1990's	10,804

   A. 1950's to 1960's
   B. 1960's to 1970's
   C. 1970's to 1980's
   D. 1980's to 1990's

8. The most reasonable estimate of the total cost of 21 new cars if each car cost $16,792 is

   F. $34,000
   G. $170,000
   H. $320,000
   J. $340,000

Short-Response Questions

9. An athlete's point total for a combined event was 457.952. To the nearest tenth, the athlete's score was _____.

10. What is the sum of five thousand six hundred forty three and two thousandths and four thousand eight hundred twenty and one tenth?

Extended-Response Questions

11. A bookstore received an order of 72 books. There were 5 more novels than biographies and 5 more mysteries than novels. How many of each type of book were there? Explain how you solved the problem.

12. According to the 2000 census, the population of New York City was 8,008,278.

    a. Round the population number to the nearest thousand.
    b. Write the rounded population in scientific notation.

13. Jason biked 20 kilometers on Friday. On Saturday, he biked 0.8 as far. On Sunday, he biked 3 kilometers more than he did on Saturday. What was the total distance Jason biked for the three days? Show your work.

14. Each skier in a race skied the course twice. The winner is the skier with the least total time for the two runs.

Race Results (times in seconds)				
	Joan	Lewis	Ada	Carlos
First Run	41.53	39.90	43.62	40.00
Second Run	39.26	40.86	41.50	41.36

a. Who had the fastest first run? The fastest second run?
b. Show the final order of the skiers from first to fourth.
c. One of the skiers broke the course record. To the nearest hundredth of a second, what is the fastest time the previous record could have been? Explain.

15. Mr. Ross ordered 2 large house salads, 1 tofu burger, 3 fruit fantasies, and 2 frozen yogurts at Healthy Eats. He gave the cashier two $20 bills. How much change should he receive?

Healthy Eats Menu	
Vegetable Soup	2.50
House Salad (small)	2.79
House Salad (large)	4.39
Tofu Burger	3.75
Fruit Fantasy	3.25
Frozen Yogurt	1.99

16. In one week, about 150 million people throughout the world were given flu shots. About how many people received flu shot during a four-hour period? Assume that the same number of shots was given each hour for 8 hours per day for 5 days.

17. Mrs. Lawson buys a case of 24 bottles of iced tea for $18.00. She sells each bottle for $1.10.

    a. How much profit does Mrs. Lawson make if she sells the whole case?
    b. How much would she have to sell each bottle for in order to make a profit of $10.80?

18. The town of Breezy Shores hosts a sailing race every other summer. The table shows the population of Breezy Shores for a 4-year period.

Summer Population of Breezy Shores

1997	4,120
1998	3,820
1999	4,220
2000	3,720
2001	

a. If the pattern continues, predict the population for 2001.

b. In what year would the predicted population be 4,620?

19. A textbook is opened at random. To which pages is it opened if the product of the facing page numbers is 5,550?

20. Annette completed the course shown in a charity walkathon. She had pledges from sponsors for $31.50 per kilometer. To the nearest cent, how much money did Annette raise for the charity? Show your work.

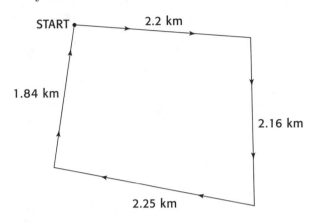

CHAPTER 3

Fractions

Chapter Vocabulary

factor
prime factorization
equivalent fractions
improper fraction
repeating decimal
least common denominator (LCD)

prime number
prime factors
lowest terms
mixed number
multiples
reciprocals
multiplicative inverses

composite number
factor pair
simplest form
terminating decimal
least common multiple (LCM)

3.1 Factors, Primes, and Composites

A **factor** of a number is an exact divisor. For example, 3 and 5 are factors of 15.

- A **prime number** is a whole number greater than 1 whose only factors are 1 and itself. 2, 3, 5, 7, 11, 13, 17, 19, and 23 are the first nine prime numbers.
- A **composite number** is a whole number greater than 1 that is not prime. 4, 6, 8, 10, 12, 14, 15, 16, 18, and 20 are examples of composite numbers.

- The numbers 0 and 1 are neither prime nor composite.
- Every composite number can be written as the product of two or more prime numbers. This product is called the **prime factorization.**

Example

The factors of 24 are: 1, 2, 3, 4, 6, 8, 12, and 24.

The **prime factors** of 24 are 2 and 3.

$2 \times 2 \times 2 \times 3 = 24$ or $2^3 \times 3 = 24$

Prime or Composite?

To tell whether a number is prime or composite, you can use the guess and check method and a calculator. Try to discover factors of the number by dividing. If you divide and there is no remainder, then the divisor is a factor of the dividend. If you find any factors other than 1 and the number itself, then the number is composite.

Model Problems

1. Tell whether each number is prime or composite.

Number	Solution	
a. 21	composite	$21 = 3 \times 7$
b. 36	composite	$36 = 9 \times 4$
c. 47	prime	$47 = 1 \times 47$
d. 51	composite	$51 = 3 \times 17$
e. 52	composite	$52 = 13 \times 4$
f. 111	composite	$111 = 3 \times 37$

2. Find the prime factorization of 72.

   Solution   Use a factor tree. Begin with a pair of factors for 72. Continue finding factors until all factors are prime. Each result has the same prime factors.

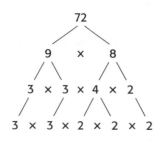

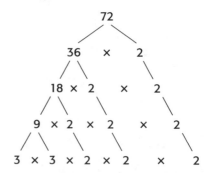

Answer Since the factors 2 and 3 are repeated, they can be expressed with exponents: $72 = 2^3 \times 3^2$.

3. Determine if 26 is a factor of 1,118.

Solution Divide to see if the remainder is 0.

$$
\begin{array}{r}
43 \\
26 \overline{)\ 1{,}118} \\
-104 \\
\hline
78 \\
-78 \\
\hline
0
\end{array}
$$

Answer 26 is a factor of 1,118, since 0 is the remainder. 43 is also a factor, since $26 \times 43 = 1{,}118$.

Practice

Multiple-Choice Questions

1. Which number has factors of 8 and 12?

   A. 16   B. 36
   C. 48   D. 60

2. For which pair is the first number a factor of the second number?

   F. 19, 727   G. 23, 667
   H. 28, 872   J. 34, 924

3. In which group are all the numbers prime?

   A. 11, 29, 51   B. 13, 37, 63
   C. 17, 49, 57   D. 19, 41, 73

4. In which group are all the numbers composite?

   F. 46, 77, 98   G. 42, 65, 97
   H. 38, 53, 89   J. 45, 59, 87

5. Which statement is false?

   A. 399 is divisible by 21.
   B. 22 is a factor of 572.
   C. 34 is a factor of 792.
   D. 26 and 32 are exact divisors of 832.

6. The prime factorization of 120 is

   F. $2 \times 3 \times 5$ G. $2^3 \times 3 \times 5$
   H. $2^3 \times 3^2 \times 5$ J. $8 \times 3 \times 5$

7. The number whose prime factorization is $3^2 \times 5^2$ is

A. 60 B. 75
C. 150 D. 225

8. The prime factorization of 180 is

F. $4 \times 3 \times 5$ G. $2 \times 2 \times 45$
H. $2^2 \times 3^2 \times 5$ J. $2^3 \times 3 \times 5^2$

9. What is the next prime number greater than 31?

A. 33 B. 35
C. 37 D. 41

10. Which is a factor pair of 288?

F. 16×18 G. 8×26
H. 12×28 J. 6×38

Short-Response Questions

11. Is the number prime or composite?

a. 27 b. 42
c. 61 d. 83

Find the prime factorization of each number using a factor tree. Write the answer using exponents.

12. 60

13. 100

14. 156

15. 162

Find each number, given the prime factorization.

16. $2^3 \times 3^2 \times 7$

17. $2^3 \times 5^2$

18. $2^2 \times 3^2 \times 7^2$

Extended-Response Questions

19. Which numbers in the box below are prime? Explain how you know that the remaining numbers are composite.

24	25	26	27	28	29
30	31	32	33	34	35
36	37	38	39	40	41
42	43	44	45		

20. Is the product of two prime numbers a prime number? Give examples that support your answer.

3.2 Divisibility and Greatest Common Factor

One number is divisible by another if the remainder is 0 when you divide. The rules below can be used as a quick check for divisibility.

A number is divisible by:

- 2 if the ones digit is 0, 2, 4, 6, or 8.
- 3 if the sum of the digits is divisible by 3.
- 4 if the number formed by the last two digits is divisible by 4.
- 5 if the ones digit is 0 or 5.
- 6 if the number is divisible by 2 and 3.
- 8 if the number formed by the last three digits is divisible by 8.
- 9 if the sum of the digits is divisible by 9.
- 10 if the ones digit is 0.

Recall that a whole number is divisible by each of its factors.

The factors of 12 are 1, 2, 3, 4, 6, and 12.
The factors of 20 are 1, 2, 4, 5, 10, and 20.
The common factors of 12 and 20 are 1, 2, and 4.
The greatest common factor (GCF) of 12 and 20 is 4.

Examples

In a division problem where the remainder is 0, the divisor and the quotient are a **factor pair** of the dividend.

1.

$$\overset{13 \ \leftarrow \ \text{quotient}}{3\overline{)39}}$$

divisor ↗   ↖ dividend

$3 \times 13 = 39$, so 3 and 13 are a factor pair of 39.

2. $104 \div 4 = 26$

$4 \times 26 = 104$, so 4 and 26 are a factor pair of 104.

 ## Model Problems

1. What are the factors of 42?

Solution

The last digit of 42 is 2, so 2 is a factor.
$4 + 2 = 6$ and 6 is divisible by 3, so 3 is a factor of 42.
Since 42 is divisible by 2 and 3, it is divisible by 6.

Find other factors by pairing.

$2 \times [\ ] = 42$ so $42 \div 2 = 21$
$3 \times [\ ] = 42$ so $42 \div 3 = 14$
$6 \times [\ ] = 42$ so $42 \div 6 = 7$

Answer The factors of 42 are 1, 2, 3, 6, 7, 14, 21, 42.

2. Find the GCF of 36 and 48.

Solution

Factors of 36: 1, 2, 3, 4, 6, 9, 12, 18, 36
Factors of 48: 1, 2, 3, 4, 6, 8, 12, 16, 24, 48
Common factors of 36 and 48 are 1, 2, 3, 4, 6, and 12.
So, the GCF of 36 and 48 is 12.

Answer 12

3. Vanna has 24 oatmeal cookies and 40 chocolate cookies. She wants all the bags to have the same number of cookies and wants only one kind of cookie in each bag. What is the greatest number of cookies Vanna can put in each bag?

Solution

Find the GCF of 24 and 40.
Factors of 24: 1, 2, 3, 4, 6, 8, 12, 24
Factors of 40: 1, 2, 4, 5, 8, 10, 20, 40

Answer   The GCF is 8. She can put 8 cookies of one kind in each bag.

Practice

Multiple-Choice Questions

1. Which set of digits could NOT be used to form a three-digit number that is divisible by 4?

   A. 0, 3, 4   B. 1, 4, 6
   C. 2, 3, 8   D. 3, 5, 8

2. Which number is divisible by 9?

   F. 3,796   G. 10,027
   H. 213,459   J. 1,407,033

3. Which number is divisible by 8?

   A. 4,124   B. 7,816
   C. 9,502   D. 10,340

4. Which number is divisible by 3, but not by 6?

   F. 18   G. 48
   H. 72   J. 93

5. What is the GCF of 48 and 54?

   A. 3   B. 6
   C. 9   D. 16

6. Which is the GCF of 18 and 108?

   F. 3   G. 6
   H. 9   J. 18

7. What is the GCF of 28, 42, and 98?

   A. 2   B. 6
   C. 14   D. 28

8. How many factor pairs does 90 have?

   F. 4   G. 5
   H. 6   J. 7

9. Which number has 4 as a factor, but not 8?

   A. 54   B. 68
   C. 72   D. 106

10. If two numbers are both primes, their GCF is

    F. 1
    G. the smaller prime
    H. the greater prime
    J. the product of the numbers

Short-Response Questions

Find the GCF for each.

11. 24, 64

12. 18, 42

13. 78, 195

14. 54, 189

15. 68, 153, 204

Extended-Response Questions

16. Fran bought some jewelry. Bracelets cost $8 and necklaces cost $20. If Fran bought 26 necklaces, how many bracelets could she have bought instead? Show the steps you take to arrive at your answer.

17. A florist has 30 carnations and 42 roses. He wants to put the flowers in vases so that each vase has the same number of like flowers.

    a. How many vases can he fill?
    b. How many flowers will be in each vase?

18. a. Give an example of two numbers whose greatest common factor is one of the numbers.
    b. In general, what is true about such pairs of numbers?

19. A tabletop is 16 inches wide and 24 inches long. David is covering the top using square tiles, all the same size. What size tiles could David use to cover the tabletop exactly? Include a diagram that illustrates your solution. Show the steps you took to arrive at your answer.

20. Sylvia's aunt Pilar is 32 years old. The greatest common factor of their ages is 8. How old is Sylvia? Explain how you found your answer.

3.3 Equivalent Fractions

Fractions that represent the same quantity are called **equivalent fractions**. $\frac{2}{4}$, $\frac{3}{6}$, and $\frac{4}{8}$ are equivalent fractions since each represents $\frac{1}{2}$ of a whole quantity.

$$\frac{2}{4} = \frac{1}{2} \qquad \frac{3}{6} = \frac{1}{2} \qquad \frac{4}{8} = \frac{1}{2}$$

- To find an equivalent fraction, multiply or divide the numerator and denominator by the same nonzero number.
- Equivalent fractions have cross products that are equal.

A fraction is in **lowest terms** or **simplest form** when the GCF of its numerator and denominator is 1.

- To write a fraction is simplest form, divide the numerator and denominator by their GCF.

Creating Equivalent Fractions

When you multiply or divide to create an equivalent fraction, you are changing the *appearance* of the fraction, but not the *value* of it. This is true because when you multiply or divide the numerator and denominator by the same nonzero number, you are actually

---

multiplying the whole fraction by 1. The identity property of multiplication (see page 45) tells you that if you multiply any number by 1, that number remains the same.

$$\frac{3}{4} \times \frac{5}{5} = \frac{15}{20} \qquad \frac{5}{5} = 1 \qquad \frac{9}{15} \div \frac{3}{3} = \frac{3}{5} \qquad \frac{3}{3} = 1$$

Model Problems

1. Find the fraction with a denominator of 24 that is equivalent to $\frac{2}{3}$.

   Solution $\frac{2}{3} = \frac{?}{24}$ $24 \div 3 = 8$

   Note: You must multiply 3 by 8 to get 24. To find the equivalent fraction, multiply the numerator by 8. Multiplying by $\frac{8}{8}$ is the same as multiplying by 1. It does not change the value of the fraction.

   Answer $\frac{2}{3} \times \frac{8}{8} = \frac{16}{24}$

2. Are $\frac{3}{20}$ and $\frac{12}{80}$ equivalent?

   Solution Use cross products.

   $\frac{3}{20} \quad \frac{12}{80} \quad \begin{array}{l} 3 \times 80 = 240 \\ 20 \times 12 = 240 \end{array}$

   Answer Yes, they are equivalent because they have the same cross product.

3. Are $\frac{21}{56}$ and $\frac{3}{7}$ equivalent?

   Solution $\frac{21}{56} \quad \frac{3}{7} \quad \begin{array}{l} 21 \times 7 = 147 \\ 56 \times 3 = 168 \end{array}$

   Answer No, they are not equivalent. Since the cross products are not equal, the fractions are not equivalent.

4. Write $\frac{18}{24}$ in simplest form.

   Solution The GCF of 18 and 24 is 6.

   $\frac{18 \div 6}{24 \div 6} = \frac{3}{4}$

   Answer The GCF of 3 and 4 is 1. $\frac{3}{4}$ is in simplest form.

 **Practice**

Multiple-Choice Questions

1. Which fraction is NOT equivalent to $\frac{4}{5}$?

   A. $\frac{8}{10}$ B. $\frac{36}{45}$

   C. $\frac{21}{25}$ D. $\frac{48}{60}$

2. Which fraction is equivalent to $\frac{3}{8}$?

   F. $\frac{9}{40}$ G. $\frac{21}{56}$

   H. $\frac{20}{48}$ J. $\frac{36}{80}$

3. Replace ? to make a true statement.
$$\frac{36}{81} = \frac{?}{9}$$

   A. 4   B. 5
   C. 8   D. 9

4. Which pair of fractions is NOT equivalent?

   F. $\frac{2}{6}, \frac{10}{30}$ G. $\frac{15}{50}, \frac{6}{20}$

   H. $\frac{3}{5}, \frac{27}{45}$ J. $\frac{16}{40}, \frac{7}{15}$

5. Which fraction is in simplest form?

   A. $\frac{3}{21}$ B. $\frac{5}{30}$

   C. $\frac{9}{25}$ D. $\frac{12}{60}$

6 Simplify $\frac{27}{64}$.

   F. $\frac{3}{8}$ G. $\frac{3}{9}$

   H. $\frac{14}{32}$ J. $\frac{27}{64}$

7. The simplest form of $\frac{68}{200}$ is

   A. $\frac{8}{25}$ B. $\frac{17}{50}$

   C. $\frac{34}{100}$ D. $\frac{136}{400}$

8. Todd made a quilt with 300 squares. There were 90 blue squares. What part of the quilt was blue?

   F. $\frac{1}{3}$ G. $\frac{2}{5}$

   H. $\frac{3}{10}$ J. $\frac{9}{50}$

9. Which fraction represents the same amount as shown shaded in the figure?

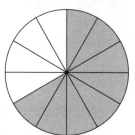

   A. $\frac{6}{10}$ B. $\frac{15}{21}$

   C. $\frac{30}{45}$ D. $\frac{40}{50}$

10. Which fraction does NOT belong?

   F. $\frac{15}{36}$ G. $\frac{35}{60}$

   H. $\frac{30}{72}$ J. $\frac{45}{108}$

Short-Response Questions

11. Replace each ( ) with = or ≠.

   a. $\frac{2}{3}$ ( ) $\frac{12}{15}$ b. $\frac{24}{27}$ ( ) $\frac{8}{9}$

   c. $\frac{7}{12}$ ( ) $\frac{49}{84}$ d. $\frac{15}{40}$ ( ) $\frac{25}{64}$

12. Write each fraction as an equivalent fraction that has the denominator shown in parentheses.

   a. $\frac{2}{3}$, (42)   b. $\frac{5}{16}$, (80)

   c. $\frac{85}{100}$, (20)   d. $\frac{198}{432}$, (24)

13. Write each fraction in simplest form.

a. $\frac{18}{30}$ b. $\frac{96}{32}$

c. $\frac{48}{72}$ d. $\frac{35}{53}$

14. Write three fractions that name each point on the number line shown.

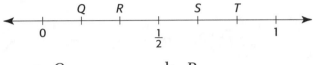

a. Q b. R
c. S d. T

15. A case of juice contains 24 bottles. What part of a case is 18 bottles? Write your answer in lowest terms.

16. Larry walked $\frac{5}{8}$ of a mile. Noreen walked $\frac{17}{24}$ of a mile. Did they walk the same distance? If not, who walked farther? How do you know?

17. A survey reported that three fourths of the people surveyed owned a computer. If 390 people in the survey owned a computer, how many people were in the survey group?

Extended-Response Questions

18. Andrew made a mosaic design. He used 65 green tiles, 95 blue tiles, 115 white tiles, and 75 yellow tiles. (All tiles were the same size).

a. Show the fractional part of the design for each color. Write answers in simplest form.
b. If Andrew decides to double the size of the mosaic, how will the fractional parts for each color change? Explain your thinking.

19. Write a fraction with a denominator of 30 that is in simplest form. (The numerator must be greater than 1.) Explain why you know that it is in simplest form.

20. Renee sewed a quilt and sold it to her grandmother for $210. Renee had spent $45 for materials.

a. What fraction of the sale price was profit for Renee?
b. Renee's grandmother sold the quilt to a quilt collector for $840. What fraction of the sale price was profit for Renee's grandmother?

3.4 Mixed Numbers and Improper Fractions

A fraction in which the numerator is greater than or equal to the denominator is called an **improper fraction.** An improper fraction has a value greater than or equal to 1 whole. Examples of improper fractions are $\frac{8}{7}, \frac{8}{8}, \frac{9}{1},$ and $\frac{6}{2}.$

A **mixed number** is a number written as a whole number and a proper fraction. Examples of mixed numbers are $2\frac{3}{8}, 4\frac{1}{5},$ and $12\frac{2}{9}.$

- To write an improper fraction as a mixed number or whole number, divide the numerator of the fraction by the denominator. The quotient is the whole number part of the mixed number. The

  $\frac{\text{remainder}}{\text{divisor}}$ is the fraction part of the mixed number.

- To write a mixed number as an improper fraction, multiply the whole number part by the denominator of the fraction part. Add this product to the numerator of the fraction part. Write the sum over the denominator of the fraction part of the original mixed number.

Model Problems

1. Write 3 as an improper fraction with a denominator of 8.

   Solution $3 = \frac{3}{1}$ ← Write 3 over a denominator of 1.

   $3 = \frac{3 \times 8}{1 \times 8} = \frac{24}{8}$ ← Multiply numerator and denominator by 8.

   So, $3 = \frac{24}{8}$.

2. Write $\frac{17}{6}$ as a mixed number.

   Solution $\frac{17}{6} = 17 \div 6$

   $$6)\overline{17}\quad 2\frac{5}{6}$$
   $$\underline{-12}$$
   $$5$$

   $\frac{17}{6} = 2\frac{5}{6}$

3. Write $4\frac{2}{3}$ as an improper fraction.

   Solution $4\frac{2}{3} = 4 + \frac{2}{3}$

   $3 \times 4 = 12$ ← Multiply

   $12 + 2 = 14$ ← Add

   $\frac{14}{3}$ ← Write over original denominator

   Here is a good way to remember the procedure:

   $4\frac{2}{3}$ ← start $4\frac{2}{3} = \frac{(3 \times 4) + 2}{3} = \frac{14}{3}$

Practice

Multiple-Choice Questions

1. Which is an improper fraction?

   A. $\frac{17}{19}$ B. $\frac{59}{58}$

   C. $\frac{91}{92}$ D. $\frac{0}{31}$

2. Write the shaded amount as a mixed number.

   F. $2\frac{3}{8}$ G. $2\frac{5}{8}$

   H. $3\frac{3}{8}$ J. $3\frac{5}{8}$

3. Which fraction is less than 1?

   A. $\frac{15}{14}$ B. $\frac{31}{30}$

   C. $\frac{77}{76}$ D. $\frac{112}{113}$

4. $\frac{81}{7}$ is equivalent to

   F. $10\frac{1}{7}$ G. $11\frac{4}{7}$

   H. $8\frac{1}{7}$ J. $12\frac{3}{7}$

5. Which represents a whole number?

   A. $\frac{23}{3}$ B. $\frac{34}{4}$

   C. $\frac{58}{6}$ D. $\frac{72}{9}$

6. $9\frac{4}{5}$ is equivalent to

   F. $\frac{39}{5}$ G. $\frac{29}{4}$

   H. $\frac{49}{5}$ J. $\frac{41}{4}$

7. $\frac{91}{8}$ is equivalent to

   A. $11\frac{1}{8}$ B. $11\frac{3}{8}$

   C. $10\frac{7}{8}$ D. $9\frac{1}{8}$

8. Granola bars come in packages of 8. If Keith has 5 full packages and one that is $\frac{5}{8}$ full, how many granola bars does he have?

   F. 45   G. 43
   H. 28   J. 23

9. Write the quotient as a mixed number.
   $4\overline{)63}$ =

   A. $6\frac{3}{4}$ B. $12\frac{1}{4}$

   C. $15\frac{3}{4}$ D. $16\frac{1}{4}$

10. Replace the [ ] with the appropriate mixed number. 2,250,000 = [ ] million

    F. $2\frac{1}{4}$ G. $2\frac{2}{5}$

    H. $20\frac{1}{4}$ J. $22\frac{1}{2}$

Short-Response Questions

11. Write each as an improper fraction.

    a. $7\frac{2}{3}$ b. $8\frac{3}{5}$

    c. $10\frac{7}{8}$ d. $6 = \frac{?}{8}$

12. Write each as a mixed number or whole number.

    a. $\frac{27}{4}$ b. $\frac{39}{8}$

    c. $\frac{49}{7}$ d. $\frac{85}{12}$

Find each value of n.

13. $5\frac{4}{4} = n$

14. $8\frac{11}{6} = 9\frac{n}{6}$

15. $10\frac{11}{3} = 13\frac{n}{3}$

16. $12\frac{n}{8} = 13\frac{3}{8}$

Extended-Response Questions

17. A chef makes 4 turkey burgers from each pound of ground meat. How many pounds of ground meat does the chef need to make 31 burgers?

18. a. Each ice cream sundae requires $\frac{1}{3}$ cup of chocolate sauce. How many sundaes were made if $8\frac{2}{3}$ cups of chocolate sauce were used?

b. Each sundae also requires $\frac{3}{4}$ oz of chopped walnuts. How many ounces of chopped walnuts are needed for the sundaes in part a?

19. Each of 104 guests will be served $\frac{1}{2}$ avocado with dinner.

a. How many boxes of avocados will be used if there are 12 avocados in a box?

b How many boxes of avocados will have to be used if the cook discovers that $\frac{1}{3}$ of the avocados in each box are too ripe to eat?

20.

a. Suppose this number line has been divided so that there are 4 equal parts between whole numbers. What number does point P represent?

b. Suppose the number line has been divided so that there are 6 equal parts between whole numbers. What number does point P represent?

3.5 Fractions and Decimals

Fractions and decimals can be converted from one form to the other.

- To write a fraction as a decimal, divide the numerator by the denominator. If the division ends, the decimal is called a **terminating decimal**. When the division does not end, the decimal is a **repeating decimal**.

Note: There are some decimals that do not terminate or repeat. Pi (π) is one example.

- To write a decimal as a fraction, say the place-value name of the decimal, then write a fraction using that place value as the denominator. Rewrite the fraction in simplest form.

Model Problems

1. Write $\frac{3}{8}$ as a decimal.

Solution   Divide numerator by denominator.

$$
\begin{array}{r}
0.375 \\
8\overline{)\ 3.000} \\
\underline{-2\,4} \\
60 \\
\underline{-56} \\
40 \\
\underline{-40} \\
0
\end{array}
$$

Write as many zeros as necessary in the dividend to complete the division.

Answer   0.375 is a terminating decimal. The remainder is 0.

2. Write $\frac{5}{3}$ as a decimal.

Solution

$$
\begin{array}{r}
1.66\overline{6} \\
3\overline{)\ 5.000} \\
\underline{-3} \\
20 \\
\underline{-18} \\
20 \\
\underline{-18} \\
20 \\
\underline{-18} \\
2
\end{array}
$$

Since $\frac{5}{3}$ is an improper fraction, the equivalent decimal has a whole number part.

$1.66\overline{6}$ is a repeating decimal. The remainder is never 0. The digit 6 in the quotient will continue to repeat. A bar is used to show the digit or digits that repeat.

Answer $\frac{5}{3} = 1.666 \ldots$ or $1.\overline{6}$

3. Write 0.625 in fraction form.

Solution   0.625 is 625 *thousandths*. Write a fraction with 1,000 as the denominator.

$$0.625 = \frac{625}{1,000}$$

Answer   Simplify the fraction. $\dfrac{625 \div 125}{1,000 \div 125} = \dfrac{5}{8}$

4. Write 5.12 as a mixed number.

Solution 5.12 is 5 and 12 *hundredths*. Write a mixed number with 100 as the denominator of the fraction. $5.12 = 5\frac{12}{100}$

Then, simplify the fraction. $\frac{12 \div 4}{100 \div 4} = \frac{3}{25}$

Answer $5.12 = 5\frac{3}{25}$

5. Write the repeating decimal 0.313131 . . . as a fraction.

Solution The following steps can be used to write any repeating decimal as a fraction.

Step 1 Since the pattern repeats every two places, multiply by 100 (10^2) to shift the decimal point two places. 100 × the number = 31.3131 . . .

Step 2 Subtract the original number.

$$\begin{array}{r} 31.3131\ldots \\ -0.3131\ldots \\ \hline 31. \end{array}$$

Step 3 99 times the original number = 31. Therefore you can use the inverse operation and *divide* by 99 to get the original number $\frac{31}{99}$.

Answer $\frac{31}{99}$

Note: If the pattern repeats every three places, the denominator would be 999. If it repeats every four places, the denominator would be 9,999, and so on.

Practice

Multiple-Choice Questions

1. The decimal form of $\frac{9}{25}$ is

   A. 0.27   B. $0.3\overline{3}$
   C. 0.36   D. 0.45

2. Which pair of numbers is equivalent?

   F. $\frac{3}{4}$, 0.65   G. $\frac{4}{5}$, 0.8

   H. $\frac{2}{3}$, 0.75   J. $\frac{7}{8}$, 0.925

3. The decimal form of $\frac{11}{4}$ is

   A. $0.3\overline{6}$ B. 1.75
   C. $2.\overline{36}$ D. 2.75

4. The decimal form of $1\frac{1}{6}$ is

   F. $1.1\overline{6}$ G. $1.\overline{16}$
   H. 1.6   J. $1.\overline{66}$

5. $0.\overline{074}$ represents

   A. 0.07444 . . .
   B. 0.0747474 . . .
   C. 0.074074074 . . .
   D. 0.074704704 . . .

6. Which fraction is equivalent to 0.28?

   F. $\frac{4}{7}$ G. $\frac{7}{50}$

   H. $\frac{7}{25}$ J. $\frac{14}{25}$

7. Which mixed number is equivalent to 2.65?

A. $2\frac{5}{6}$ B. $2\frac{3}{15}$

C. $2\frac{13}{20}$ D. $3\frac{1}{5}$

8. Daniel ran $4.\overline{3}$ kilometers. This distance is the same as

F. $4\frac{3}{10}$ km   G. $4\frac{3}{100}$ km

H. $4\frac{33}{100}$ km   J. $4\frac{1}{3}$ km

9. Which number is NOT equivalent to the others?

A. 1.18   B. $\frac{9}{5}$

C. 1.8   D. $1\frac{20}{25}$

10. $0.7 - \frac{3}{10} =$

F. $\frac{4}{7}$ G. $\frac{2}{5}$

H. 4   J. $6\frac{7}{10}$

Short-Response Questions

11. Convert each fraction to a decimal.

a. $\frac{17}{50}$ b. $\frac{4}{9}$

c. $\frac{5}{6}$ d. $\frac{7}{20}$

12. Convert each decimal to a fraction or mixed number in simplest form.

a. 6.08   b. 3.125
c. $0.135\overline{135}$ d. 0.5625

13. Compare. Use <, >, or = for each ( ).

a. 0.82 ( ) $\frac{4}{5}$ b. 1.55 ( ) $1\frac{11}{20}$

c. $\frac{2}{3}$ ( ) 0.66 d. 1.575 ( ) $1\frac{5}{8}$

Find each answer and write it as a decimal.

14. $\frac{3}{4} + 0.65 + \frac{1}{5}$

15. $\frac{4}{5} \times 0.9$

16. $3.5 - 1\frac{7}{8} + 0.155$

Extended-Response Questions

17. a. Yesterday's high for a share of Futuro stock was $57\frac{3}{4}$ dollars. Write this price as a decimal.

b. Today, 8 shares of Futuro stock cost $461. What is the price per share in decimal form and in fraction form? Is the price higher or lower than yesterday?

18. a. Lori needs $1\frac{1}{4}$ pounds of pinto beans for tacos. If she buys a package that weighs 1.65 pounds, what fractional part of a pound will she have leftover?

b. The store is having a sale on beans, so Lori considers buying another kind of beans, but is not certain which is the best value. All the packages cost $0.99, but she notices that they contain different amounts of beans. Which kind of beans is the best value? Explain how you determined your answer.

Beans on Sale for $0.99	Quantity in One Package
Black Beans	$1\frac{1}{4}$ lb
White Beans	$1\frac{1}{3}$ lb
Garbanzo Beans	1.5 lb
Red Beans	$\frac{5}{3}$ lb
Navy Beans	1 lb

19. a. Gavin practiced his guitar for a total of 11.25 hours during a 3-day weekend. Write the mixed number that shows the average time he practiced each day.

b. It takes Gavin 40 hours of practice to learn how to play a new song. If he practices an average of 7 hours per week, approximately how many new songs can he learn in one year?

20. Is 0.00734 closer to $\dfrac{73}{10,000}$ or $\dfrac{74}{10,000}$? Explain your answer.

3.6 Least Common Multiple and Least Common Denominator

To find the **multiples** of a number, multiply the number by every whole number.

Some multiples of 4 are 0, 4, 8, 12, 16, 20, 24, 28, 32, 36, . . .

Some multiples of 6 are 0, 6, 12, 18, 24, 30, 36, 42, . . .

Notice that 12, 24, and 36 appear on both lists. These numbers are common multiples of 4 and 6.

- The **least common multiple** (LCM) of two or more numbers is the least nonzero number that is a multiple of each number. The LCM of 4 and 6 is 12.
- To find the LCM of two or more numbers, list the multiples of the greatest number. The first nonzero number that is a multiple of each of the other numbers is the LCM.
- The **least common denominator** (LCD) of two or more fractions is the LCM of the denominators.

1. Find the LCM of 8 and 10.

Solution   The multiples of 10 are 0, 10, 20, 30, 40, 50, 60, . . .

40 is the first nonzero number that is also a multiple of 8. So, 40 is the LCM of 8 and 10.

2. Find the LCM of 9, 18, and 36.

Solution   The greatest number, 36, is a multiple of both 9 and 18. So, 36 is the LCM of 9, 18, and 36.

3. Marah runs a lap around the reservoir in 5 minutes. Joy runs each lap in 7 minutes. If both girls start at the same time, after how many minutes will they both be together again at the starting point?

Solution   The answer is the LCM of 5 and 7.

Multiples of 7: 0, 7, 14, 21, 28, 35, 42, . . .

The first nonzero multiple of 7 that is also a multiple of 5 is 35.

Answer   The girls will again be together at the start after 35 minutes.

4. Find the LCD of $\frac{5}{6}$ and $\frac{7}{8}$.

Solution   Multiples of 8: 0, 8, 16, 24, 32, . . .

Since 24 is the first common multiple of 6 and 8, 24 is the LCD of $\frac{5}{6}$ and $\frac{7}{8}$.

5. $0.9 - \frac{3}{5} =$

A. $\frac{6}{5}$ B. $1\frac{1}{5}$ C. $\frac{3}{10}$ D. $\frac{3}{5}$

Solution   Since the multiple-choice answers are fractions, it makes sense to turn 0.9 into a fraction and then to subtract. 0.9 is nine *tenths*, which can be written as $\frac{9}{10}$.

To complete the subtraction, $\frac{9}{10} - \frac{3}{5}$, you must find the LCD. Since 10 is the first common multiple of 5 and 10, the LCD is 10.

Rewriting $\frac{3}{5}$, we get $\frac{3}{5} \times \frac{2}{2} = \frac{6}{10}$.

$\frac{9}{10} - \frac{6}{10} = \frac{3}{10}$

Answer   C

Practice

Multiple-Choice Questions

1. Which is NOT a common multiple of 3 and 4?

   A. 0   B. 12
   C. 36   D. 42

2. The LCM of 5 and 10 is

   F. 5   G. 10
   H. 20   J. 50

3. The LCM of 6 and 16 is

   A. 32
   B. 48
   C. 96
   D. 160

4. The LCM of 10, 20, and 30 is

   F. 600
   G. 300
   H. 60
   J. 30

5. The LCM of 4, 5, and 9 is

   A. 180   B. 120
   C. 90   D. 45

6. The LCD of $\frac{2}{3}$ and $\frac{7}{12}$ is

   F. 3
   G. 12
   H. 24
   J. 36

7. The LCD of $\frac{3}{10}$ and $\frac{8}{15}$ is

   A. 30
   B. 45
   C. 60
   D. 150

8. The LCD of $\frac{1}{2}$, $\frac{5}{6}$, and $\frac{3}{7}$ is

   F. 14   G. 28
   H. 42   J. 56

9. Which statement is false?

   A. 0 is a multiple of every number.
   B. Any multiple of 9 must also be a multiple of 3.
   C. The LCM of 6 and 18 is 18.
   D. 7 and 13 have no common multiples greater than 0.

10. Which statement is true?

    F. Any multiple of 7 must also be a multiple of 14.
    G. The LCM of 3, 5, and 9 is 135.
    H. The LCM of 2, 7, and 11 is 154.
    J. There are exactly 12 common multiples of 4 and 13.

Short-Response Questions

11. Find three common multiples of 4 and 10.

12. Find three common multiples of 3, 6, and 12.

13. Find the LCM for each.

    a. 7, 9
    b. 16, 18
    c. 4, 9, 12
    d. 2, 7, 15

14. What is the LCM of 2, 4, 7, 8, 14, and 28?

15. The LCM of 5, 6, and 8 is also a multiple of what numbers?

16. Find the LCD for each group of fractions.

    a. $\frac{2}{3}$ and $\frac{7}{10}$

    b. $\frac{5}{8}$ and $\frac{11}{12}$

    c. $\frac{1}{2}$, $\frac{1}{6}$, and $\frac{4}{9}$

    d. $\frac{3}{5}$, $\frac{4}{7}$, $\frac{9}{10}$

17. Doughnuts are sold in boxes of 8. Bottles of juice are sold in cases of 12. What is the minimum number of boxes and cases needed to have the same number of each item?

18. The Cleaning Crew comes to Dr. Simon's office every 4 days. The Plant People come every 7 days. If both services show up on March 1, what will be the next date on which they both show up?

19. Tom takes 9 minutes to jog around the lake, Roy takes 10 minutes, and Franco takes 12 minutes. They all start at the same time from the same spot. They jog from noon until 2:00 P.M. How many times will they again be together at the starting point? Explain how you got your answer.

20. A group of students went on a trip to the museum. One sixth of the students rode in a van and $\frac{3}{5}$ went by bus.

  a. If 90 students went on the trip, how many went by van?
  b. How many students must have traveled by some means other than van or bus?

3.7 Comparing and Ordering Fractions

Like fractions are fractions that have a common denominator. Unlike fractions are fractions that have different denominators.

- To compare two like fractions, compare the numerators. The greater fraction has the greater numerator.
- To compare unlike fractions, write equivalent fractions using the LCD. Then compare as above.
- To order a group of fractions, compare them two at a time.

 **Model Problems**

1. Compare $\frac{9}{11}$ and $\frac{7}{11}$.

   Solution The fractions have the same denominator.

   Compare numerators. $9 > 7$

   So, $\frac{9}{11} > \frac{7}{11}$ or $\frac{7}{11} < \frac{9}{11}$.

2. Compare $\frac{2}{3}$ and $\frac{5}{8}$.

Solution   Write equivalent fractions using the LCD. The LCM of 3 and 8 is 24.

$$\frac{2 \times 8}{3 \times 8} = \frac{16}{24}$$

$$\frac{5 \times 3}{8 \times 3} = \frac{15}{24} \quad 16 > 15$$

$\frac{16}{24} > \frac{15}{24}$ so $\frac{2}{3} > \frac{5}{8}$ Give the answer in terms of the original fractions.

3. Compare $2\frac{7}{10}$ and $2\frac{3}{4}$.

Solution   First compare the whole number parts. Since 2 = 2, compare the fractions.

$$\frac{7 \times 2}{10 \times 2} = \frac{14}{20} \quad \text{The LCD is 20.}$$

$$\frac{3 \times 5}{4 \times 5} = \frac{15}{20} \quad 15 > 14$$

So, $2\frac{7}{10} < 2\frac{3}{4}$.

4. List the following fractions in order from least to greatest: $\frac{2}{3}, \frac{3}{5}, \frac{1}{6}$.

Solution   The LCD is 30.

$$\frac{2 \times 10}{3 \times 10} = \frac{20}{30} \quad \frac{3 \times 6}{5 \times 6} = \frac{18}{30} \quad \frac{1 \times 5}{6 \times 5} = \frac{5}{30}$$

$18 < 20$ and $5 < 18$ so $\frac{1}{6} < \frac{3}{5} < \frac{2}{3}$

To round a mixed number to the nearest whole number, compare its fraction part to $\frac{1}{2}$.

- A fraction is greater than $\frac{1}{2}$ if twice the numerator is greater than the denominator.
  $\frac{9}{13} > \frac{1}{2}$ because $9 \times 2 = 18$ and $18 > 13$.

- The fraction is less than $\frac{1}{2}$ if twice the numerator is less than the denominator.
  $\frac{3}{7} < \frac{1}{2}$ because $3 \times 2 = 6$ and $6 < 7$.

- If the fraction is $\frac{1}{2}$ or greater, round up to the next whole number. If the fraction is less than $\frac{1}{2}$, do not change the whole number.

Model Problem

Estimate the sum $3\frac{7}{9} + 6\frac{3}{11}$.

Solution   Round each number to the nearest whole.

$14 > 9$ so $\frac{7}{9} > \frac{1}{2}$ Round $3\frac{7}{9}$ to 4.

$6 < 11$ so $\frac{3}{11} < \frac{1}{2}$ Round $6\frac{3}{11}$ to 6.

Add the rounded numbers: $4 + 6 = 10$. The sum is about 10.

Practice

Multiple-Choice Questions

1. Select the greatest fraction.

A. $\frac{7}{29}$ B. $\frac{14}{29}$

C. $\frac{23}{29}$ D. $\frac{6}{29}$

2. Select the least fraction.

F. $\frac{3}{4}$ G. $\frac{1}{2}$

H. $\frac{7}{8}$ J. $\frac{3}{8}$

3. Select the greatest fraction.

A. $\frac{2}{3}$ B. $\frac{5}{6}$

C. $\frac{7}{12}$ D. $\frac{3}{4}$

4. Which number is the least?

F. $3\frac{9}{10}$ G. $3\frac{1}{2}$

H. $3\frac{5}{8}$ J. $3\frac{3}{4}$

5. Which comparison is false?

A. $\frac{19}{24} < \frac{11}{12}$

B. $\frac{7}{8} > \frac{5}{6}$

C. $\frac{4}{15} < \frac{1}{4}$

D. $\frac{7}{10} > \frac{2}{3}$

6. The sum of $8\frac{11}{16} + 5\frac{5}{7} + 1\frac{4}{9}$ is

F. less than 17
G. exactly 17
H. more than 17
J. more than 18

7. The best estimate of $12\frac{9}{15} - 7\frac{4}{13}$ is

A. 4   B. 5
C. 6   D. 7

8. Which set of numbers is ordered from least to greatest?

F. $2\frac{3}{8}, 2\frac{5}{12}, 2\frac{2}{3}, 2\frac{1}{6}$

G. $2\frac{1}{6}, 2\frac{2}{3}, 2\frac{3}{8}, 2\frac{5}{12}$

H. $2\frac{2}{3}, 2\frac{1}{6}, 2\frac{3}{8}, 2\frac{5}{12}$

J. $2\frac{1}{6}, 2\frac{3}{8}, 2\frac{5}{12}, 2\frac{2}{3}$

9. Which number is the greatest?

A. $1\frac{8}{19}$ B. $1\frac{24}{45}$

C. $1\frac{31}{65}$ D. $1\frac{27}{59}$

10. Which comparison is true?

F. $\frac{8}{3} > \frac{40}{15}$ G. $\frac{11}{7} > \frac{11}{8}$

H. $\frac{5}{5} < \frac{8}{8}$ J. $\frac{13}{24} < \frac{21}{44}$

Short-Response Questions

11. Compare. Use $<$, $>$, or $=$.

a. $\frac{5}{9}$ ( ) $\frac{8}{9}$ b. $\frac{11}{12}$ ( ) $\frac{9}{12}$

c. $\frac{18}{25}$ ( ) $\frac{23}{25}$ d. $\frac{79}{100}$ ( ) $\frac{71}{100}$

12. Compare. Use $<$, $>$, or $=$.

a. $\frac{6}{10}$ ( ) $\frac{3}{5}$ b. $\frac{7}{16}$ ( ) $\frac{3}{4}$

c. $\frac{11}{15}$ ( ) $\frac{2}{3}$ d. $\frac{13}{6}$ ( ) $\frac{29}{12}$

13. Compare. Use $<$, $>$, or $=$.

a. $\frac{3}{4}$ ( ) $\frac{3}{5}$ b. $\frac{7}{16}$ ( ) $\frac{5}{12}$

c. $\frac{5}{6}$ ( ) $\frac{8}{9}$ d. $\frac{13}{10}$ ( ) $\frac{11}{8}$

14. Compare. Use $<$, $>$, or $=$.

a. $5\frac{2}{3}$ ( ) $5\frac{7}{8}$ b. $3\frac{3}{7}$ ( ) $3\frac{1}{4}$

c. $2\frac{4}{9}$ ( ) $\frac{14}{6}$ d. 7 ( ) $\frac{77}{12}$

15. Write each group of numbers in order from least to greatest.

a. $\frac{1}{6}, \frac{3}{4}, \frac{2}{5}$ b. $\frac{2}{5}, \frac{1}{2}, \frac{4}{9}$

c. $1\frac{2}{3}, 1\frac{7}{9}, 1\frac{3}{4}, 1\frac{5}{12}$

d. $3\frac{9}{10}, 3\frac{7}{8}, 3\frac{5}{6}, 3\frac{3}{4}$

16. Estimate each sum or difference.

a. $3\frac{7}{12} + 1\frac{5}{8}$

b. $9\frac{11}{15} - 2\frac{7}{20}$

c. $13\frac{5}{9} + 4\frac{8}{11} + 6\frac{10}{17}$

d. $14\frac{3}{13} - 3\frac{9}{16}$

Extended-Response Questions

17. To get to a convention, Mr. Sims must drive $1\frac{1}{4}$ hours and then fly for $5\frac{2}{3}$ hours. If he leaves at 8:00 A.M., will he arrive by 4:00 P.M.? Explain your thinking.

18. Ms. Lopez found that 17 out of 20 students in her first period class passed the math test. In her second period class, 22 out of 25 students passed. In which class did a larger part pass the test? Show your work.

19. Tayo estimated that he spends $\frac{1}{3}$ of his day sleeping and $\frac{7}{24}$ of his day at school. At which activity does he spend more time? Explain your answer.

20. Janet worked $7\frac{1}{4}$ hours on Monday, $8\frac{3}{4}$ hours on Tuesday, $9\frac{2}{3}$ hours on Wednesday, $7\frac{1}{2}$ hours on Thursday, and $9\frac{1}{3}$ hours on Friday. She receives overtime pay for every hour she works over 40 hours. Estimate how many hours of overtime she will be paid for. How did you arrive at your estimate?

3.8 Adding Fractions and Mixed Numbers

Remember that fractions must have like denominators before they can be added.

To add fractions, do the following:

- If necessary, write equivalent fractions using the LCD.
- Add the numerators. Use the like denominator as the denominator in your answer. Write the answer in simplest form.
- For mixed numbers, add the fractions, then add the whole numbers.

Model Problems

1. Add $\frac{5}{12} + \frac{11}{12}$.

Solution $\frac{5}{12} + \frac{11}{12} = \frac{5+11}{12} = \frac{16}{12}$. Rewrite as a mixed number in lowest terms.

Answer $\frac{16}{12} = 1\frac{4}{12} = 1\frac{1}{3}$

2. Add $\frac{2}{5} + \frac{2}{3}$.

Solution Rewrite using the LCD of 5 and 3, which is 15.

$$\frac{2}{5} \times \frac{3}{3} = \frac{6}{15} \quad \frac{2}{3} \times \frac{5}{5} = \frac{10}{15}$$

Note: Multiplying the fractions by $\frac{3}{3}$ and $\frac{5}{5}$ does not change the *value* of the fractions because it is the same as multiplying by 1.

Answer Adding the rewritten fractions, you get

$\frac{6}{15} + \frac{10}{15} = \frac{6+10}{15} = \frac{16}{15}$, which can be written as the mixed number $1\frac{1}{15}$.

3. Add $7\frac{3}{4} + 5\frac{7}{8} + 8\frac{9}{16}$.

Solution Rewrite using the LCD of 16.

$$7\frac{3}{4} = 7\frac{12}{16}$$

$$5\frac{7}{8} = 5\frac{14}{16}$$

$$+ 8\frac{9}{16} = 8\frac{9}{16}$$
$$\overline{\phantom{+ 8\frac{9}{16} = }\,20\frac{35}{16}}$$

Answer Simplify. $20\frac{35}{16} = 20 + 2\frac{3}{16} = 22\frac{3}{16}$

4. Ben hiked $4\frac{4}{5}$ miles on Saturday. He hiked $1\frac{3}{4}$ miles more on Sunday than Saturday. How many miles did he hike in all?

Solution   First, find Sunday's distance.

$$4\frac{4}{5} = 4\frac{16}{20}$$
$$+ 1\frac{3}{4} = 1\frac{15}{20}$$
$$5\frac{31}{20} = 5 + 1\frac{11}{20} = 6\frac{11}{20}$$

Then find the total distance.
$$4\frac{4}{5} = 4\frac{16}{20}$$
$$+ 6\frac{11}{20} = 6\frac{11}{20}$$
$$10\frac{27}{20} = 10 + 1\frac{7}{20} = 11\frac{7}{20}$$

Answer   Ben hiked $11\frac{7}{20}$ miles.

Practice

Multiple-Choice Questions

1. Which sum is greater than 1?

  A. $\frac{3}{8} + \frac{4}{8}$ B. $\frac{2}{6} + \frac{3}{6}$

  C. $\frac{5}{9} + \frac{4}{9}$ D. $\frac{7}{10} + \frac{7}{10}$

2. Add $\frac{8}{9} + \frac{5}{6}$.

  F. $1\frac{5}{9}$ G. $1\frac{1}{3}$

  H. $1\frac{13}{18}$ J. $2\frac{1}{18}$

3. Add $\frac{5}{5} + \frac{7}{8}$.

  A. $\frac{3}{5}$ B. $\frac{4}{5}$

  C. $1\frac{7}{8}$ D. $1\frac{2}{15}$

4. Add $\frac{5}{6} + \frac{3}{2} + \frac{2}{3}$.

  F. $\frac{10}{11}$ G. $1\frac{3}{4}$

  H. $1\frac{23}{24}$ J. 3

5. Add $3\frac{7}{10} + 2\frac{3}{4}$.

  A. $5\frac{5}{7}$ B. $6\frac{9}{20}$

  C. $6\frac{5}{7}$ D. $7\frac{1}{10}$

6. Advantage Industries' stock was priced at $36\frac{5}{8}$. The stock gained $1\frac{3}{4}$ points. What was the new price of the stock?

  F. $37\frac{1}{8}$ G. $37\frac{2}{3}$

  H. $37\frac{3}{4}$ J. $38\frac{3}{8}$

7. Which sum is a whole number?

A. $\dfrac{4}{10} + \dfrac{4}{10}$ B. $\dfrac{5}{9} + \dfrac{5}{9}$

C. $\dfrac{7}{12} + \dfrac{7}{12}$ D. $\dfrac{8}{16} + \dfrac{8}{16}$

8. Which statement is true?

F. The sum of two mixed numbers is always a whole number.

G. The sum of two mixed numbers is never a whole number.

H. The sum of two fractions is always greater than 1.

J. The sum of two mixed numbers is always greater than 1.

Short-Response Questions

9. Add. Write answers in simplest form.

a. $\dfrac{3}{7} + \dfrac{2}{7}$ b. $\dfrac{11}{29} + \dfrac{6}{29}$

c. $\dfrac{8}{16} + \dfrac{4}{16}$ d. $\dfrac{13}{15} + \dfrac{8}{15}$

10. Add. Write answers in simplest form.

a. $\begin{aligned}&\ \ \dfrac{4}{5}\\ +&\ \dfrac{7}{10}\\ \hline\end{aligned}$ b. $\begin{aligned}&\ \ \dfrac{2}{3}\\ +&\ \dfrac{3}{4}\\ \hline\end{aligned}$

c. $\begin{aligned}&\ \ \dfrac{1}{6}\\ &\ \ \dfrac{3}{8}\\ +&\ \dfrac{1}{2}\\ \hline\end{aligned}$ d. $\begin{aligned}&\ \ \dfrac{5}{9}\\ &\ \ \dfrac{3}{4}\\ +&\ \dfrac{7}{12}\\ \hline\end{aligned}$

11. Add. Write answers in simplest form.

a. $\begin{aligned}&\ 8\dfrac{2}{3}\\ +&\ 3\dfrac{5}{6}\\ \hline\end{aligned}$ b. $\begin{aligned}&\ 3\dfrac{3}{5}\\ +&\ 2\dfrac{2}{3}\\ \hline\end{aligned}$

c. $\begin{aligned}&\ 4\dfrac{5}{12}\\ &\ 3\dfrac{1}{6}\\ +&\ 7\dfrac{7}{8}\\ \hline\end{aligned}$ d. $\begin{aligned}&\ 6\dfrac{3}{10}\\ &\ 9\dfrac{2}{3}\\ +&\ 4\dfrac{3}{4}\\ \hline\end{aligned}$

Extended-Response Questions

12. Grace mixed $\dfrac{3}{4}$ pound cashews, $\dfrac{7}{8}$ pound almonds, and $1\dfrac{1}{2}$ pounds of peanuts. What was the total weight of the nut mixture?

13. A package of ground beef weighs $2\dfrac{4}{5}$ pounds. A package of ground pork weighs $1\dfrac{7}{10}$ pounds more than the beef. What is the combined weight of the packages?

14. On his bike, Daryl rode $\dfrac{2}{3}$ mile from his house to the library and then $\dfrac{5}{8}$ mile to the market. His return trip home was $\dfrac{1}{4}$ mile longer than going. What was the total distance Daryl rode?

15. Tina spent $1\dfrac{1}{3}$ hours on math homework, 45 minutes on history, $\dfrac{1}{2}$ hour on English, and 25 minutes on art. How many hours did Tina spend on homework?

16. a. Show that
$$\left(1\dfrac{1}{2} + 2\dfrac{1}{5}\right) + 4\dfrac{1}{3} = 1\dfrac{1}{2} + \left(2\dfrac{1}{5} + 4\dfrac{1}{3}\right).$$

b. What property have you demonstrated?

3.9 Subtracting Fractions and Mixed Numbers

Addition and subtraction are inverse operations that have the same rules. When subtracting fractions, you must follow the same rules as when adding fractions.

- To subtract fractions with like denominators, subtract the numerators. The denominator remains the same. Simplify if possible.
- To subtract fractions with unlike denominators, first write equivalent fractions using the LCD. Then use the procedure for like fractions.

Model Problems

1. Chris had a piece of wire $\frac{7}{8}$ inch long. He cut off $\frac{3}{16}$ inch. How long is the original piece now?

   Solution Subtract $\frac{7}{8} - \frac{3}{16}$. The LCD is 16.

   $$\frac{7}{8} = \frac{14}{16} \quad \text{Subtract numerators. } 14 - 3 = 11$$
   $$-\frac{3}{16} = \frac{3}{16}$$
   $$\overline{\qquad \frac{11}{16}}$$

   Answer The length of the wire is $\frac{11}{16}$ inch.

2. Subtract $\frac{7}{4} - \frac{2}{3}$.

   Solution The LCD is 12.
   $$\frac{7}{4} = \frac{21}{12}$$
   $$-\frac{2}{3} = \frac{8}{12}$$
   $$\overline{\qquad \frac{13}{12}}$$

   Answer The difference is $\frac{13}{12}$, which can also be written as the mixed number $1\frac{1}{12}$.

When you subtract whole numbers, it is sometimes necessary to use renaming. The same situation may occur when a mixed number is subtracted from a whole number or another mixed number.

- To subtract mixed numbers, first write equivalent fractions if necessary. Subtract the fractions, then subtract the whole numbers. Simplify if possible.
- If the fraction part of the bottom number is greater than the fraction part of the top number, rename the top number. Then follow the steps for subtraction.

---

3. Lee worked $7\frac{2}{3}$ hours. Kim worked $4\frac{1}{2}$ hours. How many more hours did Lee work?

Solution   Subtract $7\frac{2}{3} - 4\frac{1}{2}$.

$$7\frac{2}{3} = 7\frac{4}{6}$$
$$-4\frac{1}{2} = 4\frac{3}{6}$$
$$\rule{3cm}{0.4pt}$$
$$3\frac{1}{6}$$

Answer   Lee worked $3\frac{1}{6}$ hours more.

4. Inez bought 8 yards of fabric. She used $3\frac{3}{4}$ yards for a dress. How much fabric does she have left?

Solution   Subtract $8 - 3\frac{3}{4}$.
Rename 8 as a mixed number with a denominator of 4.

$$8 = 7 + 1$$
$$8 = 7 + \frac{4}{4} = 7\frac{4}{4}$$
$$-3\frac{3}{4} = 3\frac{3}{4}$$
$$\rule{3cm}{0.4pt}$$
$$4\frac{1}{4}$$

Answer   Inez has $4\frac{1}{4}$ yards left.

5. Subtract $11\frac{5}{6} - 5\frac{7}{8}$.

Solution   The LCD of 6 and 8 is 24.

$$11\frac{5}{6} = 11\frac{20}{24} = 10\frac{44}{24} \quad \text{Since } \frac{21}{24} > \frac{20}{24}, \text{ rename } 11\frac{20}{24} \text{ as } 10\frac{44}{24}.$$

$$-5\frac{7}{8} = \ \ 5\frac{21}{24} = \ \ 5\frac{21}{24}$$
$$\rule{4cm}{0.4pt}$$
$$5\frac{23}{24}$$

 **Practice**

Multiple-Choice Questions

1. Find the missing denominator.
   $$\frac{11}{12} - \frac{5}{12} = \frac{1}{[\ ]}$$
   A. 2   B. 3
   C. 6   D. 12

2. Subtract $\frac{4}{5} - \frac{1}{20}$.
   F. $\frac{11}{20}$ G. $\frac{3}{10}$
   H. $\frac{3}{4}$ J. $\frac{17}{20}$

3. Subtract $\frac{9}{3} - \frac{2}{7}$.
   A. 1   B. $1\frac{6}{7}$
   C. $2\frac{5}{7}$ D. $2\frac{2}{3}$

4. Beth used $3\frac{7}{8}$ yards of cotton to make a
   dress and $2\frac{3}{8}$ yards to make a jacket.
   How much more fabric did she use for
   the dress than the jacket?
   F. $1\frac{1}{4}$ yd   G. $1\frac{1}{2}$ yd
   H. 6 yd   J. $6\frac{1}{4}$ yd

5. Walter promised to do 12 hours of vol-
   unteer work this week. So far, he has
   done $5\frac{3}{5}$ hours. How many more hours
   must he work this week?
   A. $5\frac{4}{5}$ B. $6\frac{2}{5}$
   C. $6\frac{4}{5}$ D. $7\frac{1}{5}$

6. A jeweler cut $\frac{1}{2}$ of an inch off a bracelet
   that was $6\frac{3}{16}$ inches long. How long is
   the bracelet now?
   F. $5\frac{3}{4}$ in.   G. $5\frac{3}{16}$ in.
   H. $5\frac{9}{16}$ in.   J. $5\frac{11}{16}$ in.

7. Find the missing addend:
   $2\frac{5}{6} + [\ ] = 7\frac{2}{3}$.
   A. $4\frac{1}{6}$ B. $4\frac{5}{6}$
   C. $5\frac{1}{6}$ D. $10\frac{1}{2}$

8. Subtract $4\frac{9}{10}$ from $7\frac{1}{6}$.
   F. $2\frac{1}{10}$ G. $2\frac{1}{5}$
   H. $2\frac{4}{15}$ J. $3\frac{11}{15}$

9. What number must be subtracted
   from $12\frac{1}{8}$ so that the result is $2\frac{1}{2}$?
   A. $14\frac{5}{8}$ B. $9\frac{5}{8}$
   C. $9\frac{3}{8}$ D. $8\frac{7}{8}$

10. From a 10-foot board, a carpenter
    sawed off pieces that were $2\frac{3}{4}$ feet and
    $3\frac{2}{3}$ feet long. How much of the board
    is left?
    F. $3\frac{7}{12}$ ft   G. $3\frac{11}{12}$ ft
    H. $4\frac{1}{6}$ ft   J. $6\frac{1}{3}$ ft

Short-Response Questions

11. Subtract. Write answers in lowest terms.

a. $\dfrac{9}{10}$
$-\dfrac{7}{10}$

b. $\dfrac{11}{12}$
$-\dfrac{3}{8}$

c. $\dfrac{5}{3}$
$-\dfrac{7}{15}$

d. $\dfrac{7}{9}$
$-\dfrac{7}{12}$

12. Subtract. Write answers in lowest terms.

a. $8\dfrac{3}{8} - 3\dfrac{7}{8} =$

b. $9 - 4\dfrac{3}{10} =$

c. $6\dfrac{5}{8} - 2 =$

d. $12\dfrac{1}{6} - 4\dfrac{5}{6} =$

13. Subtract. Write answers in lowest terms.

a. $5\dfrac{1}{5} - 1\dfrac{3}{4} =$

b. $14\dfrac{1}{8} - 6\dfrac{9}{10} =$

c. $11\dfrac{1}{2} - 3\dfrac{2}{7} =$

d. $10\dfrac{2}{15} - 4\dfrac{1}{6} =$

14. Write each answer in lowest terms.

a. $\dfrac{7}{2} - \dfrac{3}{4} - \dfrac{5}{8} =$

b. $1\dfrac{2}{3} + \dfrac{5}{6} - \dfrac{3}{10} =$

c. $3\dfrac{1}{4} - 1\dfrac{2}{3} + \dfrac{1}{6} =$

d. $\dfrac{16}{9} - \dfrac{4}{3} + \dfrac{11}{1} =$

15. Find the missing numbers in each pattern.

a. $8\dfrac{1}{3}, 7\dfrac{2}{3}, 7, \underline{?}, \underline{?}$

b. $12, 11\dfrac{3}{4}, 11\dfrac{1}{4}, 10\dfrac{1}{2}, \underline{?}, \underline{?}$

c. $9\dfrac{4}{5}, 9\dfrac{7}{10}, 9\dfrac{1}{2}, 9\dfrac{1}{5}, \underline{?}, \underline{?}$

16. Mr. Logan spends $\dfrac{3}{8}$ of his income on rent, $\dfrac{1}{4}$ of his income on food, and $\dfrac{1}{5}$ of his income on clothing. What part of his income is left to spend on other items?

Extended-Response Questions

17. Andrea's weekly time sheet is shown below.

Day	Mon.	Tue.	Wed.	Thur.	Fri.	Sat.
Hours worked	$5\dfrac{1}{2}$	$5\dfrac{2}{3}$	$7\dfrac{3}{4}$	$6\dfrac{1}{4}$	$5\dfrac{1}{3}$	

a. How many more hours did Andrea work on Wednesday than Friday?

b. If Andrea wants to work a total of 35 hours this week, how many hours must she work on Saturday?

18. A pack of flour contains 10 cups. Stan used $2\dfrac{7}{8}$ cups to make bread and $1\dfrac{3}{4}$ cups to make cookies. How many cups of flour are left?

19. Study the Input/Output table.

a. What rule is being used to change the input to the output?

b. Complete the table.

Input	Output
$1\dfrac{1}{3}$	$\dfrac{7}{12}$
$2\dfrac{3}{5}$	$1\dfrac{17}{20}$
$3\dfrac{5}{6}$	[ ]
[ ]	$3\dfrac{5}{8}$

20. Sheila and Dan each travel to school by bicycle. Sheila's house is $1\dfrac{5}{8}$ miles from school. Dan's house is $1\dfrac{1}{3}$ miles from school. How much longer is Sheila's round-trip than Dan's?

3.10 Multiplying Fractions and Mixed Numbers

The figure shows that $\frac{1}{2}$ of $\frac{3}{4}$ is $\frac{3}{8}$. Remember that the word *of* often indicates that the operation to use is multiplication.

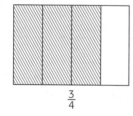

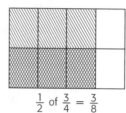

$\frac{3}{4}$

$\frac{1}{2}$ of $\frac{3}{4} = \frac{3}{8}$

$\frac{1}{2}$ of $\frac{3}{4}$ means $\frac{1}{2} \times \frac{3}{4} = \frac{3}{8}$

To multiply two fractions:

- Try to simplify before multiplying because it is easier to work with smaller numbers. If possible, divide a numerator and a denominator by a common factor.
- Multiply the numerators. Then multiply the denominators. Write the product in lowest terms.

Model Problems

1. Multiply: $\frac{2}{5} \times \frac{7}{12}$.

   Solution   Divide 2 and 12 by their common factor of 2. Then multiply.

   $$\frac{\overset{1}{2}}{5} \times \frac{7}{\underset{6}{12}} = \frac{1 \times 7}{5 \times 6} = \frac{7}{30} \quad \begin{matrix} \leftarrow \text{Multiply numerators} \\ \leftarrow \text{Multiply denominators} \end{matrix}$$

2. Multiply: $\frac{3}{4} \times \frac{8}{9}$.

   Solution $\dfrac{\overset{1}{3}}{\underset{1}{4}} \times \dfrac{\overset{2}{8}}{\underset{3}{9}} = \dfrac{1 \times 2}{1 \times 3} = \dfrac{2}{3}$

   Note: If you do not simplify first, you will still get the same answer eventually.

   $$\frac{3}{4} \times \frac{8}{9} = \frac{24}{36} = \frac{2}{3}$$

For problems 3 and 4, to multiply with mixed numbers or whole numbers, write these numbers as improper fractions, and then multiply the fractions.

3. Five eighths of people surveyed owned a computer. If there were 120 people in the survey, how many owned a computer?

Solution Find $\frac{5}{8}$ of 120.

Write 120 as $\frac{120}{1}$ and multiply. $\quad \frac{5}{8_1} \times \frac{\cancel{120}^{15}}{1} = \frac{5 \times 15}{1 \times 1} = 75$

Answer There were 75 computer owners.

4. A recipe calls for $1\frac{1}{3}$ cups tomato sauce. If Julia needs $2\frac{1}{2}$ times more servings than the recipe makes, how many cups of tomato sauce should she use?

Solution Find $2\frac{1}{2}$ times the amount of tomato sauce.

$2\frac{1}{2} \times 1\frac{1}{3} = \frac{5}{2_1} \times \frac{\cancel{4}^2}{3} = \frac{5 \times 2}{1 \times 3} = \frac{10}{3} = 3\frac{1}{3}$

Answer $3\frac{1}{3}$ cups of tomato sauce will be needed.

Practice

Multiple-Choice Questions

1. Which situation is shown in the figure?

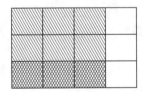

A. $\frac{1}{2}$ of $\frac{2}{3}$ B. $\frac{1}{3}$ of $\frac{3}{4}$

C. $\frac{1}{4}$ of $\frac{5}{6}$ D. $\frac{1}{3}$ of 9

2. Multiply $\frac{5}{6} \times \frac{3}{10}$.

F. $\frac{1}{12}$ G. $\frac{1}{4}$

H. $\frac{1}{3}$ J. $\frac{1}{2}$

3. Find the missing number. $\frac{5}{7} \times \frac{}{5} = 1$

A. 35   B. 25
C. 7   D. 5

4. Multiply $\frac{3}{8} \times 14$.

F. $5\frac{1}{4}$ G. $2\frac{5}{8}$

H. $18\frac{2}{3}$ J. $10\frac{1}{2}$

5. $\left(\frac{2}{3} \times 15\right) \times \frac{1}{5} =$

A. $\frac{1}{3}$ B. 5

C. $10\frac{2}{5}$ D. 2

6. Which is NOT equivalent to $\frac{3}{4} \times 8$?

F. 6

G. $\frac{3}{2} \times 4$

H. $\frac{16}{4}$

J. $\frac{3}{4} + \frac{3}{4} + \frac{3}{4} + \frac{3}{4} + \frac{3}{4} + \frac{3}{4} + \frac{3}{4} + \frac{3}{4}$

7. Alicia bought $1\frac{3}{4}$ pounds of chicken salad at $6.80 per pound. How much did she spend?

   A. $7.95   B. $9.80
   C. $11.90   D. $13.30

8. Each bag of peanuts weighs $1\frac{7}{8}$ pounds. How much do $1\frac{1}{3}$ bags of peanuts weigh?

   F. $2\frac{1}{2}$ lb   G. $3\frac{1}{3}$ lb
   H. $4\frac{1}{6}$ lb   J. 5 lb

9. Which expression can be used to find $2\frac{5}{6}$ of 30?

   A. $(2 \times 30) \times \left(\frac{5}{6} \times 30\right)$

   B. $(2 \times 30) + \frac{5}{6}$

   C. $\left(2 \times \frac{5}{6}\right) + \left(30 \times \frac{5}{6}\right)$

   D. $(2 \times 30) + \left(\frac{5}{6} \times 30\right)$

10. A tailor used $3\frac{7}{8}$ yards of wool that cost $12 a yard. The best estimate for the total cost of the wool is

   F. less than $36
   G. exactly $42
   H. less than $48
   J. more than $54

Short-Response Questions

11. Multiply. Write each product in lowest terms.

   a. $\frac{2}{3} \times \frac{5}{6}$ b. $\frac{7}{9} \times \frac{3}{10}$

   c. $\frac{5}{8} \times \frac{4}{15}$ d. $\frac{3}{4} \times \frac{7}{12}$

12. Multiply. Write each product in lowest terms.

   a. $\frac{2}{3} \times 16$ b. $30 \times \frac{11}{12}$

   c. $156 \times \frac{3}{4}$ d. $\frac{5}{6} \times 222$

13. Multiply. Write each product in lowest terms.

   a. $\frac{2}{3} \times 5\frac{3}{4}$ b. $\frac{5}{8} \times 7\frac{3}{5}$

   c. $2\frac{2}{3} \times \frac{3}{8}$ d. $2\frac{2}{7} \times 4\frac{9}{10}$

14. Complete. Write *less than, greater than,* or *equal to.*

   a. The product of a whole number greater than 1 and a proper fraction is always _____ the whole number.

   b. The product of a whole number greater than 1 and a mixed number is always _____ the whole number.

   c. The product of two improper fractions is always _____ 1.

15. Find each answer. Do all work in parentheses first.

   a. $\left(2 \times \frac{3}{8}\right) \times \frac{2}{9}$

   b. $\left(8 \times 1\frac{1}{4}\right) - 3$

   c. $\left(\frac{4}{5} \times 12\frac{1}{2}\right) \times \left(6\frac{1}{4} \times \frac{2}{5}\right)$

   d. $\left(\frac{3}{4} \times 2\frac{2}{5}\right) - \left(\frac{5}{6} \times 1\frac{11}{25}\right)$

Extended-Response Questions

16. Rita wants to make Savory Mushrooms for 10 people.

   a. By what factor must she increase the recipe?

   b. Complete the table to show how much she needs of each ingredient.

Savory Mushrooms		
Ingredient	Serves 4	Serves 10
Mushrooms	$1\frac{1}{2}$ lb	
Olive Oil	$\frac{1}{4}$ c	
Seasoned bread crumbs	$\frac{2}{3}$ c	
Parsley flakes	2 tbsp	
Garlic salt	$\frac{1}{2}$ tbsp	

17. Carlos earns $360 per month working part-time as an usher at the movie theater. The graph shows how Carlos spends his earnings. How much does he spend on each item?

Carlos's Budget

Savings $\frac{1}{10}$
Food $\frac{1}{4}$
Entertainment $\frac{3}{10}$
All Other Expenses $\frac{3}{20}$
Clothing $\frac{1}{5}$

a. Food
b. Clothing
c. Entertainment
d. Savings
e. Other
f. What should be the sum of your answers for a–e?

18. If Carina takes the bus to her aunt's house, the trip takes $3\frac{1}{2}$ hours. By train, the trip takes $\frac{2}{3}$ as long. How many hours does the trip take by train?

19. Mr. Chang worked $12\frac{3}{4}$ hours overtime last week. If he earns $20.60 for each hour of overtime, how much overtime pay did he earn?

20. With 10 pounds of jelly beans, Mike could fill $6\frac{2}{3}$ jars.

a. Complete the table.

Pounds of Jelly Beans	5	10	20	30
Jars Filled		$6\frac{2}{3}$		

b. How many jars could Mike fill with 25 pounds of jelly beans?

3.11 Dividing Fractions and Mixed Numbers

You have seen that 1 is a very useful number. As a fraction (like $\frac{3}{3}$ or $\frac{8}{8}$) 1 makes it possible to change the appearance of numbers without changing their value.

It is also useful to make any number equal to 1. This can be done easily by multiplying a number by its reciprocal. Two numbers whose product is 1 are **reciprocals** (or **multiplicative inverses**). Exchanging the numerator and denominator of a fraction forms its reciprocal.

Examples

1. To make 7 into 1, first write 7 as a fraction, and then multiply it by its reciprocal.

$$7 = \frac{7}{1} \qquad \frac{7}{1} \times \frac{1}{7} = \frac{7}{7} = 1$$

2. The reciprocal of $\frac{4}{5}$ is $\frac{5}{4}$. $\qquad \frac{4}{5} \times \frac{5}{4} = 1$

3. Since $2\frac{2}{4} = \frac{11}{4}$, the reciprocal of $2\frac{3}{4} = \frac{4}{11}$. $\qquad \frac{11}{4} \times \frac{4}{11} = 1$

One instance of when reciprocals are useful is when you need to divide fractions. Division is the inverse operation of multiplication. So, if you divide by a number, it is the same as multiplying by its reciprocal.

$$20 \div 4 = 5 \qquad 20 \times \frac{1}{4} = 5$$

- To divide by a fraction, multiply by its reciprocal.
- To divide with mixed numbers or whole numbers, first write these numbers as improper fractions. Then divide the fractions.

Model Problems

1. A piece of wood $\frac{3}{4}$ yard long is to be cut into 6 equal pieces. What is the length of each piece?

    Solution Divide: $\frac{3}{4} \div 6$

    The divisor is 6 or $\frac{6}{1}$. Its reciprocal is $\frac{1}{6}$.

    Multiply by the reciprocal: $\frac{3^1}{4} \times \frac{1}{6_2} = \frac{1}{8}$

    Answer Each piece is $\frac{1}{8}$ yd long.

2. Divide. $\frac{3}{8} \div \frac{3}{5}$

    Solution The reciprocal of $\frac{3}{5}$ is $\frac{5}{3}$.

    $$\frac{3}{8} \div \frac{3}{5} = \frac{3^1}{8} \times \frac{5}{3_1} = \frac{5}{8}$$

3. It took Sarah $4\frac{1}{2}$ hours to paint one room. What part of the job did she complete in 1 hour?

    Solution Divide 1 room by $4\frac{1}{2}$ hours.

    $$4\frac{1}{2} = \frac{9}{2} \qquad \text{The multiplicative inverse of } \frac{9}{2} \text{ is } \frac{2}{9}.$$

    $$1 \div 4\frac{1}{2} = 1 \div \frac{9}{2}$$

$$1 \div \frac{9}{2} = 1 \times \frac{2}{9} = \frac{2}{9}$$

Answer   Sarah completed $\frac{2}{9}$ of the job in 1 hour.

4. A utility worker cut a wire that was $40\frac{1}{2}$ feet long into pieces that were $3\frac{3}{8}$ feet long. How many pieces were there?

Solution   Divide. $40\frac{1}{2} \div 3\frac{3}{8}$

Write the reciprocal of the divisor: $\frac{8}{27}$

Multiply: $\frac{\cancel{81}3}{2} \times \frac{\cancel{8}4}{\cancel{27}} = 12$

Answer   There were 12 pieces of wire.

Practice

Multiple-Choice Questions

1. Which expression is equivalent to $\frac{3}{12} \div \frac{2}{5}$?

   A. $\frac{3}{12} \times \frac{2}{5}$ B. $\frac{3}{12} \times \frac{5}{2}$

   C. $\frac{12}{3} \div 5$ D. $\frac{12}{3} \times \frac{2}{5}$

2. Which expression is equivalent to $\frac{5}{6} \times \frac{1}{2}$?

   F. $\frac{5}{6} \div 2$ G. $\frac{5}{6} \times 2$

   H. $\frac{6}{5} \div 2$ J. $\frac{6}{5} \div 2$

3. $\frac{5}{12} \div \frac{5}{6} =$

   A. $\frac{25}{72}$ B. $2\frac{1}{2}$

   C. $\frac{1}{2}$ D. 2

4. Lana has 12 yards of fabric for pillows. Each pillow takes $\frac{2}{3}$ of a yard. How many pillows can she make?

   F. 8   G. 10
   H. 18   J. 24

5. Mr. Nolan is making 6 sandwiches from $\frac{3}{4}$ pounds of turkey. How much turkey is on each sandwich?

   A. $\frac{1}{8}$ lb   B. $\frac{1}{3}$ lb
   C. $\frac{9}{2}$ lb   D. 4.5 lb

6. Divide. $8\frac{2}{5} \div 1\frac{1}{2}$

   F. $12\frac{3}{5}$ G. $2\frac{4}{5}$
   H. $5\frac{2}{3}$ J. $5\frac{3}{5}$

7. The best estimate of $31\frac{3}{4} \div 4\frac{1}{7}$ is

   A. 5   B. 6
   C. 8   D. 9

8. Which quotient is about 7?

   F. $23\frac{3}{4} \div 4\frac{1}{6}$ G. $48\frac{7}{8} \div 7\frac{3}{10}$
   H. $35\frac{8}{9} \div 3\frac{6}{7}$ J. $56\frac{1}{5} \div 6\frac{1}{9}$

9. Divide. $\frac{5}{12} \div \frac{5}{9}$

   A. $\frac{25}{108}$ B. $\frac{3}{4}$
   C. $1\frac{1}{3}$ D. $3\frac{3}{4}$

10. $0 \div 1\frac{7}{8} =$

    F. 0   G. $\frac{8}{15}$

    H. $\frac{4}{7}$ J. $\frac{15}{8}$

Short-Response Questions

11. Divide. Write each answer in lowest terms.

    a. $\frac{2}{9} \div \frac{1}{3}$ b. $\frac{2}{5} \div \frac{1}{6}$

    c. $\frac{5}{8} \div \frac{5}{7}$ d. $\frac{2}{7} \div \frac{5}{9}$

12. Divide. Write each answer in lowest terms.

    a. $10 \div \frac{5}{8}$ b. $6 \div \frac{5}{6}$

    c. $4\frac{4}{5} \div 9$ d. $1\frac{7}{8} \div 4$

13. Divide. Write each answer in lowest terms.

    a. $6\frac{7}{8} \div 2\frac{3}{4}$ b. $5\frac{3}{4} \div \frac{5}{8}$

    c. $3\frac{1}{2} \div 4\frac{1}{5}$ d. $2\frac{7}{10} \div 4\frac{4}{5}$

Extended-Response Questions

14. Ted can mow the lawn in 3 hours. Carol takes 4 hours to mow the lawn.

    a. What part of the lawn can Ted mow in 1 hour?

    b. What part of the lawn can Carol mow in 1 hour?

    c. If Ted and Carol work together, what part of the lawn can they mow in 1 hour?

15. Gina has 7 yards of ribbon for bows. Each bow takes $\frac{2}{3}$ yard. How many bows can Gina make? Explain.

16. Find each quotient.

    a. $\frac{1}{2} \div \frac{1}{2}$

    b. $\frac{1}{2} \div \frac{1}{2} \div \frac{1}{2}$

    c. $\frac{1}{2} \div \frac{1}{2} \div \frac{1}{2} \div \frac{1}{2}$

    d. Find $\frac{1}{2} \div \frac{1}{2} \div \frac{1}{2} \div \frac{1}{2} \div \frac{1}{2}$ without dividing.

    e. Write an expression equal to 64 using $\frac{1}{2}$'s.

17. If $3\frac{1}{2}$ pounds of mushrooms cost $8.40, what is the cost of $1\frac{1}{2}$ pounds of mushrooms?

18. Find the missing factor: $6\frac{1}{2} \times ? = 23\frac{2}{5}$

19. If 2 is divided by the fraction $\frac{a}{b}$, what is the quotient?

20. A company packs its fruit salad in containers that hold $1\frac{5}{8}$ pounds. How many containers does it need to hold 585 pounds of fruit salad?

Chapter 3 Review

Multiple-Choice Questions

1. The prime factorization of 135 is

    A. 27×5 B. $3 \times 3 \times 5$
    C. $3^3 \times 5$ D. $3^2 \times 5^2$

2. Which number is NOT divisible by 6?

    F. 558   G. 772
    H. 846   J. 978

3. What is the greatest common factor of 52, 91, and 143?

    A. 2   B. 3
    C. 7   D. 13

4. What is the least common multiple of 4, 12, and 20?

    F. 48   G. 60
    H. 120   J. 240

5. Which set of numbers is in order from least to greatest?

    A. $5.375, 5\frac{1}{3}, 5\frac{5}{12}, 5.5$

    B. $5\frac{1}{3}, 5\frac{5}{12}, 5.375, 5.5$

    C. $5\frac{1}{3}, 5.375, 5\frac{5}{12}, 5.5$

    D. $5.375, 5\frac{5}{12}, 5\frac{1}{3}, 5.5$

6. Which number is NOT equivalent to the others?

    F. $\frac{16}{6}$ G. 2.666

    H. $2\frac{2}{3}$ J. $2.\overline{6}$

7. Which sum is the greatest?

    A. $5\frac{2}{3} + 3\frac{1}{2} + 2\frac{3}{8}$

    B. $2\frac{5}{6} + 9\frac{4}{15}$

    C. $4\frac{8}{11} + 4\frac{5}{8} + 4\frac{9}{10}$

    D. $5\frac{11}{12} + 5\frac{7}{8}$

8. When $1\frac{4}{5}$ is added to a number, the sum is $5\frac{3}{10}$. Find the number.

    F. $4\frac{1}{2}$ G. $4\frac{1}{5}$

    H. $3\frac{7}{10}$ J. $3\frac{1}{2}$

9. A hiking trail is $3\frac{3}{4}$ miles long. Amy hiked $\frac{2}{3}$ of the way, then stopped to rest. How much farther must she walk to complete the trail?

    A. $1\frac{1}{4}$ miles   B. $1\frac{3}{4}$ miles

    C. $2\frac{1}{2}$ miles   D. $2\frac{1}{12}$ miles

10. A bottle of perfume containing $10\frac{1}{2}$ ounces is used to fill $\frac{3}{4}$-ounce sample bottles. How many sample bottles can be filled?

    F. $7\frac{7}{8}$ G. $9\frac{1}{3}$

    H. 14   J. 28

Short-Response Questions

11. Find the answer in simplest form.
    $3\frac{3}{5} \div \left(\frac{2}{5} \times \frac{3}{4}\right)$

12. What number multiplied by $1\frac{1}{4}$ is equal to $3\frac{1}{4}$? Write the answer in simplest form.

13. Find $\frac{1}{3} \div \frac{1}{3} \div \frac{1}{3} \div \frac{1}{3} \div \frac{1}{3}$.

Extended-Response Questions

14. When she began painting her room, Vicki had $2\frac{2}{3}$ gallons of paint. She used $\frac{3}{4}$ of the paint to complete the room. How much paint does she have left?

15. The Wu family budget allows $\frac{7}{20}$ for rent and utilities, $\frac{3}{10}$ for food, $\frac{1}{4}$ for clothing and recreation, and the rest for savings.

a. What part of the Wu budget is allowed for savings?

b. If the family's monthly income is $3,000, how much money does the budget allow for rent and utilities? For food?

16. Ari skated $3\frac{7}{12}$ miles, Bryan skated 3.65 miles, and Courtney skated $3\frac{5}{8}$ miles.

a. Write the skaters in order from greatest to least distance skated.

b. What was the difference between the distances for Ari and Courtney?

17. At the Country Fresh Food Market, every 12th customer gets $1.00 off his or her purchase. Every 30th customer gets a free bag of apples.

a. Which customer will be the first to save $1.00 *and* get the apples?

b. If there are 500 customers between 9 A.M. and noon, how many will get both items?

18. In a shipment of 108 T-shirts, $\frac{2}{3}$ of them were size large. In another shipment of 152 T-shirts, $\frac{3}{4}$ were size large.

a. What is the total number of large T-shirts?

b. What fraction of the total was NOT size large? Write the answer in simplest form.

19. Martha works 8 hours each day at *Surfer* magazine. The table shows how she spends her time. How much time does Martha spend surfing during a 5-day week? Show your work.

Activity	Time (in hours)
Research	$2\frac{1}{2}$
Writing	$1\frac{3}{4}$
Lunch	$\frac{3}{4}$
Editing	$1\frac{1}{4}$
Meetings	1
Surfing	?

20. A developer plans to build homes on $\frac{5}{8}$ of a 112-acre plot of land. On $\frac{2}{3}$ of the remaining land, there will be stores and service buildings. The rest of the land will be used for trees and gardens. How many acres will be used for trees and gardens? Show your work.

Multiple-Choice Questions

1. What is the standard form of the number $7 \times 10^4 + 4 \times 10^2 + (8 \times 1) + \left(6 \times \frac{1}{10^2}\right) + \left(5 \times \frac{1}{10^3}\right)$?

   A. 7,408.065   B. 7,408.65
   C. 70,048.605   D. 70,408.065

2. $26.35 \times 7.34 =$

   F. 108.035   G. 165.04
   H. 174.0405   J. 193.409

3. What is the value of n in the sentence below?
   $n \div 0.06 = 1,310.4 \div 6$

   A. 13.104   B. 21.84
   C. 131.04   D. 218.4

4. In 1924, Harold Osborn of the United States scored 7,710.77 points in the Olympic decathlon. In 1928, Paavo Yrjola of Finland scored 8,053.29 points. What was the difference between their scores?

   F. 98.52 points   G. 298.52 points
   H. 342.52 points J. 1,242.52 points

5. Write 27,630,000 in scientific notation.

   A. 2.763×10^8 B. 27.63×10^6
   C. 2.763×10^7 D. 2.763×10^{-7}

6. Read the problem in the box and think about the steps that you would take to solve it. Do not solve the problem.

   About 1.5 million people visit the observation deck at the Empire State Building each year. To get to the observation deck, visitors must take an elevator. If an elevator holds 15 people at a time, how many elevator trips are needed to take 1.5 million people to the observation deck?

   Which of the problems below would you solve the same way?

   F. Dolly has $100. Henry has $45. How much more money does Dolly have than Henry?
   G. Victor has two water bottles that hold 15 ounces each and another that holds 24 ounces. How many total ounces do his water bottles hold?
   H. Jean wants to buy a scooter that costs $90. If he makes $15 per day bagging groceries, how many days does he have to work to buy the scooter?
   J. Erika makes $7 for shoveling a driveway. How much would she make for shoveling 14 driveways?

7. Which is equivalent to 3.56?

   A. $3\frac{7}{20}$ B. $3\frac{14}{25}$
   C. $3\frac{7}{8}$ D. $3\frac{17}{50}$

8. $48 \div 3\frac{3}{5} =$

   F. $6\frac{2}{3}$ G. $13\frac{1}{3}$
   H. $26\frac{2}{3}$ J. $172\frac{4}{5}$

9. Which is a prime number?

   A. 51   B. 68
   C. 77   D. 103

Short-Response Questions

10. Put the numbers below in order from least to greatest.
    $1\frac{2}{3}$, 1.53, $1\frac{4}{7}$, 1.645

11. Jeremy has $39. This is $\frac{3}{4}$ of the money he needs to buy a jacket. How much does the jacket cost?

Extended-Response Questions

12. The senior class had a spring flower sale. They made 250 small bunches and 150 large bunches of daffodils to sell for the prices shown. They sold all the small bunches and some of the large bunches. If they took in $1,612.50, how many large bunches were left over? Show your work.

Daffodils	
Small Bunch	$4.25
Large Bunch	$5.50

13. During a sale, Sneak Sportique sold 120 pairs of sneakers at $\frac{4}{5}$ the regular price. How many pairs of sneakers at the regular price would Sneak Sportique have to sell to take in the same amount of the money?

14. Paul makes a snack mix by combining $\frac{3}{4}$ of a pound of gummy bears with 1 pound of peanuts. He has $3\frac{1}{2}$ pounds of peanuts.

  a. How many pounds of gummy bears does he need to make the snack mix?

  b. How many pounds of the mix can he make?

15. Kate makes and sells jewelry. She charges $5.50 for pairs of earrings and $3.50 for bracelets. At a craft fair, she sold 30 items and collected $139.00. How many of each item did she sell?

16. Five cages are being used to transport 15 puppies. Two cages are each only large enough for 2 puppies. No cage can hold more than 4 puppies. No puppy will be by itself. How will the puppies be arranged?

17. Ivan wants to improve his vocabulary. He decides to learn a new word on the first day of his plan, and then double the number of new words learned each day.

  a. How many new words will he have learned after 4 days?

  b. How many days will it take him to learn 127 new words?

18. Deals on Wheels rents cars for $38 per day plus 38¢ per mile for each mile over 120 miles per day. How much would the company charge for a one-day rental if the car were driven 206 miles? Show your work.

19. The table shows the population of California for the last four census counts.

Population of California	
1970	7,032,075
1980	7,477,657
1990	8,863,164
2000	9,519,338

  a. Between which two counts did the population have the greatest increase?

  b. Round the 2000 population to the nearest ten thousand.

20. One third of the books returned to the library this morning were mysteries. One half were nonfiction. The remaining 19 books were novels. How many books were returned this morning? Show your work.

CHAPTER 4

Rational Numbers

Chapter Vocabulary

signed numbers
integers
square
irrational number

opposites
rational number
square root
real numbers

additive inverse
absolute value
perfect square

4.1 Using Signed Numbers

Signed Numbers can be used to represent changes such as the rise and fall of temperatures, distances above and below sea level, or other quantities in situations where there are two opposite directions.

Positive and Negative Signs

Positive numbers are written with a + sign or with no sign.
Examples: +1, 1, +25, 25, +392, 392

Negative numbers are always written with a − sign.
Examples: −1, −25, −392

Zero is neither positive nor negative. It has no sign.

The set of signed numbers can be shown as points on a number line.

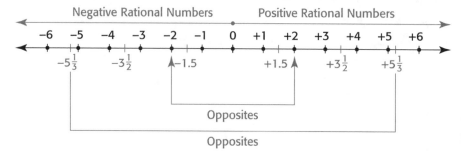

Two numbers that are the same distance from 0 on a number line but in different directions are called **opposites**. Zero is its own opposite. The opposite of a number is also called its **additive inverse**. The sum of a number and its opposite, or additive inverse, is 0.

The numbers . . . , -2, -1, 0, $+1$, $+2$, . . . , are **integers**. A **rational number** is any number that can be expressed as the quotient of two integers where the divisor is not 0. Rational numbers include all the positive fractions, their opposites, and 0. Any decimal that can be expressed as a fraction is, therefore, a rational number.

Examples

1. 7 can be expressed as $\frac{7}{1}$ and therefore is a rational number.

2. 0.4 can be expressed as $\frac{4}{10}$ and therefore is a rational number.

The distance of a signed number from 0 on the number line is its **absolute value**. Since distance is never negative, absolute value is never negative.

$|+8| = +8$ The absolute value of positive 8 equals 8.

$|-8| = +8$ The absolute value of negative 8 equals 8.

Note: The sign is often omitted for positive numbers, so 8 is understood to mean $+8$.

 Model Problems

1. Use a signed number to represent each situation.

		Solution
a.	13° below 0°	-13
b.	a gain of 8 yards	$+8$
c.	a withdrawal of $4.25	-4.25
d.	$10\frac{1}{2}$ meters above sea level	$+10\frac{1}{2}$
e.	owing $40	-40

2. Name the rational number that is represented by each of the letters shown on the number line.

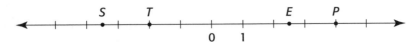

Solution

S is $3\frac{1}{2}$ units to the left of 0 $-3\frac{1}{2}$ or -3.5

T is 2 units to the left of 0 -2

E is $2\frac{1}{2}$ units to the right of 0 $2\frac{1}{2}$ or 2.5

P is 4 units to the right of 0 4

3. Find the absolute value of each number.

    a. $|{-6.2}|$ b. $\left|9\frac{1}{2}\right|$ c. $|{-17}|$ d. $|0|$

Solution

    a. $|{-6.2}| = 6.2$ b. $\left|9\frac{1}{2}\right| = 9\frac{1}{2}$ c. $|{-17}| = 17$ d. $|0| = 0$

Practice

Multiple-Choice Questions

1. The opposite of $-\frac{3}{4}$ is

    A. $-\frac{4}{3}$ B. $\frac{4}{3}$

    C. $\frac{3}{4}$ D. $-1\frac{3}{4}$

2. The number 1.25 written as a quotient of two integers is

    F. $\frac{1.25}{1}$ G. $\frac{5}{4}$

    H. $\frac{4}{5}$ J. $1\frac{1}{4}$

3. What is the absolute value of -105?

    A. -105 B. -150

    C. 150 D. 105

4. What integer is 5 units to the right of 3 on a number line?

    F. -2 G. 2

    H. -8 J. 8

5. What integer is 7 units to the left of 2 on a number line?

    A. -5 B. 5

    C. -9 D. 9

6. What number has the same absolute value as 4.3?

    F. -8.6 G. 14.3

    H. -4.3 J. -14.3

7. The numbers X and $-X$ are additive inverses. What number is midway between them on a number line?

    A. 1 B. 0

    C. -1 D. $\frac{1}{2}$

8. What number is most likely represented at point P on the number line?

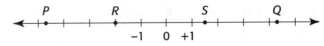

    F. -2.7 G. -3.7

    H. -4.7 J. -5.7

9. Which point on the number line most likely represents $-2\frac{5}{8}$?

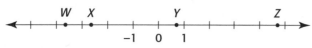

A. W B. X
C. Y D. Z

10. Which could be represented as $+12$?

  F. a loss of 12 pounds
  G. a drop of 12 degrees in temperature
  H. a profit of $12
  J. a withdrawal of $12

Short-Response Questions

11. Write the additive inverse of each number.

  a. 15   b. $-7\frac{1}{5}$

  c. 12.38   d. 0

12. Find the absolute value of each number.

  a. $|-23|$ b. $\left|16\frac{2}{3}\right|$

  c. $|-41.9|$ d $|0|$

13. Graph each number on the number line given below.

  a. -4 b. 3.2

  c. $-2\frac{7}{8}$ d. $+\frac{19}{10}$

14. How are -14 and $+14$ the same? How are they different?

15. Name the integer that is 12 units to the right of -5 on the number line.

16. Name a pair of integers located 27 units from zero on the number line.

17. Find the value of $-|-18|$.

Extended-Response Questions

18. The table shows the price of MDC stock at different times of the day.

Time	MDC Stock Price
10 A.M.	$25\frac{1}{2}$
11 A.M.	$29\frac{1}{2}$
12 noon	$26\frac{3}{4}$
1 P.M.	$24\frac{1}{2}$
2 P.M.	30
3 P.M.	$31\frac{7}{8}$
4 P.M.	$30\frac{1}{2}$

  a. Complete the table below, using signed numbers to show the change in price for each hour.

Time	10 A.M. to 11 A.M.	11 A.M. to 12 noon	12 noon to 1 P.M.
Change			
Time	1 P.M. to 2 P.M.	2 P.M. to 3 P.M.	3 P.M. to 4 P.M.
Change			

  b. Between which hours did the price have the greatest change?
  c. Between which hours did the price have the least change?
  d. Between which hours did the price have the greatest decrease?
  e. What signed number represents the overall change in price for the day?

19. Describe the opposite of each situation and write an integer to represent it.

  a. 8 seconds before liftoff
  b. 15 years from now
  c. walking 4 blocks east

20. A mountain peak is 3.45 kilometers above sea level. The ranger station at the base of the mountain is 1.95 kilometers above sea level.

a. Franklin hiked down from the peak to the ranger station to report a mountain lion sighting. Express his change in elevation as a signed number.

b. The mountain lion was spotted $\frac{2}{5}$ of the way from the ranger station to the peak. What is the difference between the elevation where the mountain lion was spotted and the elevation at the peak?

4.2 Comparing and Ordering Signed Numbers

To compare two signed numbers, think of their location on a number line. The number that is farther to the right is the greater number.

Note: A negative number is always less than a positive number.

Model Problems

1. Replace each ( ) with $<$ or $>$ to make a true comparison.
   a. -2 ( ) -7
   b. -5 ( ) 3

   Solution
   a. -2 is to the right of -7 on a number line. $-2 > -7$
   b. -5 is to the left of 3 on a number line. $-5 < 3$

2. Write the numbers in order from least to greatest.
   a. $-9, -11, 3, 0$
   b. $6, -13, -25, 21, -8$
   c. $\frac{1}{2}, -1\frac{4}{5}, 1\frac{3}{4}, \frac{3}{5}$

   Solution
   a. $-11 < -9, -9 < 0, 0 < 3$ The order is $-11, -9, 0, 3$.

b. $-25 < -13$, $-13 < -8$, $-8 < 6$, $6 < 21$
The order is $-25, -13, -8, 6, 21$.

c. Write equivalent fractions with a common denominator of 20.

$$\frac{1}{2} \times \frac{10}{10} = \frac{10}{20}$$

$$-1\frac{4}{5} \times \frac{4}{4} = -1\frac{16}{20} = -\frac{36}{20}$$

$$-1\frac{3}{4} \times \frac{5}{5} = -1\frac{15}{20} = -\frac{35}{20}$$

$$\frac{3}{5} \times \frac{4}{4} = \frac{12}{20}$$

Ordered from least to greatest, these fractions are $-\frac{36}{20}, -\frac{35}{20}, \frac{10}{20}, \frac{12}{20}$.
The order is $-1\frac{4}{5}, -1\frac{3}{4}, \frac{1}{2}, \frac{3}{5}$.

Practice

Multiple-Choice Questions

1. Which value of n makes the comparison $-16 < n$ true?

   A. -29
   B. -23
   C. -19
   D. -14

2. Which statement is false?

   F. $-8 > -12$
   G. $0 > -6$
   H. $-9 > -4$
   J. $13 > -5$

3. Which number is the greatest?

   A. $-6\frac{1}{3}$

   B. $-9\frac{1}{2}$

   C. $-12\frac{3}{4}$

   D. -15

4. Which number is farthest to the left on a number line?

   F. -56 G. 95
   H. -118 J. -99

5. The following is true for the numbers represented by W, X, Y, Z:

   $$Y < X$$
   $$Z < Y$$
   $$W > X$$

   In what order from left to right would the points appear on a number line?

   A. Y, X, Z, W
   B. Z, Y, X, W
   C. Z, X, Y, W
   D. W, X, Z, Y

6. Which group of numbers is in order from greatest to least?

   F. $0.4, -0.44, 0.44, -0.04$
   G. $0.44, 0.4, -0.04, -0.44$
   H. $-0.44, -0.04, 0.4, 0.44$
   J. $0.4, 0.44, -0.004, -0.44$

7. Which statement is false?

  A. -3 is between -4.1 and -1.8

  B. $-1\frac{1}{2}$ is between -2 and -1.8

  C. $-5\frac{1}{2}$ is between 0 and 11

  D. 4.3 is between -2 and 6.9

8. In Denver, the afternoon temperature rose $12°$ from the morning temperature of $-3°F$. What was the afternoon temperature?

  F. $-9°F$

  G. $9°F$

  H. $15°F$

  J. $-15°F$

Short-Response Questions

9. Replace each ( ) with $<$, $>$, or $=$ to make a true comparison.

  a. -9 ( ) -19

  b. $-7\frac{1}{2}$ ( ) $-3\frac{1}{2}$

  c. -26 ( ) -44

  d. 2.8 ( ) -1.5

10. Tell whether the given statement is *true* or *false*.

  a. $-7 < -4 < -10$

  b. $9 > -3 > -5$

  c. $-12 > -8 > 0$

  d. $-14 < 0 < 2$

11. Write the numbers in order from least to greatest.

  a. $-5, 0, 3, -9$

  b. $-8, 9, -10, 11, -5$

  c. $3\frac{1}{4}, -7\frac{1}{8}, -7\frac{3}{4}, 4\frac{1}{2}, -4\frac{5}{8}$

  d. $0.03, -0.33, 0.303, -3.03, -3.3$

Extended-Response Questions

12. Use the number line and the information given below. Match the points with the correct letters.

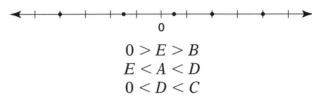

$$0 > E > B$$
$$E < A < D$$
$$0 < D < C$$

13. a. What is the least positive integer?

  b. What is the greatest negative integer?

  c. What is the greatest integer less than -4?

  d. What is the least integer greater than -15?

14. Kareem recorded the deposits and debits in his checking account.

  a. When did Kareem deposit the greatest amount of money?

  b. When did he write a check for the greatest amount?

  c. Which transaction had a greater absolute value, the one on 3/9 or the one on 3/25?

  d. Place all the deposits and debits on a number line in order from greatest decrease to greatest increase.

  e. On the number line, which change in Kareem's account is between the one on 3/14 and the one on 3/30?

Date	Deposit/Debit (in dollars)
3/5	+116
3/9	−47
3/14	−39
3/22	+174
3/25	−83
3/30	+155
4/3	−122

4.3 Adding Signed Numbers

One way to add signed numbers is with a number line. Begin at 0. Then move to the right to add a positive number or to the left to add a negative number.

Examples

1. Add $-2 + (-4)$. Begin at 0. Move 2 units to the left. Then move 4 more units to the left.

   $-2 + (-4) = -6$

   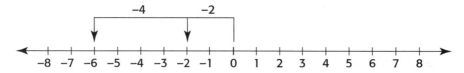

2. Add $8 + (-3)$. Begin at 0. Move 8 units to the right. Then move 3 units to the left.

   $8 + (-3) = 5$

   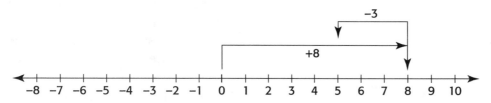

To add two signed numbers without using a number line, follow these rules.

- The sum of two positive numbers is positive. Add the absolute values.
- The sum of two negative numbers is negative. Add the absolute values.
- When one number is positive and the other is negative, find the absolute value of each. Then subtract the lesser from the greater. The sum has the same sign as the number with the greater absolute value.

Model Problems

1. Add $-7 + (-5)$.

   Solution   Both signs are negative, so the sum is negative.

   $|-7| = 7$ Add the absolute values.

   $|-5| = 5$ $7 + 5 = 12$

   So, $-7 + (-5) = -12$. Use the negative sign.

   Answer -12

2. Add $14\frac{3}{5} + \left(-9\frac{1}{5}\right)$.

   Solution   The signs are different. The sum will have the sign of the number with the greater absolute value.

   $\left|14\frac{3}{5}\right| = 14\frac{3}{5}$

   $14\frac{3}{5} > 9\frac{1}{5}$ Therefore the sum is positive.

   $\left|-9\frac{1}{5}\right| = 9\frac{1}{5}$

   $14\frac{3}{5} - 9\frac{1}{5} = 5\frac{2}{5}$ Subtract the lesser from the greater.

   So, $14\frac{3}{5} + \left(-9\frac{1}{5}\right) = 5\frac{2}{5}$

   Answer $5\frac{2}{5}$

3. Add $-21 + 8$.

   Solution   Since $|-21| > |8|$, the sum is negative.

   $|-21| = 21$

   $|8| = 8$

   $21 - 8 = 13$

   Answer $-21 + 8 = -13$

4. Add $-15\frac{1}{2} + 15\frac{1}{2}$.

   Solution   The numbers $-15\frac{1}{2}$ and $15\frac{1}{2}$ are opposites or additive inverses.

   Answer $\left|-15\frac{1}{2}\right| = \left|15\frac{1}{2}\right|$, so the sum is 0.

Multiple-Choice Questions

1. Which addition is shown on the number line?

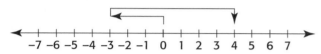

   A. $3 + (-7)$ B. $-3 + 7$
   C. $-3 + 4$ D. $4 + (-7)$

2. Which addition is shown on the number line?

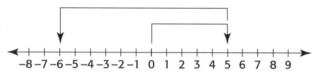

   F. $-6 + 5$ G. $-6 + 11$
   H. $5 + (-6)$ J. $5 + (-11)$

3. Which number line shows $9 + (-5)$?

   A.

   B.
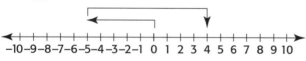

   C.
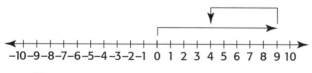

   D.
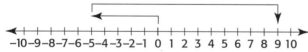

4. Find $-6\frac{1}{2} + \left(-4\frac{1}{4}\right)$.

   F. $-10\frac{3}{4}$ G. $-2\frac{1}{4}$

   H. $10\frac{3}{4}$ J. $2\frac{1}{4}$

5. Find $-5 + 17$.

   A. -12 B. 12
   C. 22 D. -22

6. Find the missing number: $5\frac{3}{8} + \left[\ \right] = 0$.

   F. $5\frac{3}{8}$ G. $-10\frac{5}{8}$

   H. $-5\frac{3}{8}$ J. $-4\frac{5}{8}$

7. Add: $-16\frac{2}{3} + 16\frac{2}{3} = \left[\ \right]$.

   A. $32\frac{1}{3}$ B. $-33\frac{1}{3}$

   C. $33\frac{1}{3}$ D. 0

8. What number must be added to 75 so that the sum is 56?

   F. 19 G. -19
   H. -131 J. 131

9. The sum of -35 and what number is 23?

   A. 12 B. -12
   C. -58 D. 58

10. Find the sum of $29 + (-16) + (-8)$.

   F. -21 G. 21
   H. 5 J. -5

Short-Response Questions

Write an addition expression that matches each description. Complete the addition.

11. Start at 0. Move left 7 units. Then move right 26 units.

12. Start at 0. Move right 13 units. Then move left 25 units.

13. Start at 0. Move left 18 units. Then move right 18 units.

14. Find each sum.

a. $-8 + (-17)$ b. $-12\frac{1}{2} + 5$

c. $20 + (-7)$ d. $-6\frac{2}{3} + 6\frac{2}{3}$

e. $-3.8 + 10.3$

15. Find each sum.

a. $-52 + 27 + (-33)$
b. $75 + (-19) + 8$
c. $-13.6 + (-9.4) + 40$
d. $7\frac{3}{4} + \left(-14\frac{1}{2}\right) + \left(-9\frac{1}{4}\right)$

Write an addition sentence for each. Solve.

16. The temperature dropped 15°F and then rose 8°F. What was the change in temperature?

17. John received 60 points for his correct answers, lost 18 points for his incorrect answers, and received 10 points for a bonus question. What was John's score?

18. Mr. Landis left home and drove 49 miles west and stopped for lunch. Then he drove another 62 miles west, where he made a sales call. He then drove back east 70 miles and checked into a motel for the night. Where was Mr. Landis in relation to his home?

19. On May 5, the balance in Laura's checking account was $187.50. On May 6, she wrote a check for $130.25. On May 8, she deposited $92.85. What was her balance on May 8?

20. A stock gained $2\frac{7}{8}$ points, lost $1\frac{3}{4}$ points, gained $\frac{1}{2}$ point, and then lost 5 points. What was the total change in the stock's price?

4.4 Subtracting Signed Numbers

To subtract a signed number, add its opposite, or additive inverse.

Model Problems

1. Using a number line, subtract $3 - 5$.

Solution Subtracting a number means adding its opposite.

The opposite of 5 is -5. Change subtraction to addition.
$3 - 5 = 3 + (-5)$

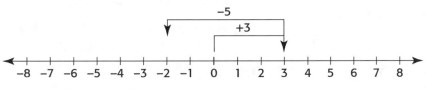

Answer $3 - 5 = -2$

2. What is the difference between a temperature of $-8°F$ and a temperature of $-17°F$?

Solution Subtract to find the difference.

$-17 - (-8) = -17 + 8$ The opposite of -8 is 8.

$-17 + 8 = -9$

Answer A temperature of $-17°F$ is $9°$ colder than $-8°F$. Note that a positive answer, such as $5°$, would indicate that the temperature had gotten $5°$ warmer.

3. Find the change in time between $4\frac{1}{2}$ minutes before a rocket liftoff and $1\frac{1}{2}$ minutes after liftoff.

Solution $1\frac{1}{2} - \left(-4\frac{1}{2}\right) = 1\frac{1}{2} + 4\frac{1}{2} = 6$ minutes

Answer The change in time is 6 minutes later. Note that a negative answer, such as $-3\frac{1}{2}$, would indicate $3\frac{1}{2}$ minutes earlier.

Practice

Multiple-Choice Questions

1. The number line model corresponds to which subtraction problem?

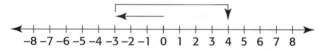

A. $3 - 7$ B. $-3 - 7$
C. $-3 - (-7)$ D. $3 - (-7)$

2. Which expression is equivalent to $-15 - (-18)$?

F. $-15 + (-18)$ G. $15 + (-18)$
H. $15 + 18$ J. $-15 + 18$

3. Which expression is equivalent to $25 - 17$?

A. $25 + 17$ B. $-25 + 17$
C. $-25 + (-17)$ D. $25 + (-17)$

4. Subtract: $8 - 18$.

F. 26   G. -26
H. 10   J. -10

5. Find the difference: $-20 - (-12)$.

A. 8   B. -8
C. -32 D. 32

6. Find the difference: $-9.6 - 7.3$.

F. 2.3   G. -2.3
H. -16.9 J. 16.9

7. Find the missing number: $10 - [\ ] = 13$.

A. -3 B. 3
C. -23 D. 23

8. Find the missing number: $[\ ] - 5 = 19$.

F. 24   G. -24
H. 14   J. -14

9. Find the change in temperature between a morning reading of $-10°F$ and an afternoon reading of $13°F$.

A. $3°F$ B. $-3°F$
C. $-23°F$ D. $23°F$

10. Which difference is 0?

F. $-2 - (-2)$ G. $-3 - 3$
H. $4 - (-4)$ J. $-5 - 5$

11. Use the number line below to show $-6 - (-2)$. Write the answer.

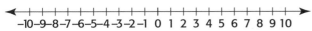

12. Write an equivalent addition expression for each.

a. $12 - 21$ b. $-15.3 - 8.9$

c. $6\frac{4}{5} - \left(-3\frac{1}{2}\right)$ d. $-100 - (-65)$

13. Find each difference.

a. $-19 - (-7)$ b. $12.5 - (-3.2)$

c. $-2\frac{1}{2} - 1\frac{1}{4}$ d. $-600 - (-200)$

14. Find each missing number.

a. $14 - [\ ] = -4$
b. $[\ ] - (-5) = -8$
c. $-6.9 - [\ ] = 2.4$
d. $[\ ] - 20 = 50$

15. From the sum of -43 and 28, subtract -15.

16. Add -31 to the difference of 80 and -56.

17. The highest point in Australia, Mount Kosciusko, is 7,310 feet above sea level. The lowest point, Lake Eyre, is 52 feet below sea level. What is the difference in altitude from the lake to the mountain?

18. The temperature is 68°F inside and -4°F outside. What is the difference in temperature?

19. Last year, a toy company had a profit of $248,000. This year, the company had a loss of $53,000. What was the change in the company's earnings from last year to this year?

20. Ryan's test score was 8 points above the class average. Ava's test score was 11 points below the class average. What is the difference between Ryan's score and Ava's score?

4.5 Multiplying Signed Numbers

The temperature dropped 3°F each hour for 3 hours. To find the total change in temperature, multiply signed numbers using these rules.

- The product of two positive or two negative numbers is always positive.
- The product of a positive and negative number is always negative.

Multiplication Symbols

Multiplication can be written several ways.

$\frac{1}{2} \times 16 = 8$ $\frac{1}{2} \cdot 16 = 8$ $\left(\frac{1}{2}\right)(16) = 8$

Model Problems

1. The temperature dropped 3°F each hour for 3 hours. Find the total change in temperature.

   Solution   Multiply $3 \cdot -3$.

   The signs are different so the product is negative.

   $3 \cdot -3 = -9$

   Answer   The temperature dropped 9°F.

2. How much more money did Jeff have 5 days ago if he spent $4.50 each day for lunch?

   Solution   Represent 5 days ago as -5 and the money spent each day as $-\$4.50$. Multiply to find the total.

   Answer $-5 \times (-\$4.50) = \$22.50$

   Note: The positive answer makes sense because 5 days ago Jeff had $22.50 *more* than he has today.

3. Multiply $\left(-2\frac{1}{2}\right)(-6)\left(-1\frac{4}{5}\right)$.

   Solution   First find the product of the first two numbers.

   $\left(-2\frac{1}{2}\right)(-6) = \dfrac{-5}{2} \times \dfrac{-6}{1} = \dfrac{30}{2} = 15$ The first product is positive.

   Multiply this product by the third number.

   $(15)\left(-1\frac{4}{5}\right) = \dfrac{^{3}\cancel{15}}{1} \times \dfrac{-9}{\cancel{5}_{1}} = -27$

   Answer -27

   Note: The product of an odd number of negative factors is always negative.

Practice

Multiple-Choice Questions

1. Which product is positive?

   A. $-1 \cdot -2 \cdot -3$ B. $5 \cdot -6 \cdot 7$
   C. $-8 \cdot 4 \cdot -9$ D. $-10 \cdot 12 \cdot 15$

2. Find the product of -7 and -8.

   F. -56
   G. 56
   H. -15
   J. 15

3. Alicia drank 8 ounces of orange juice from a large container each morning for the last 4 days. Which expression represents how much more juice was in the container 4 days ago?

  A. -4×8
  B. $4 \times (-8)$
  C. $-4 \times (-8)$
  D. $-4 + (-4) + (-4) + (-4)$

4. Multiply $10 \times \left(-1\frac{3}{5}\right)$.

  F. $8\frac{2}{5}$ G. -6
  H. 16 J. -16

5. Multiply $-2.4 \times (-1.3)$.

  A. -3.12 B. 3.12
  C. -0.312 D. 31.2

6. Find the missing number: $-9 \times [\ ] = 126$.

  F. -14 G. 14
  H. -13 J. 13

7. Multiply $(-5)\,(-7)\,(-6)$.

  A. 180 B. -180
  C. 210 D. -210

8. Which product is negative?

  F. $-4 \cdot -4 \cdot 4 \cdot 4$
  G. $-4 \cdot 4 \cdot -4 \cdot -4$
  H. $-4 \cdot -4 \cdot -4 \cdot -4$
  J. $4 \cdot 4 \cdot -4 \cdot -4$

9. $n \times (-7) = -119$. Find the value of n.

  A. 17 B. 19
  C. -17 D. -19

10. Find the missing number: $-6 \times [\ ] = 12 \times (-4)$.

  F. -9 G. 9
  H. -8 J. 8

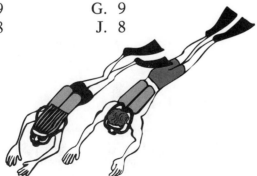

Short-Response Questions

11. Represent each situation as the product of signed numbers.

  a. a stock drops $\frac{3}{4}$ point every day for 4 days
  b. a salesperson earns $58 a day for 6 days
  c. a skydiver descends 2.5 meters per second for 45 seconds
  d. how much heavier a sack of rice was 8 days ago, if $\frac{1}{2}$ pound has been used each day

12. Find each product.

  a. $-9\,(8)$ b. $10\left(-\frac{2}{5}\right)$
  c. $-12.7 \times (-1.4)$ d. -51×36

13. Find each product.

  a. $-6 \times (-5) \times (-9)$
  b. $3 \times (-11) \times (-10)$
  c. $-7 \times 8 \times (-14)$
  d. $15 \times 16 \times (-3)$

14. Replace each $[\ ]$ by a number to make a true statement.

  a. $-4 \times [\ ] = -28$
  b. $6 \times [\ ] = -48$
  c. $-9 \times [\ ] = 3$
  d. $[\ ] \times (-7) \times (-10) = 35$

15. Find each product.

  a. $(-2)^5$ b. $(-3)^4$
  c. $(-1)^{10}$ d. $(-5)^2$

Extended-Response Questions

16. Howard and Ivanna went scuba diving. Howard descended to a depth of 32 feet. Ivanna dove 2.5 times deeper than Howard.

  a. Write a multiplication expression for the depth Ivanna dove.
  b. Find the product.

17. Complete the grid so that the product of numbers in each row, column, and diagonal is 1.

−1		
	1	1

18. A clock loses 25 seconds each day. If the clock is not reset, by how many seconds will the clock's time differ from the correct time after one week?

19. A clock that loses 5 seconds each hour has not been reset for 48 hours.

  a. By how many seconds will the clock's time differ from the correct time?

  b. What time does the clock show if the correct time is 8:49?

20. New Orleans, Louisiana, has an altitude in relation to sea level of −5 feet. The altitude of Death Valley, California, is about 56.4 times that of New Orleans. What is the altitude of Death Valley?

4.6 Dividing Signed Numbers

Dividing signed numbers is similar to multiplying signed numbers. First you figure out what sign the quotient will be, then you divide.

- The quotient of two positive or two negative numbers is positive.
- The quotient of a positive and negative number is negative.

Model Problems

1. Mr. Colon's new car cost $15,886. After 3 years, the car's value will decrease $5,580. What is the average change in value each year?

   Solution Represent the decrease in value as −$5,580.

   Divide to find the average change.

   $$
   \begin{array}{r}
   -1{,}860 \\
   3\overline{)-5{,}580} \\
   \underline{-3\phantom{{,}000}} \\
   2\,5 \\
   \underline{-2\,4} \\
   18 \\
   \underline{-18} \\
   0
   \end{array}
   $$

   $-\$5{,}580 \div 3 = -\$1{,}860$

   Answer The average change is −$1,860, which is a decrease of $1,860 each year.

2. Divide the sum of $-6\frac{3}{4}$ and $30\frac{1}{2}$ by $\frac{-5}{8}$.

Solution First find the sum. $-6\frac{3}{4} + 30\frac{1}{2} =$

$-6\frac{3}{4} + 30\frac{2}{4} = 23\frac{3}{4}$

Then, find the quotient. $23\frac{3}{4} \div \frac{-5}{8} =$

$\frac{95}{4} \div \frac{-5}{8} = \frac{\cancel{95}^{19}}{\cancel{4}} \times \frac{-\cancel{8}^2}{\cancel{5}} = -38$

Answer -38

Practice

Multiple-Choice Questions

1. Divide -60 by -5.

  A. 12
  B. -12
  C. 300
  D. -300

2. Which quotient is equal to 1?

  F. $-14 \div 14$
  G. $5.5 \div (-5.5)$
  H. $0 \div (-2)$
  J. $-6\frac{1}{3} \div \left(-6\frac{1}{3}\right)$

3. A climber descended 102 meters in 3 hours. On average, how much did the climber's elevation change each hour?

  A. 34 meters
  B. 306 meters
  C. -34 meters
  D. -305 meters

4. Find the missing number: $-24 \div [\ ] = -8$.

  F. 192
  G. -192
  H. -3
  J. 3

5. Find the missing number: $[\ ] \div (-9) = 18$.

  A. -162
  B. 162
  C. -2
  D. 2

6. Which quotient is positive?

  F. $-1 \div 1 \div 1$
  G. $1 \div (-1) \div (-1)$
  H. $-1 \div (-1) \div (-1)$
  J. $1 \div 1 \div (-1)$

7. Which quotient is negative?

  A. $125 \div (-5) \div (-5)$
  B. $-125 \div 5 \div (-5)$
  C. $125 \div 5 \div (-5)$
  D. $-125 \div (-5) \div 5$

8. $(-2 - 6) \div (-4) =$

  F. 1
  G. -1
  H. 2
  J. -2

9. When the sum of -1 and -5 is divided by 2, the quotient is

  A. 2
  B. -2
  C. 3
  D. -3

10. $369 \div (-9 \times 5) = [\ ]$

    F. 8.2   G. -8.2
    H. -8.6 J. 8.6

Short-Response Questions

11. Divide.

    a. $-90 \div 10$
    b. $144 \div (-12)$
    c. $-120 \div (-30)$
    d. $78 \div (-13)$

12. Divide.

    a. $7.25 \div (-5)$
    b. $-0.169 \div (-13)$
    c. $-72.3 \div 5$
    d. $0 \div (-5.8)$

13. Divide.

    a. $-2\frac{5}{8} \div (-3)$

    b. $\frac{3}{5} \div \left(-1\frac{1}{2}\right)$

    c. $-6\frac{1}{4} \div \frac{-1}{2}$

    d. $\frac{-4}{5} \div \frac{3}{10}$

14. Replace each [ ] by a number to make a true statement.

    a. $[\ ] \div (-8) = -10$
    b. $-115 \div [\ ] = 23$
    c. $(-25 \div 5) \div [\ ] = -1$
    d. $(-8 \div [\ ]) \div (-2) = -2$

15. Write *true* or *false* for each.

    a. The quotient of two integers is always an integer.
    b. The quotient of any number and its opposite is always -1.

Extended-Response Questions

16. Determine the rule and then complete the table. Write an explanation of the rule.

Input	Output
12	-3
-20	5
32	[ ]
[ ]	-10
-100	[ ]

17. The low temperature in Rochester, New York, was $-6°F$ on Monday, $-10°F$ on Tuesday, and $-5°F$ on Wednesday. Find the average low temperature for the 3-day period.

18. A piece of machinery that cost \$12,000 will be worth only \$8,000 after 5 years. What will be the average change in the value of the machinery each year?

19. At the base of a mountain, the temperature was $64°F$. At a height of 5,000 feet, the temperature was $50°$. What was the average change in temperature for each 1,000 feet up the mountain?

20. Between 10:00 A.M. and 1:00 P.M., the price of INC stock went from $91\frac{1}{2}$ to $85\frac{7}{8}$. Between 1:00 P.M. and 2:00 P.M., the stock lost another $1\frac{3}{8}$ points. What was the average hourly change in the stock's price between 10:00 A.M. and 2:00 P.M.?

4.7 Squares and Square Roots

A square room has an area of 144 square feet. What do you think is the length of its sides? Since the area of a square is found by multiplying the length of a side by itself, and $12 \times 12 = 144$, the length of a side must be 12 feet.

- The **square** of a number is that number multiplied by itself. Since this is equivalent to saying that the number is used as a factor twice, the square of a number is the number raised to the second power.

  $6 \times 6 = 6^2$, which can be read as *6 squared*.

  36 is the square of 6.

- The **square root** of a given number is a number whose square is the given number. The symbol is $\sqrt{}$ is called the radical and is read as *the square root of*.

  $6 = \sqrt{36}$ 6 is the square root of 36.

There are two square roots for every positive rational number, one positive and one negative.

  $\sqrt{36} = \pm 6$ because:

  $6 \times 6 = 36$, so $\sqrt{36}$ equals 6, and

  $-6 \times -6 = 36$, so $\sqrt{36}$ also equals -6.

The symbol $\pm$ is read as *plus or minus*. The symbol $-\sqrt{}$ means *the negative square root*.

- A whole number with a square root that is also a whole number is a **perfect square**. Some perfect squares are 1, 4, 9, 16, 25, 36, 49, 64, and so on.

- The square root of a number that is not a perfect square is a nonterminating, nonrepeating decimal. This type of decimal is an **irrational number**. Its value can only be approximated. (π is an example of an irrational number.)

Model Problems

1. Find each square. a. 14 b. −1.5

   Solution

   a. The square of 14 means 14^2.
      $14^2 = 14 \times 14 = 196$
   b. $(-1.5)^2 = -1.5 \times (-1.5) = 2.25$

2. Find each square root. a. $-\sqrt{121}$ b. $\sqrt{\dfrac{9}{16}}$

   Solution

   a. Since $11^2 = 11 \times 11 = 121$, then $-\sqrt{121} = -11$.
   b. $\dfrac{3}{4} \times \dfrac{3}{4} = \dfrac{9}{16}$, so $\sqrt{\dfrac{9}{16}} = \dfrac{3}{4}$

3. $\sqrt{100} - \sqrt{64} =$

   Solution
   $10^2 = 10 \times 10 = 100$, so $\sqrt{100} = 10$
   $8^2 = 8 \times 8 = 64$, so $\sqrt{64} = 8$
   $10 - 8 = 2$

   Answer 2

4. Between what two consecutive integers is $\sqrt{29}$?

   Solution

   $5^2 = 25$ and $6^2 = 36$
   $25 < 29 < 36$, therefore $\sqrt{25} < \sqrt{29} < \sqrt{36}$.
   Answer $\sqrt{29}$ is between 5 and 6.

   A calculator displays the approximate value 5.3851648 for $\sqrt{29}$. Rounded to the nearest thousandth, $\sqrt{29} = 5.385$.

Practice

Multiple-Choice Questions

1. Which number is NOT a perfect square?

   A. 49
   C. 110
   B. 81
   D. 144

2. What is the square of 16?

   F. 196
   G. 256
   H. 296
   J. 324

3. Find $\left(\frac{4}{9}\right)^2$.

   A. $\frac{2}{3}$

   B. $\frac{16}{9}$

   C. $\frac{4}{81}$

   D. $\frac{16}{81}$

4. Find $(-9)^2$.

   F. -3 G. 3
   H. 81 J. -81

5. Which number is irrational?

   A. $\sqrt{25}$ B. $-\sqrt{64}$
   C. $\sqrt{90}$ D. $-\sqrt{121}$

6. Find $-\sqrt{225}$.

   F. -15 G. 15
   H. -25 J. 25

7. Between which two consecutive integers is $\sqrt{75}$?

   A. 4 and 5   B. 6 and 7
   C. 8 and 9   D. 9 and 10

8. Which statement is false?

   F. $\sqrt{2} + \sqrt{4} = 3$
   G. $\sqrt{9} + \sqrt{16} = 7$
   H. $\sqrt{36} + \sqrt{64} = 14$
   J. $\sqrt{81} + \sqrt{100} = 19$

9. Solve: $16^2 \div \sqrt{64}$.

   A. 0.0625   B. 0.5
   C. 4   D. 32

10. Which statement is always true?

   F. The square of a number is greater than the number.
   G. The square of a number is non-negative.
   H. The square root of a negative number is negative.
   J. Perfect squares have exactly one square root.

Short-Response Questions

11. Find each square.

   a. 13^2

   b. $(-17)^2$

   c. $\left(\frac{1}{4}\right)^2$

   d. $(1.2)^2$

12. Which numbers are perfect squares?

   a. 1,000
   b. 400
   c. 625
   d. 160

13. Find each square root.

   a. $\sqrt{324}$

   b. $\sqrt{1.69}$

   c. $-\sqrt{900}$

   d. $\sqrt{\frac{25}{49}}$

14. Between what two consecutive integers is each square root?

   a. $\sqrt{19}$ b. $-\sqrt{60}$
   c. $\sqrt{102}$ d. $\sqrt{200}$

15. Solve.

   a. $\sqrt{36} + \sqrt{100}$
   b. $\sqrt{25} + \sqrt{49}$
   c. $-\sqrt{169} + \sqrt{225}$
   d. $50^2 \div 5^2$

Extended-Response Questions

16. a. What are the next three terms in the series 1, 4, 9, 16, 25, 36, . . . ?
   b. How would you write the above series of numbers using exponents?

17. When is the square of a number less than the number? Give examples.

18. The area of a square tabletop is 441 square inches. What is the length of a side of the tabletop?

19. For what numbers is it true that the number, its square, and its square root are all equal?

20. A square garden has an area of 484 square feet. If fencing costs $10.75 per foot, how much would it cost to erect a fence around the lot? Show your work.

4.8 Sets of Numbers

Recall these important sets of numbers that have been used in problems:

- The natural or counting numbers are the numbers {1, 2, 3, 4, 5, . . .}.
- The whole numbers are the numbers {0, 1, 2, 3, 4, 5, . . .}.
- The integers are the numbers {. . . −3, −2, −1, 0, 1, 2, 3, . . .}.
- The rational numbers are the numbers that can be expressed as a quotient $\frac{a}{b}$ where a and b are integers and b is not zero. Some examples of rational numbers are $\frac{1}{4}$, $1.\overline{27}$, 11.5, −19, and $\sqrt{4}$.

- Any number that is not rational is irrational. Some examples of irrational numbers are $\sqrt{3}$, π, $\sqrt{7}$ and 1.515225333 . . . . In decimal form, an irrational number is nonterminating and non-repeating.

- The rational numbers and the irrational numbers together form the **real numbers**. All of the numbers used in everyday life are real numbers. Each real number corresponds to exactly one point on the number line, and every point on the number line represents exactly one real number.

The sets of natural numbers, whole numbers, and integers are all subsets of rational numbers. The diagram shows the relationships among the different sets of numbers.

REAL NUMBERS

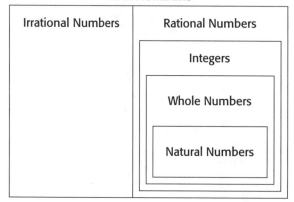

Model Problems

1. Name the sets of numbers to which each value belongs.

   a. 9   b. $\dfrac{-2}{3}$ c. $7.\overline{36}$ d. $\sqrt{12}$

   Solution

   a. $9 = \dfrac{9}{1}$ natural, whole, integers, rational, real

   b. $-\dfrac{2}{3}$ rational, real

   c. $7.\overline{36}$ rational, real

   d. $\sqrt{12} = 3.4641016\ldots$ rational, real

2. Find a real number between 2.3 and 2.4.

   Solution   One way to find a real number between two real numbers is to use an average or mean.

   $$\dfrac{2.3 + 2.4}{2} = \dfrac{4.7}{2} = 2.35$$

   Answer   2.35 is a real number between 2.3 and 2.4. Other real numbers between 2.3 and 2.4 are $2.\overline{3}$, 2.39, and 2.368.

Practice

Multiple-Choice Questions

1. Which number is an integer?

   A. 1.5

   B. $-17\dfrac{1}{2}$

   C. $\sqrt{14}$

   D. 33

2. Which number is irrational?

   F. $\sqrt{21}$

   G. $1.\overline{45}$

   H. $-\sqrt{36}$

   J. 10,000.29

3. The number -8 belongs to each of the sets below except the

   A. real numbers
   B. rational numbers
   C. integers
   D. whole numbers

4. Which set includes $\sqrt{15}$?

   F. rational numbers
   G. natural numbers
   H. irrational numbers
   J. integers

5. Which of the choices shows all the rationals in this group of numbers?

$-6, 0, \frac{4}{5}, 1.7, \sqrt{3}, -\sqrt{16}, 4.763$

  A. $-6, 0, \frac{4}{5}, 1.7$

  B. $-6, 0, \frac{4}{5}, 1.7, -\sqrt{16}, 4.763$

  C. $-6, \frac{4}{5}, 1.7, -\sqrt{16}, 4.763$

  D. $0, \frac{4}{5}, 1.7, 4.763$

6. The numbers $\frac{6}{7}$, $\sqrt{30}$, π, -5, 55.83, and $-8.\overline{63}$ are all

  F. real numbers
  G. rational numbers
  H. irrational numbers
  J. integers

7. It is possible for a number to belong to which two sets?

  A. rational numbers and irrational numbers
  B. irrational numbers and integers
  C. rational numbers and natural numbers
  D. whole numbers and irrational numbers

8. Which number is rational?

  F. $\sqrt{18}$
  G. $\sqrt{49}$
  H. $-\sqrt{15}$
  J. $-\sqrt{40}$

9. The number π is

  A. real and rational
  B. irrational and whole
  C. real and irrational
  D. rational and natural

10. Every integer is

  F. a whole number
  G. an irrational number
  H. a natural number
  J. a rational number

Short-Response Questions

11. Name all the sets of numbers to which each value belongs.

  a. 31
  b. $-\frac{8}{9}$
  c. $\sqrt{61}$
  d. $1.\overline{72}$
  e. -93

12. Find a real number between each two numbers.

  a. 3.2 and 3.3
  b. -1.5 and -1.6
  c. 0.74 and 0.75

13. Consider this group of numbers.

$-7, 0, \frac{5}{8}, 3.9, \sqrt{2}, -\sqrt{6}, \frac{21}{16}, 32, -71, \pi$

List the numbers that belong to each set.

  a. natural numbers
  b. whole numbers
  c. integers
  d. rational numbers
  e. irrational numbers
  f. real numbers

14. Tell whether each of the following can be found between the numbers 7 and 8. If the answer is yes, give an example.

  a. a real number
  b. a rational number
  c. an irrational number
  d. an integer

15. Write *true* or *false* for each statement.

  a. All natural numbers are integers.
  b. All rational numbers are whole numbers.
  c. Some integers are irrational numbers.
  d. All irrational numbers are real numbers.

16. Write *sometimes*, *always*, or *never*, to make each statement true.

   a. Rational numbers are _____ natural numbers.

   b. Terminating decimals are _____ irrational numbers.

   c. Integers are _____ rational numbers.

   d. Negative numbers are _____ integers.

Extended-Response Questions

17. Give an example of a number that is negative and irrational.

18. Give an example of a number that is rational, but NOT a whole number.

19. Put the numbers below into the appropriate section of the Venn diagram.
$$-3, -2, -1, 0, 1, 2, 3$$

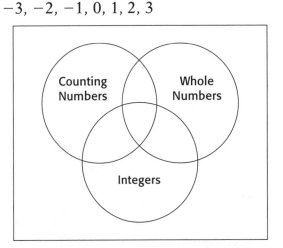

20. a. Order the following rational numbers from least to greatest.

$$-2\frac{1}{2}, \ \frac{3}{5}, \ \frac{100}{265}, \ \frac{2}{3}, \ \frac{5}{3}$$

   b. Order the *reciprocals* of the rational numbers above from least to greatest.

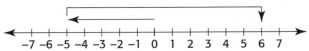

Chapter 4 Review

Multiple-Choice Questions

1. Which number is the greatest?

   A. $\frac{1}{10}$

   B. -2.2

   C. $-4\frac{5}{6}$

   D. -6

2. Which comparison is NOT true?

   F. $|-9| > |-2|$

   G. $|0.8| = |-0.80|$

   H. $|-11| < |-7|$

   J. $\left|\frac{-3}{4}\right| > \left|\frac{1}{2}\right|$

3. What operation is shown on the number line?

$$
\begin{array}{c}
\xleftarrow{\hspace{2cm}}\Big| \\
\text{—7 —6 —5 —4 —3 —2 —1 \ 0 \ 1 \ 2 \ 3 \ 4 \ 5 \ 6 \ 7}
\end{array}
$$

   A. $6 + (-5)$ B. $-11 + 6$

   C. $5 + 6$ D. $-5 + 11$

4. $-(-8) = [\ \ ]$

   F. 8 G. 0

   H. -8 J. 16

5. $-12 - (-7) = [\ \ ]$

   A. -19 B. 19

   C. -5 D. 5

6. $9 \div \left(-\frac{1}{3}\right) =$

   F. 3
   G. −3
   H. 27
   J. −27

7. Which group of integers is in order from least to greatest?

   A. −2, −4, 0, 3, −8
   B. −8, −4, −2, 0, 3
   C. 0, −2, 3, −4, −8
   D. 3, 0, −2, −4, −8

8. Which group of numbers are all perfect squares?

   F. 1, 16, 27
   G. 4, 39, 64
   H. 9, 49, 81
   J. 10, 25, 36

9. Between which two integers is $-\sqrt{72}$?

   A. −8 and −9
   B. −71 and −73
   C. 8 and 9
   D. 71 and 73

10. Which statement is false?

   F. All integers are real numbers.
   G. Some rational numbers are whole numbers.
   H. No irrational numbers are natural numbers.
   J. No negative numbers are real numbers.

Short-Response Questions

11. Find all the values of n that make the following statement true:

$$|n| = 23$$

12. From 4:00 P.M. to 9:00 P.M., the outdoor temperature went from 12°F to −3°F. What was the average change in temperature per hour? Show your work.

13. Use the number line. Name the point for each rational number.

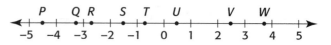

   a. −2.6   b. $3\frac{4}{5}$
   c. $-4\frac{1}{2}$ d. −0.75

14. Label the diagram using the following letters.
   R real numbers
   I integers
   T rational numbers
   W whole numbers
   N natural numbers
   S irrational numbers

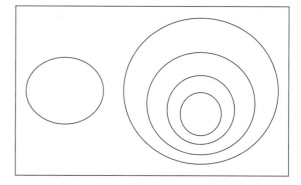

15. Find the answer:
$$\left(\frac{2}{3} - \left(-\frac{1}{3}\right)\right) \times \left(-\frac{1}{2} + \left(-\frac{1}{2}\right)\right).$$

16. Find each value.

   a. $(-1)^{10}$
   b. $(-2)^7$
   c. $(-3)^5$

Extended-Response Questions

17. A square playground has an area of 1225 square meters.

   a. What is the length of one side of the playground?
   b. How many meters of fencing are needed to enclose the playground? Show your work.

18. Without a calculator, estimate $\sqrt{83}$ to the nearest whole number. Explain your reasoning.

19. Find the next two numbers in each pattern. Explain the rule.

a. 3, −6, 12, −24, ____, ____
b. −400, 80, −16, 3.2, ____, ____

20. Airline passengers were told to fasten their seat belts at 2.5 minutes before takeoff. At 7.8 minutes after takeoff, the passengers were allowed to unfasten their seat belts. Write and solve a number sentence to find the total time seat belts were fastened.

Chapter 4 Cumulative Review

Multiple-Choice Questions

1. What exponent will make the following statement true: $125 = 5^{[\ ]}$?

A. 1
B. 2
C. 3
D. 4

2. What is the value of $(2 \times 10^2) + (6 \times 10^3)$?

F. 620
G. 1,840
H. 6,200
J. 62,000

3. Subtract 96.3008 from 285.1006.

A. 188.7098
B. 188.7998
C. 198.8998
D. 199.8006

4. Andrea has enough daffodils to divide them into 2, 3, 4, or 9 groups of the same number. What is the least number of daffodils she can have?

F. 54
G. 36
H. 24
J. 18

5. Which group of numbers are all divisible by 3?

A. 96, 128, 252
B. 102, 233, 372
C. 124, 249, 405
D. 177, 291, 438

6. Divide. Write the answer in lowest terms.

$$\frac{7}{10} \div \frac{3}{8}$$

F. $\frac{21}{80}$

G. $1\frac{13}{15}$

H. $1\frac{3}{5}$

J. $3\frac{11}{15}$

7. $6,679 \div 46 =$

A. 1,145 r 9
B. 1,144 r 45
C. 1.945 r 9
D. 145 r 9

8. Pierre works $5\frac{3}{4}$ hours each day, Monday through Saturday. How many hours does he work in all?

F. $11\frac{3}{4}$

G. $30\frac{3}{4}$

H. $32\frac{1}{4}$

J. $34\frac{1}{2}$

9. Which number is farthest to the right on a number line?

A. $-\frac{13}{2}$ B. -4.88

C. -7 D. $-\frac{21}{4}$

Short-Response Questions

10. $-3.6 \div 0.4 =$

11. Vernon read 3 books last week for a total of 706 pages. This week he read 2 books for a total of 549 pages. How many more pages did he read the first week?

12. If the pattern continues, how many dots will there be in Fig. 7?

```
 •       •       • •     • • • •
 •       •       • •     • • • •
         •               •     •
Fig. 1  Fig. 2  Fig. 3   Fig. 4
```

13. The sun has a diameter of 864,000 miles. The interior temperature of the sun is thought to be about 27,000,000°F. Write these numbers in scientific notation.

14. A box contained $2\frac{1}{2}$ pounds of raisins. Cathy used $\frac{7}{8}$ of the box to make oatmeal cookies. How many pounds of raisins did she use?

Extended-Response Questions

15. Eric had a piece of wire 96 inches long. He cut the wire into 3 pieces. The second piece was 6 inches longer than the first piece. The third piece was 9 inches longer than the second. What was the length of each piece of wire?

16. If you wrote the integers from 0 to 100, how many times would you write the digit 8? Explain your reasoning.

17. One 8-ounce glass of regular milk contains 150 calories. The same glass of low-fat milk has 120 calories. What is the least number of glasses of each kind of milk you will have to drink to consume an equal number of calories? Explain your answer.

18. For 10 weeks, Pilar recorded the weight change of a laboratory rat given different types of food.

Week	1	2	3	4	5
Weight Change (in ounces)	+1.8	+3.0	−1.6	−4.2	0

Week	6	7	8	9	10
Weight Change (in ounces)	−0.7	+2.9	+1.3	+0.5	−2.5

If the rat weighed 11 ounces at the start of the experiment, how much did it weigh at the end? Show your work.

19. Find all the numbers that give 7.29 when squared. Show your work.

20. At the end of August, Pronto Appliance had 14 air conditioners left in stock. The store had sold 59 in June, 47 in July, and 25 in August. If 60 were delivered in June, 44 were delivered in July, and 36 were delivered in August, how many air conditioners were in stock at the beginning of June? Show your work.

Writing and Solving Equations

Chapter Vocabulary

numerical expression	evaluate	order of operations
variables	value	equation
inequality	solution	formula

5.1 Numerical Expressions and Order of Operations

A **numerical expression** contains numbers and operation symbols. To **evaluate** a numerical expression means to perform the operations shown to obtain a single numerical value. To evaluate an expression with more than one operation, use this **order of operations**.

- Perform all calculations within parentheses, brackets, or above or below the division bar.
- Do all calculations involving exponents or roots.
- Multiply or divide, in order, from left to right.
- Add or subtract in order, from left to right.

1. Simplify.

a. $9 - 3 \times 2 + 7$ b. $12 \div 3 + 5 \times 8$

Solution

a. $9 - 3 \times 2 + 7$ b. $12 \div 3 + 5 \times 8$

$= 9 - 6 + 7$ $ = 4 + 5 \times 8$

$= 3 + 7$ $ = 4 + 40$

$= 10$ $ = 44$

2. Simplify $6 + 4(8 + 3^2)$.

Solution

$6 + 4(8 + 3^2)$ Simplify exponent.

$= 6 + 4(8 + 9)$ Add in parentheses.

$= 6 + 4(17)$ Multiply.

$= 6 + 68$ Add.

Answer 74

3. Find the value of $31 + \dfrac{28}{7 - \sqrt{9}} - 2 \times 10$.

Solution

$31 + \dfrac{28}{7 - \sqrt{9}} - 2 \times 10$ Simplify square root.

$= 31 + \dfrac{28}{7 - 3} - 2 \times 10$ Subtract below division bar.

$= 31 + \dfrac{28}{4} - 2 \times 10$ Divide and multiply.

$= 31 + 7 - 20$ Add.

$= 38 - 20$ Subtract.

Answer 18

4. The cost of a regular admission to the museum is $6.00. The cost of a student admission is $4.00. Write an expression for the amount of money that will be collected from 28 regular admissions and 53 student admissions. Evaluate the expression.

Solution

Write an expression for the regular admissions. 28×6
Write an expression for the student admissions and add it to the product above.
$28 \times 6 + 53 \times 4$
Use the order of operations to evaluate.
$28 \times 6 + 53 \times 4$ Multiply first.
$= 168 + 212$ Add.
$= 380$

Answer The amount collected is $380.

Multiple-Choice Questions

1. Evaluate $23 + 15 - 6 \times 3$.

   A. 20   B. 42
   C. 50   D. 96

2. Evaluate $72 \div 2^3 + \dfrac{45 - 17}{7}$.

   F. 6   G. 13
   H. 15   J. 40

3. The value of $6^2 \div 2 \times 9 + 3$ is equal to

   A. 1.5   B. 5
   C. 57   D. 165

4. Simplify the expression:
   $40 \div 2 + \left(5 - \sqrt{4}\right)$.

   F. $13\frac{1}{3}$ G. 21
   H. 23   J. 43

5. Evaluate $12^2 - 9 \times 6 + \dfrac{18}{10 - 7}$.

   A. -39 B. 63
   C. 84   D. 96

6. Which expression is equal to 2?

   F. $24 \div (2 + 2) \times 5$
   G. $(24 \div 2) + (2 \times 5)$
   H. $(24 \div 2 + 2) \times 5$
   J. $24 \div (2 + 2 \times 5)$

7. Which expression is equal to 1?

   A. $64 \div 8 \div 2 + 6$
   B. $64 \div 8 \div (2 + 6)$
   C. $64 \div (8 \div 2) + 6$
   D. $(64 \div 8 \div 2) + 6$

8. Admission to a state fair is $5 and each ride costs an additional $2. Which expression shows the cost of admission to the fair and 7 rides?

   F. $(5 + 7) \times 2$ G. $7 + 2 \times 5$
   H. $7 \times 2 + 5$ J. $7 \times (2 + 5)$

9. Each box of candy contains 24 caramels and 18 creams. Which expression shows the total number of candies in 12 boxes?

   A. $12 + 24 \times 18$
   B. $12 \times (24 + 18)$
   C. $(12 \times 24) + 18$
   D. $(12 + 18) \times 24$

10. For which expression would the first step be to multiply?

   F. $100 \div 2 \times 5 + 21$
   G. $(8 + 7) \times 3 + 19$
   H. $\sqrt{25} + 4 \times 6 \div 3$
   J. $\dfrac{5 \times 9}{3} - 8 \div 2$

Short-Response Questions

11. Write the operation you would perform first.

   a. $9 + 7 \times 15 - 4$
   b. $14 \div 2 + 16 - 9$
   c. $(11 + 7) \times 6 \div 3$
   d. $100 \div 5^2 + 15 \times 4$

12. Simplify.

   a. $20 - 8 \div 2 \times 3$
   b. $36 \div 4 - 5 + 3$
   c. $19 + 4 \times 3 \times 5$
   d. $14 \times 5 + 6 \div 6$

13. Find the value of each expression.

   a. $9 \times (8 + 2) \div 6$
   b. $(8 - 12) \times 4 + 6$
   c. $45 \div (3 \times 5) + 5$
   d. $(4 + 3)^2 \times 3 - 1$

14. Evaluate each expression.

   a. $48 \div \sqrt{64} - \dfrac{23 - 13}{2}$
   b. $22 \div \sqrt{121} \times (4 - 7)$
   c. $\dfrac{53 + 5^2}{2^4 - 2 \times 5} \times 3$
   d. $\dfrac{7^2 + 9 \times 3}{24 - 11 \times 2}$

15. Rewrite each sentence using parentheses to make it true.

   a. $40 \div 5 - 2 \times 3 = 18$
   b. $72 \div 7 + 2 \times 5 = 40$
   c. $8 - 3 \times 9 + 2 = 55$
   d. $18 + 9 \times 3 \div 9 - 2^3 = 1$

Extended-Response Questions

16. Write a numerical expression for the sum of six and fourteen, multiplied by the difference of ten and seven. Evaluate.

Write and evaluate an expression to solve each problem.

17. The Paramount Theater charges $8 for adults and $4.50 for children under 12. The cost of parking is $6 per car. What is the total cost for a group of 3 adults and 5 children that came in 2 cars?

18. Shavel has $65. She wants to purchase a pair of jeans for $32, three tank tops for $14.50 each, and one pair of socks priced at 3 for $7.53 (prices include tax). How much more money does she need?

19. Isaac had $280 in his checking account. He wrote 3 checks for $20 each and 2 checks for $75 each. Then he made a deposit of $110. How much money was in his account then?

20. Angela practices piano for 5 hours on Sundays and for 3 hours on all other days of the week. How many hours does Angela practice in one year?

5.2 Writing Algebraic Expressions

In algebraic expressions, letters called **variables** are used to represent numbers.

$x - 4$
x is a variable

$(a + b) - 10$
a and b are variables

- Verbal phrases can be translated into algebraic expressions by using variables, numbers, and operation signs.
- The unknown quantity can be represented by any variable. First letters of key words are often used.
- A dot, parentheses, or no symbol are used for multiplication to avoid confusing the variable x with the multiplication sign $\times$.
- Complicated expressions may require parentheses or other grouping symbols.

Examples

1. Translate each verbal phrase into an algebraic expression.

   a. the sum of a number (n) and 25 $n + 25$

   b. 5 points less than your grade (g) $g - 5$

   c. 7.5 centimeters more than the plant's height (h) $h + 7.5$

   d. Adam's age (a) divided by 3 $a \div 3$

   e. the dog's weight (w) decreased by 2 pounds $w - 2$

   f. 3 times the cost (c) of an item $3 \cdot c$ or $3(c)$ or $3c$

2. Translate each verbal phrase into an algebraic expression.

   a. twice the sum of y and 3 $2(y + 3)$

   b. the difference of 100 and n, divided by 3 $\dfrac{100 - n}{3}$ or $(100 - n) \div 3$

   c. 8 more than 3 times a number x, doubled $2(3x + 8)$

The use of commas can change the meaning of an expression.

the sum of 60 and a number multiplied by 2 $60 + 2n$

the sum of 60 and a number, multiplied by 2 $2(60 + n)$

Practice

Multiple-Choice Questions

1. Which is an algebraic expression?

   A. $3 + 8 \div 4 - 2$
   B. $6^2 \div (3 + 9)$
   C. $2(y + 14) - 1$
   D. 2.81×10^5

2. Which expression means *9 more than a number*?

   F. $9n$
   G. $n + 9$
   H. $9 - n$
   J. $n \div 9$

3. Which expression means *a number decreased by 10*?

   A. $10 - n$
   B. $n \div 10$
   C. $n - 10$
   D. $\dfrac{10}{n}$

4. Which expression means *the sum of twice a number and 5*?

   F. $2(5) + n$
   G. $2(n + 5)$
   H. $5(2n)$
   J. $2n + 5$

5. Which expression means *7 less than a number, divided by 4*?

  A. $7 - \frac{n}{4}$ B. $(n - 7) \div 4$

  C. $n - (7 \div 4)$ D. $(7 - n) \div 4$

6. The expression $3n - 2$ could be used to translate which expression?

  F. 2 less than the product of 3 and a number
  G. the product of 3 and 2 less than a number
  H. a number minus the product of 3 and 2
  J. the product of a number and 2 less than 3

7. The expression $-50 \div (n + 4)$ could be used to translate which expression?

  A. the quotient of -50 and a number, increased by 4
  B. -50 plus 4, divided by a number
  C. -50 divided by the sum of a number and 4
  D. the sum of a number and 4 divided by -50

8. Gina sells bracelets for $4 apiece. Which expression shows how much Gina will earn if she sells b bracelets?

  F. $4 + b$
  G. $4b$
  H. $b \div 4$
  J. $4(b + 1)$

Short-Response Questions

9. Write an expression for each.

  a. 15 times a number n
  b. a weight w divided by 8
  c. 3 years less than Kate's age a
  d. 11 points more than Don's grade g

10. Use the variable n to translate each phrase into an algebraic expression.

  a. $\frac{2}{3}$ of a number
  b. 70 divided by a number
  c. a number divided by 16
  d. the product of a number and -10

11. Write each expression algebraically. Use the variable x.

  a. 7 times the sum of a number and 3
  b. 1 less than 2 times a number
  c. the difference between a number and 5, divided by 8
  d. the sum of a number and 4, multiplied by the difference between the number and 6

12. Write a verbal phrase for each.

  a. $n + 12$ b. $20 - b$
  c. $-5(w + 4)$ d. $100 \div (x + y)$

13. Nicole is y years old. Write an expression for each.

  a. twice as old as Nicole
  b. 3 years older than Nicole
  c. half Nicole's age
  d. 5 years younger than Nicole

14. Javon sells plants for d dollars apiece. Write an expression for each.

  a. the cost of 5 plants
  b. one third the cost of 10 plants
  c. 3 dollars more than the cost of 4 plants
  d. double the cost of 9 plants

15. The length of a room is l feet. Write an expression for each.

  a. 1 foot less than twice the length
  b. 8 feet more than the length, divided by 5
  c. 40 feet minus 3 times the length
  d. 6 feet more than $\frac{1}{4}$ the length

5.3 Evaluating Algebraic Expressions

Each number a variable can represent is a **value** of the variable. To evaluate or find the value of an algebraic expression, replace the variables by their numerical values. Then perform the operations that are shown using the correct order of operations as explained in Section 5.1.

Model Problems

1. Find the value of $20 - x$ when $x = 9$.

   Solution   Replace x with 9 to find a value of $20 - x$.

   $20 - 9 = 11$

   Answer   When $x = 9$, $20 - x = 11$.

2. Evaluate $3 + 4a$ when $a = -7$.

   Solution   Replace a with -7.

$3 + 4(-7)$	Multiply first. $4(-7)$ means $4 \times (-7)$.
$= 3 + (-28)$	Then add.
$= -25$	

   Answer   When $a = -7$, $3 + 4a = -25$.

3. Evaluate $3s - 2t + 1$ when $s = 10$ and $t = 4$.

   Solution   Replace both variables in the expression with the values given.

$3(10) - 2(4) + 1.$	Multiply first.
$= 30 - 8 + 1$	Add and subtract from left to right.
$= 22 + 1$	
$= 23$	

   Answer   When $s = 10$ and $t = 4$, $3s - 2t + 1 = 23$.

4. A machine makes 72 copies per minute. Write an expression for the number of copies in m minutes. Then find how many copies can be made in 3.5 minutes.

Solution

The machine makes 72 copies for each minute. So, in 1 minute it makes 72(1) copies. In 2 minutes it makes 72(2) copies. To find the number of copies, you multiply 72 by the number of minutes. This can be seen in the table below, where the variable m represents the number of minutes.

Minutes	1	2	3	3.5	m
Number of Copies	72(1)	72(2)	72(3)	72(3.5)	72(m)

Evaluate the expression 72(m) for $m = 3.5$.

$$72(3.5) = 252$$

Answer The machine can make 252 copies in 3.5 minutes.

Multiple-Choice Questions

1. When $n = 18$, the value of $3n - 11$ is

   A. 54   B. 43
   C. 33   D. 15

2. When $c = 3$, the value of $c^2 + 5c + 1$ is

   F. 15   G. 19
   H. 25   J. 55

3. Evaluate $4x - 3y$ if $x = 16$ and $y = 12$.

   A. 0   B. 4
   C. 16   D. 28

4. Find the value of $2s + 5t + 1$ when $s = 7$ and $t = -2$.

   F. 5   G. 25
   H. 32   J. 50

5. Evaluate $m + \frac{1}{2}(3n + 4)$ when $m = 5$ and $n = 12$.

   A. 15   B. 23
   C. 25   D. 45

6. When $x = 13$ and $y = 7$, which expression is equal to 27?

   F. $2x + y$
   G. $x + 2y$
   H. $3x - 3y$
   J. $2(x + y) - 1$

7. When $m = 16$ and $n = 4$, which expression is equal to 7?

   A. $(m \div n) + 6$
   B. $m - 2n + 1$
   C. $(2m + 3) \div (n + 1)$
   D. $m - n^2 + 3$

8. When $w = 5$, $x = 4$, $y = -15$, and $z = 7$, which expression cannot be evaluated?

   F. $\dfrac{5 - w}{x + z}$
   G. $2(y - 15)$
   H. $y^2 - 10w - 20x$
   J. $\dfrac{3 + x}{z - 7}$

Short-Response Questions

9 Evaluate each expression, using the given values of the variables.

  a. $2n + 5$ when $n = 2.6$
  b. $100 - 7x$ when $x = 15$
  c. $14 - 3y + 9$ when $y = -4$
  d. $3z \div 5$ when $z = 8$

10. Evaluate each expression using the values $y = 11$ and $z = -5$.

  a. $y + 2z$ b. $4z - y + 3$

  c. $y^2 - z^2$ d. $\dfrac{10(y + z)}{z + 1}$

11. Find the value of $3x - 2y$ using the given values of the variables.

  a. $x = 2, y = -3$
  b. $x = 6, y = 10$
  c. $x = 3\frac{2}{3}, y = 4\frac{1}{2}$
  d. $x = 7.2, y = 0$

Extended-Response Questions

12. For what value(s) of p could the expression $\dfrac{100}{p^2 - 16}$ NOT be evaluated? Explain how you arrived at your answer.

13. Jacob is h inches tall. Write an expression for each height. Then evaluate each expression if Jacob is 63 inches tall.

  a. 10 inches more than $\frac{1}{3}$ Jacob's height
  b. 19 inches less than twice Jacob's height
  c. 1.5 times Jacob's height
  d. Jacob's height plus 17 inches, divided by 5

14. Complete the table.

	x	y	$y - (x + 3)$	$2(x - y)$
a.	4	9		
b.	12	7		
c.	-8	-5		
d.	20	-11		

15. At the Sport Quench drink factory a machine fills 40 bottles per minute.

  a. How many bottles does the machine fill in m minutes?
  b. How many bottles does the machine fill in 3 minutes?
  c. How many in $1\frac{1}{2}$ hours?
  d. How many in 24 seconds?

5.4 Writing Equations and Inequalities

Equations and inequalities are algebraic sentences. An **equation** is a sentence with an equal sign ($=$). An equation states that two quantities are equal. An **inequality** is a sentence with an inequality symbol ($>$, $<$, $\geq$, $\leq$, $\neq$). An inequality states that two quantities are unequal.

Inequality Symbols

>	greater than	<	less than
≥	greater than or equal to	≤	less than or equal to
≠	not equal to		

A **solution** to an equation or inequality is a value of the variable that makes the sentence true.

Model Problems

1. Tell whether the given value is a solution of the equation $x - 9 = 21$.

a. 30   b. 12

Solution   Replace x with the given value.

a. $x - 9 = 21$
    $30 - 9 = 21$
        $21 = 21$ is true
    30 is a solution to the equation $x - 9 = 21$.

b. $x - 9 = 21$
    $12 - 9 = 21$
        $3 = 21$ is false
    12 is not a solution to the equation $x - 9 = 21$.

2. Tell whether 4 is a solution of each inequality.

a. $y + 3 < 6$

b. $9 - y \geq 3$

Solution

a. $y + 3 < 6$
    $4 + 3 < 6$
        $7 < 6$ false
    4 is not a solution to the inequality $y + 3 < 6$.

b. $9 - y \geq 3$
    $9 - 4 \geq 3$
        $5 \geq 3$ true
    4 is a solution to the inequality $9 - y \geq 3$.

To write an equation or inequality to replace a relationship stated in words, choose a variable to represent the unknown quantity. Translate the words of the sentence into algebraic expressions.

Examples

1. A number n increased by 13 equals 47. $n + 13 = 47$
2. The sum of a number m plus 8 is greater than 25. $m + 8 > 25$
3. $5 more than the cost c is $33. $c + 5 = 33$
4. 6 pounds less than twice the weight w equals 40 pounds. $2w - 6 = 40$
5. 20 miles more than the distance d is less than or equal to 115 miles. $d + 20 \leq 115$
6. Three times Jack's age a subtracted from 100 is 16. $100 - 3a = 16$
7. 9 more than half a number n is greater than 12. $\frac{n}{2} + 9 > 12$

Practice

Multiple-Choice Questions

1. Which number is a solution of $x + 17 = 39$?

   A. 12   B. 22
   C. 56   D. 66

2. Which number is a solution of $2y - 3 > 5$?

   F. 0   G. 2
   H. 4   J. 6

3. Which equation has a solution of 8?

   A. $x - 5 = 13$ B. $2x + 2 = 21$
   C. $3x - 9 = 15$ D. $20 - 2x = -4$

4. Which equation says *12 more than twice a number n is 38*?

   F. $2n + 12 = 38$
   G. $2n = 38 + 12$
   H. $38 + 2n = 12$
   J. $2n - 38 = 12$

5. Which equation says *the cost of sweater s is $10 more than the cost of vest v*?

   A. $s + v = 10$
   B. $s + 10 = v$
   C. $s = v + 10$
   D. $10 - v = s$

6. Which sentence says *20 miles more than twice the distance d is less than 150 miles*?

   F. $2d - 20 \leq 150$
   G. $2d + 20 = 150$
   H. $20 < 150 + 2d$
   J. $2d + 20 < 150$

7. The sentence $x + 1 > x$ is

   A. true only for the value 0
   B. false only for the value 1
   C. true for all whole numbers
   D. false for all whole numbers

8. Which equation could be used to solve this problem?

Ted's age 5 years ago was 13. What is Ted's age t now?

F. $t + 5 = 13$
G. $t - 5 = 13$
H. $t + 13 = 5$
J. $t = 13 - 5$

Short-Response Questions

9. Tell whether the given value is a solution of the given sentence.

a. $x + 12 = 17, x = 5$
b. $y + 3 \geq 10, y = 7$
c. $2x - 3 = 21, x = 12$
d. $100 - 3n < 10, n = 30$

10. Tell which of the given values is a solution of the given sentence.

a. $2x + 11 = 29; x = \{9, 10, 11\}$
b. $3y - 12 \leq 33; y = \{14, 15, 16\}$
c. $101 = m - 47; m = \{147, 148, 149\}$
d. $\frac{c}{4} + 1 > 0; c = \{-4, -3, -2\}$

11. Write an algebraic sentence for each.

a. A number n decreased by 7 is 25.
b. When 8 is subtracted from twice a number x, the result is 44.
c. Ten more than 3 times a number m is less than 61.
d. One half of the product of a number n and 5 more than the number equals 33.

Extended-Response Questions

12. On a test, Joan scored 23 points more than Vincent did. Joan scored 91 points.

a. Write an equation that can be used to find the number of points r that Vincent scored.
b. Which of the values 64, 68, or 78 is Vincent's score?

13. A stack of quarters is worth $9.25. Use the equation $0.25n = 9.25$ and these values for n: 35, 36, 37. Find n, the number of quarters in the stack.

14. A jacket was on sale for $\frac{3}{4}$ of its regular price. The sale price was $69.

a. Write an equation that can be used to find the regular price p.
b. Which of the values $84, $88, $92 is the regular price?

15. The sum of money Lance and Kyra collected for charity equals the sum collected by Marcus and Noelle. Kyra collected $57, Marcus collected $46, and Noelle collected $63.

a. Write an equation that can be used to find l, the amount that Lance collected.
b. Use any method you wish. Find how much Lance collected.

5.5 Solving Equations with One Operation

Recall that addition and subtraction are inverse operations and multiplication and division are inverse operations. An operation followed by its inverse operation always results in the original number or variable.

Start with $x \rightarrow$ Add 5 to get $x + 5 \rightarrow$ Subtract 5 to get $x + 5 - 5 = x$

Start with $y \rightarrow$ Multiply by 3 to get $3y \rightarrow$ Divide by 3 to get $\frac{3y}{3} = y$

To solve an equation, look at what operation has been used. Then perform the inverse operation to get the variable by itself on one side. *Always use the same inverse operation on both sides of the equation to keep the equation balanced.* Check the solution by substituting in the original equation.

 **Model Problems**

1. Solve $x + 13 = 32$.

   Solution The equation shows a number added to x. Use the inverse of adding 13 to solve.
   $$x + 13 = 32$$
   $$x + 13 - 13 = 32 - 13 \quad \text{Subtract 13 from both sides of the equation.}$$

   Answer $x = 19$

   Check Replace x in the original equation with 19.
   $$x + 13 = 32$$
   $$19 + 13 = 32$$
   $$32 = 32$$

2. Solve $27 = y - 18$.

   Solution Since y has 18 subtracted from it, use the inverse of subtracting 18.
   Add 18 to both sides.
   $$27 + 18 = y - 18 + 18$$

   Answer $45 = y$

   Check
   $$27 = y - 18$$
   $$27 = 45 - 18$$
   $$27 = 27$$

3. Four times a number is equal to 66. Find the number.

Solution Translate the word problem into an equation, and then solve.

$4n = 66$ Use the inverse of multiplying by 4.

$$\frac{4n}{4} = \frac{66}{4}$$

Answer $n = 16.5$

Check
$$4n = 66$$
$$4(16.5) = 66$$
$$66 = 66$$

4. A florist divided a shipment of roses equally among 8 vases. There were 18 roses in each vase. How many roses were in the shipment? Write and solve an equation.

Solution Let r represent the number of roses in the shipment.

$\frac{r}{8} = 18$ Use the inverse of dividing by 8.

$8 \cdot \frac{r}{8} = 18 \cdot 8$ Multiply both sides by 8.

Answer $r = 144$

Check $\frac{r}{8} = 18$

$$\frac{144}{8} = 18$$

$$18 = 18$$

5. Copies of a math workbook are $\frac{3}{4}$ inch thick. How many can be stored on a shelf that is 21 inches wide? Write and solve an equation.

Solution Let c = number of copies.

$$\frac{3}{4}c = 21$$

When an equation involves a variable that is multiplied by a fraction, the reciprocal of the fraction can be used to solve the equation.

$\frac{3}{4}c = 21.$ The reciprocal of $\frac{3}{4}$ is $\frac{4}{3}$. Multiply both sides by the reciprocal.

$\frac{4}{3}\left(\frac{3}{4}c\right) = \frac{4}{3_1}\left(\frac{21^7}{1}\right)$ Recall that the product of a fraction and its reciprocal is 1.

$(1)c = 28$

Answer $c = 28$

Check $\frac{3}{4}(28) = 21$

$$\frac{3}{4_1}\left(\frac{28^7}{1}\right) = \frac{21}{1} = 21$$

Multiple-Choice Questions

1. $6 \times 37 \div 6 =$

A. $6\frac{1}{6}$
B. 36
C. 37
D. 222

2. What is the solution to $34 + n = 109$?

F. 65
G. 75
H. 143
J. 153

3. Solve: $y - 28 = -49$.

A. -21
B. -77
C. 21
D. 77

4. Find the solution: $y \div 5 = 65$.

F. 13
G. 60
H. 70
J. 325

5. Nine times a number n equals 423. Which shows the equation for this problem and its solution?

A. $9 + n = 423$; $n = 414$
B. $n \div 423 = 9$; $n = 3,807$
C. $9n = 423$; $n = 47$
D. $n - 9 = 423$; $n = 431$

6. Solve: $D + 9.6 = 3.7$.

F. -13.3
G. -5.9
H. 5.9
J. 13.3

7. Solve: $16y = -12$.

A. -192
B. 4
C. $-1\frac{1}{3}$
D. $-\frac{3}{4}$

8. Which of the following is true about Bart's solution?

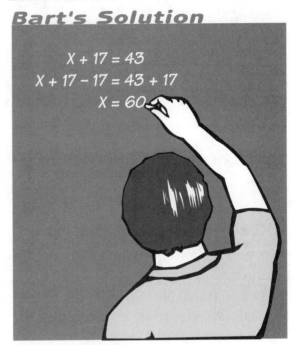

Bart's Solution

X + 17 = 43
X + 17 − 17 = 43 + 17
X = 60

F. The solution is correct.
G. He should have added 17 to the left side.
H. He should have subtracted 17 from the right side.
J. He should have subtracted 43 from both sides.

9. Solve: $\frac{3}{8}t = 24$.

A. 9
B. 64
C. 72
D. 192

10. Which equation has no solution?

F. $x - 1 = 0$
G. $x + 1 = 1$
H. $1 - x = 1$
J. $x + 1 = x$

Short-Response Questions

11. Solve and check each equation.

    a. $x + 29 = 62$ b. $y - 23 = 23$

    c. $4t = 10.8$ d. $\frac{m}{3} = 3.9$

12. Write an equation for each sentence. Use n as the variable. Then find the number.

    a. A number increased by 38 equals 84.

    b. $5\frac{7}{8}$ less than a number is $4\frac{1}{2}$.

    c. When a number is multiplied by 7, the result is -105.

    d. One sixth of a number is 11.8.

13. Solve and check each equation.

    a. $\frac{2}{5}x = 8$ b. $\frac{5}{6}c = 30$

    c. $2\frac{1}{2}n = 45$ d. $\frac{10}{3}w = 50$

Extended-Response Questions

For problems 14–19:
a. Assign a variable to the unknown quantity.
b. Write an equation you can use to find the unknown quantity.
c. Solve the equation you wrote to find the unknown quantity. Show your work.

14. A certain number decreased by 8 is equal to the product of -2 and 6. Find the number.

15. After Helena spent $46 for a pair of sneakers, she had $19 left. How much money did she have to begin with?

16. Steve's friend gave him 35 baseball cards, so Steve had 113 cards. How many cards did Steve have before his friend's gift?

17. Monica earned $98.60 working for an hourly wage of $6.80. How many hours did Monica work?

18. So far, Nick has addressed 73 envelopes. This is one eighth of the number he must complete. How many envelopes will Nick address in all?

19. Takeya has read 302 pages of the novel *Oliver Twist*. That is 58 pages more than César has read. How many pages has César read?

20. Solve each equation.

    a. $x + 9 = 15$ b. $2x + 18 = 30$

    c. $\frac{1}{3}x + 3 = 5$

What do you notice about the solutions? What do you notice about the equations? State any conclusions you might draw.

5.6 Solving Equations with Two Operations

Some equations, like $5x + 11 = 46$, require two inverse operations for their solution. If you look closely at $5x + 11 = 46$, you can see that the x is multiplied by 5 and has 11 added to it. Because there are two operations being performed on x, you will have to perform two inverse operations to find the solution. To solve such equations, first work on the addition or subtraction. Then work on the multiplication or division.

Model Problems

1. Solve and check: $5x + 11 = 46$.

   Solution $5x + 11 = 46$ Use the inverse of the addition.
   $$5x + 11 - 11 = 46 - 11$$ Subtract 11 from both sides.
   $$5x = 35$$ Then use the inverse of multiplication.
   $$\frac{5x}{5} = \frac{35}{5}$$ Divide both sides by 5.

   Answer $x = 7$

   Check $5x + 11 = 46$
   $$5(7) + 11 = 46$$
   $$35 + 11 = 46$$
   $$46 = 46$$

2. If one fourth of a number is decreased by 9, the result is 5. Find the number.

   Solution   Let n represent the number.
   $$\frac{1}{4}n - 9 = 5$$

   $$\frac{1}{4}n - 9 + 9 = 5 + 9$$ Use the inverse of subtraction.

   $$\frac{1}{4}n = 14$$ Use the inverse of multiplying by $\frac{1}{4}$.

   Remember that dividing by $\frac{1}{4}$ is the same as

   multiplying by the reciprocal of $\frac{1}{4}$, that is, 4.

   $$4\left(\frac{1}{4}n\right) = 4(14)$$

   Answer $n = 56$

   Check $\frac{1}{4}n - 9 = 5$

   $$\frac{1}{4}(56) - 9 = 5$$

   $$14 - 9 = 5$$
   $$5 = 5$$

3. Solve and check: $8(x + 7) = 24$.

Solution The first step is to use the distributive property to simplify the left side.

$8(x + 7) = 24$	Multiply each term in parentheses by 8.
$8x + 56 = 24$	
$8x + 56 - 56 = 24 - 56$	Subtract 56 from both sides.
$8x = -32$	
$\dfrac{8x}{8} = \dfrac{-32}{8}$	Divide both sides by 8.

Answer $x = -4$

$$\begin{aligned} Check \quad 8(x + 7) &= 24 \\ 8(-4 + 7) &= 24 \\ 8(3) &= 24 \\ 24 &= 24 \end{aligned}$$

Practice

Multiple-Choice Questions

1. Which equation has the same solution as $3x - 5 = 16$?

A. $2x - 4 = 15$
B. $3x = 21$
C. $3x = 11$
D. $\frac{1}{3}x = 21$

2. Which equation has the same solution as $\frac{c}{4} + 5 = 34$?

F. $4c + 10 = 34$
G. $\frac{c}{4} = 39$
H. $c - 20 = 136$
J. $\frac{c}{4} = 29$

3. Solve: $8x + 3 = 51$.

A. 6
B. 7
C. -6
D. -9

4. Solve: $11z + 9 = -79$.

F. 8
G. -6
H. 7
J. -8

5. Solve: $4(x - 3) = 72$.

A. 15
B. 18
C. 21
D. $\frac{75}{4}$

6. Solve: $5(3x + 1) = -55$.

F. -4
G. -11
H. -18
J. -92

7. Which equation has a solution of -2?

A. $3(x - 2) = 0$
B. $5(2x - 1) = 0$
C. $7(x + 2) = 0$
D. $9(x + 4) = 0$

8. What is true about Lisa's solution?

Lisa's Solution

$$5x + 12 = -13$$
$$5x + 12 - 12 = -13 - 12$$
$$5x = -1$$
$$\frac{5x}{5} = \frac{1}{5}$$
$$x = \frac{1}{5}$$

F. The solution is correct.
G. She should have subtracted -12 on the right.
H. She should have gotten -25 on the right.
J. She should have multiplied $5x$ by 5.

Short-Response Questions

9. Solve each equation and check.

a. $4w + 9 = 77$
b. $8c - 12 = 40$
c. $\frac{1}{2}x + 6 = -10$
d. $\frac{2}{3}m - 1 = 15$

10. Solve each equation and check.

a. $5(z + 3) = 30$
b. $\frac{1}{4}(y - 8) = 12$
c. $3(2y + 5) = 51$
d. $7(2x - 3) = -49$

11. Write an equation to solve each problem. Use n for the variable. Then find the number.

a. If 3 times a number is decreased by 7, the result is 32.
b. If one half of a number is increased by 10, the result is -5.
c. The sum of 8.7 and 4 times a number is equal to 33.1.
d. When 2 is added to a number and the sum is multiplied by -5, the result is 40.

Extended-Response Questions

For problems 12–16:
a. Assign a variable to the unknown quantity.
b. Write an equation you can use to find the unknown quantity.
c. Solve the equation you wrote to find the unknown quantity. Show your work.

12. It costs $28 per day plus $0.16 per mile to rent a car. Julia's charge for a one-day rental was $69.60. How many miles did Julia drive?

13. Rosa, Mel, and Kathy work as a cleaning crew. They shared the money they earned last week equally. Mel spent $26, leaving him with $45. How much money did the crew earn last week?

14. Peter bought a coat at an end-of-winter half-price sale. Sales tax of $6.24 was added to the sale price, bringing the cost of the coat to $84.24. What was the original price of the coat?

15. This week, the price of a round-trip train ticket to the city rose by $1.50. Mr. Logan bought 6 tickets and paid $64.50. How much did the round-trip ticket cost before the increase?

16. Together, Zoë and Susan were planning to make 72 bracelets for a craft sale. Zoë made 26, then cut her finger and could not work anymore. Susan had to make twice as many bracelets as she originally promised to make. How many bracelets had Susan promised to make?

5.7 Graphing and Solving Inequalities

Most of the equations you have solved have had only one solution. An inequality may have an infinite number of solutions. Since these solutions are real numbers, they can be graphed on a number line. Some typical graphs are shown below.

Examples

Inequality	Description	Graph of Solution
$x > 3$	All numbers greater than 3	
$x \geq 3$	All numbers greater than or equal to 3	
$x < 3$	All numbers less than 3	
$x \leq 3$	All numbers less than or equal to 3	
$-1 \leq x < 4$	All the numbers between -1 and 4, including -1 but not including 4	

To solve an inequality means to describe all the numbers that make the inequality true. To solve an inequality with several operations, follow the same rules as for solving equations, with one important exception:

- When multiplying or dividing both sides of an inequality by a negative number, *reverse* the direction of the inequality.

Example

$-4 < 2$, but if both sides of the inequality are multiplied by -3, the inequality symbol must be reversed.

$$-3(-4) > -3(2)$$

$$12 > -6$$

 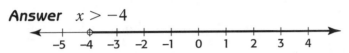

1. Solve and graph $x + 7 > 3$.

Solution $x + 7 > 3$
  $x + 7 - 7 > 3 - 7$

Answer $x > -4$

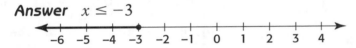

Check Substitute a value greater than -4 in the original inequality.
  $x + 7 > 3$
  $0 + 7 > 3$
    > 3

2. Solve and graph $-2y \geq 6$.

Solution $-2y \geq 6$
    $\dfrac{-2y}{-2} \leq \dfrac{6}{-2}$ Divide each side by -2. Reverse the direction of the inequality.

Answer $x \leq -3$

Check Substitute a value less than -3 in the original inequality.
    $-2y \geq 6$
  $-2(-5) \geq 6$
    $10 \geq 6$

3. Ilona wants to buy some hair barrettes that cost $3 each. She has also decided to buy a scarf for $8. She does not want to spend more than $22 total. Write and solve an inequality to express the number of barrettes she can buy.

Solution The phrase *not more than $22* means *less than or equal to $22*. Let b = number of barrettes.
      $3b + 8 \leq 22$
  $3b + 8 - 8 \leq 22 - 8$ Use the inverse of adding 8 first.
      $3b \leq 14$
      $\dfrac{3b}{3} \leq \dfrac{14}{3}$ Then use the inverse of multiplying by 3. Do not reverse the inequality.
      $b \leq 4\dfrac{2}{3}$

Answer Ilona can buy, at most, 4 barrettes.

Practice

Multiple-Choice Questions

1. Which value is NOT a solution of $4x + 6 > 50$?

    A. 11   B. 12
    C. 20   D. 50

2. What is the inequality that is represented by the graph?

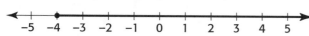

    F. $x \leq -4$ G. $x > -4$
    H. $-4 < x < 5$ J. $x \geq -4$

3. What is the inequality that is represented by the graph?

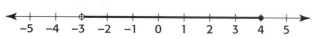

    A. $x > -3$ B. $-3 < x \leq 4$
    C. $x \leq 4$ D. $-3 \leq x < 4$

4. Which graph represents the inequality $x \geq -1$?

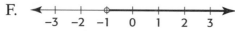

5. Which graph represents the inequality $-3 < x \leq 2$?

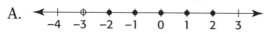

6. What is the solution to $m + 6 > -1$?

    F. $m < -7$
    G. $m > -7$
    H. $m = -7$
    J. $-7 < m < 0$

7. What is the solution to $4x \geq -20$?

    A. $x \geq -5$
    B. $x \leq -5$
    C. $x \geq 5$
    D. $-5 \leq x \leq 5$

8. What is the solution to $-5y \leq 35$?

    F. $y \leq -7$
    G. $y \geq 7$
    H. $y \geq -7$
    J. $y = 7$

9. The solution to $\frac{n}{4} - 6 > -2$ is

    A. $n > -2$ B. $n < -56$
    C. $n > 16$ D. $n < 16$

10. Which graph represents the solution to $-3w + 4 > 19$?

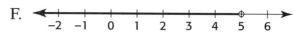

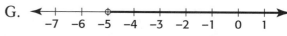

Short-Response Questions

11. For each inequality, tell whether the given value of the variable is a solution.

    a. $x + 8 > 15; x = 4$
    b. $6x \leq -30; x = -8$
    c. $4y - 7 < 12; y = -1$
    d. $-2z + 2 \geq 4; z = 2$

12. Solve each inequality and graph its solution.

  a. $2x > -12$

  b. $y + 5 \geq 2$

  c. $\frac{x}{4} \geq -1$

  d. $z - 3 < 4$

13. Solve each inequality and graph its solution.

  a. $4x + 5 \leq 17$

  b. $-2y + 3 < 11$

  c. $\frac{n}{5} - 2 \geq 3$

  d. $3w - 7 \leq -2$

14. What integers belong to the graph of $-4 \leq x < 6$?

15. Describe the solution of $x + 1 > x$.

Extended-Response Questions

For problems 16–19:

a. Assign a variable to the unknown quantity.

b. Write an inequality you can use to find the unknown quantity.

c. Solve the inequality. Show your work.

16. Sasha needs to score at least 360 points on four tests to earn an A. On her first three tests she scored 86, 93, and 88. What must she score on the fourth test to earn an A?

17. The Drama Club wants to collect at least $1,500 from their performance of *A Raisin in the Sun*. They estimate that they will sell 400 tickets. How much should they charge for tickets?

18. The greatest load that a freight elevator can carry is 2,000 pounds. Ricky has cartons of paper that each weigh 40 pounds. Ricky must ride with the cartons, and he weighs 140 pounds. What is the greatest number of cartons Ricky can safely transport at one time?

19. When 26 is subtracted from 3 times a number, the result is greater than 10. What numbers are possible?

20. a. Mr. Lopez must drive 340 miles to attend an origami convention. His car averages 28 miles per gallon and so far he has traveled 102 miles. How much gasoline must be left in his tank if he wants to complete the trip without stopping?

  b. At the convention, Mr. Lopez made origami swans, cranes, and dragons. He made three times as many swans as cranes. He made five more cranes than dragons. In all, he made 25 origami figures. How many of each did he make?

5.8 Working with Formulas

A **formula** is an equation that shows how two or more variables are related to each other. Formulas are often used to represent scientific, business, or other real-life relationships.

To evaluate a formula means to find the value of one of the variables when the values for the other variables are given.

Model Problems

1. The distance formula $d = rt$ says that the distance an object travels is equal to the product of its rate of travel and the time it travels. How far does an airplane traveling at 550 miles per hour go in 3.5 hours?

    Solution   Decide what values are given for the formula $d = rt$.
    $r = 550$ mph $t = 3.5$ hours $d = rt$
    $d = 550\,(3.5)$
    $d = 1{,}925$ miles

    Answer   The plane travels a distance of 1,925 miles.

2. The formula for the Celsius temperature equivalent to a given Fahrenheit temperature is $C = \frac{5}{9}(F - 32)$. What Celsius temperature is equivalent to 95°F?

    Solution   Substitute 95 as the value of F into the formula $C = \frac{5}{9}(F - 32)$.

    $C = \frac{5}{9}(95 - 32)$ Work inside the parentheses first.

    $C = \frac{5}{9}(63)$ Multiply.

    $C = 35$

    Answer   A temperature of 35°C is equivalent to 95°F.

3. Moira's salary is equal to $10.80 times the number of hours worked.
    a. Write a formula expressing the relationship between salary and hours worked.
    b. How much is Moira's salary if she works 31 hours?

    Solution

    a. Let S represent salary and h represent hours worked.
        Salary = $10.80(hours worked)
            $S = \$10.80h$
    b. Evaluate $S = \$10.80h$ for $h = 31$.
        $S = \$10.80(31) = \$334.80$

    Answer   Moira earns $334.80.

    Note: The next problem shows that sometimes it is convenient to rewrite a formula in terms of a different variable.

4. The formula for the area of a rectangle is $A = lw$ (length times width). Find the width of a rectangle whose area is 204 square inches and whose length is 12 inches.

Solution   To get w by itself on one side of the equal sign, solve the formula for w like you would any equation.

$A = lw$ w is multiplied by l. Use the inverse operation.

$\dfrac{A}{l} = \dfrac{lw}{l}$ Divide both sides by l.

$\dfrac{A}{l} = w$ or $w = \dfrac{A}{l}$

Substitute in the given values for A and l:

$w = \dfrac{A}{l} = \dfrac{204}{12}$

$w = 17$

Answer   The width is 17 inches.

Practice

Multiple-Choice Questions

1. Which formula is NOT equivalent to the others?

A. $\dfrac{d}{r} = t$ B. $t - r = d$

C. $d = rt$ D. $r = \dfrac{d}{t}$

2. Use the formula $C = \dfrac{5}{9}(F - 32)$. What Celsius temperature is equivalent to 59°F?

F. -15°C   G. 48.6°C

H. 15°C   J. 55°C

3. In chemistry, the formula $P = \dfrac{nRT}{V}$ is used to find the pressure of a gas. Find P for $n = 10$, $R = 50$, $T = 12$, and $V = 480$.

A. 1.1   B. 1.25

C. 11.52   D. 12.5

4. Which formula expresses the relationship between the cost C of a bag of bagels and the number of bagels n if each bagel costs $0.60?

F. $C = \$0.60n$

G. $n = \$0.60C$

H. $C \cdot n = \$0.60$

J. $C = \$0.60 + n$

5. Which formula expresses the relationship between x and y shown in the table?

x	3	6	10	17
y	1	4	8	15

A. $x = y + 2$

B. $y = x + 2$

C. $y = 2x - 5$

D. $x = y \div 3$

6. Which formula expresses the relationship between y and m shown in the table?

y	2	3.5	5	7
m	24	42	60	84

    F. $y = 12m$
    G. $m = y + 20$
    H. $m = 12y$
    J. $y = \frac{1}{2}m - 10$

7. Use the formula $d = rt$, where d = distance, r = rate, and t = time. How long would it take a car traveling 80 kilometers per hour to travel a distance of 336 kilometers (if it made no stops)?

    A. 4.2 hours   B. 4.8 hours
    C. 5.6 hours   D. 44.8 hours

8. The formula $C = \$1 + \$0.07(m - 20)$ gives the cost of a long distance telephone call of m minutes when $m \geq 20$. Find the cost of a 35-minute call.

    F. $1.05   G. $2.05
    H. $2.40   J. $3.45

Short-Response Questions

Evaluate each formula for the given values of the variables.

9. $A = \pi r^2$ (Use $\pi = 3.14$)

    a. $r = 5$
    b. $r = 30$

10. $C = P + 0.06P$

    a. $P = \$50$
    b. $P = \$120$

11. $P = 2l + 2w$

    a. $l = 26, w = 19$
    b. $l = 98.5, w = 106.3$

12. $I = prt$

    a. $p = 2,000, r = 0.07, t = 3$
    b. $p = 8,500, r = 0.06, t = 4$

13. $s = -16t^2 + 64t + h$

    a. $t = 1, h = 144$
    b. $t = 5, h = 200$

Extended-Response Questions

14. The formulas for converting Fahrenheit temperatures to Celsius and Celsius temperatures to Fahrenheit are given below. Choose the correct formula and find each temperature. Show your work.

$$°C = \frac{5}{9}(°F - 32)$$
$$°F = \frac{9}{5}°C + 32$$

    a. A temperature of 20°C corresponds to what Fahrenheit temperature?
    b. A temperature of 50°F corresponds to what Celsius temperature?

For problems 15 and 16, write a formula that expresses the relationship between the variables in the table of values.

15.

m	1	2.5	3.8	7
c	100	250	380	700

16.

x	0	2	5	8
y	1	5	11	17

17. A student's score on a math test was determined by multiplying the number of correct answers (c) by 2.5, then subtracting the number of incorrect answers (w).

    a. Write a formula for a student's score (S).
    b. Jasmine had 36 correct answers and 4 incorrect answers. What was her score?
    c. Tony had 32 correct answers and 8 incorrect answers. What was his score?

18. Write a formula for each relationship.

   a. days and the number of hours in them

   b. a car's speed and its traveling time on a 280-mile trip

   c. the total cost if admission is $5 and each ride costs $2

19. Use the formula $I = prt$ to write a formula for t.

20. The formula $C = \$8.00 + 0.45(h - 15)$ gives the cost, C, of monthly Internet access for h hours a month when $h \geq 15$. Find C when h equals

   a. 15 hours

   b. 26 hours

   c. 78 hours

Chapter 5 Review

Multiple-Choice Questions

1. Which sentence means *16 is greater than an unknown number plus 5*?

  A. $16 + n > 5$ B. $5n > 16$
  C. $16 > n + 5$ D. $16 > n \div 5$

2. Which expression means *the product of 4 and 3 less than a number*?

  F. $n - 12$ G. $4(n - 3)$
  H. $4(3 - n)$ J. $-12n$

3. When $x = -6$, the value of $5 + 3x$ is

  A. -13 B. -2
  C. 8 D. 23

4. Evaluate $2g - 4h + 8k$ when $g = 10$, $h = 3$, and $k = -1$.

  F. 26 G. 24
  H. 16 J. 0

5. Which equation is paired with its solution?

  A. $x + 5 = 18; x = 23$
  B. $y - 11 = 15; y = 26$
  C. $3z = 42; z = 13$
  D. $w \div 5 = 6; w = 1$

6. In 3 years, Joel will be 18. Which equation could be solved to find out how old Joel is now?

  F. $a - 3 = 18$ G. $3a = 18$
  H. $a + 3 = 18$ J. $a + 18 = 3$

7. Solve: $\frac{x}{4} + 2 = 5$.

  A. 3 B. 12
  C. 18 D. 28

8. Evaluate the formula $I = prt$ when $p = 3{,}000$, $r = 0.05$, and $t = 6$.

  F. $1{,}800$ G. $1{,}500$
  H. 900 J. 90

9. What is the inequality that is represented by the graph?

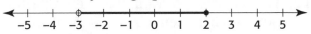

  A. $x \leq 2$
  B. $-3 < x < 2$
  C. $-3 > x \geq 2$
  D. $-3 < x \leq 2$

10. How many integers are there that make this statement true?
$-5 \leq x < 4$

  F. 8
  G. 9
  H. 40
  J. an infinite number

Short-Response Questions

11. If $m = 20$ and $n = 30$, evaluate each expression.

  a. $\frac{80}{m} + \frac{15}{n}$ b. $4(m - n)$

12. Leo bought a radio for $\frac{3}{4}$ of the original price. He paid $36.

   a. Write an equation that can be used to find the original price.
   b. Solve and check the equation you wrote.

13. If $x + 5 = 17$, what does $2x - 3$ equal? Show your work.

Extended-Response Questions

14. On the Budget Plan, the monthly cost of a health club membership is $C = \$35 + 4(v - 5)$ where v is the number of visits and $v \geq 5$. Find the monthly cost for each number of visits:

   a. 5   b. 9   c. 20

15. Solve and check: $\frac{x}{5} - 6 = -2$.

16. Solve and graph the solution: $-12x + 15 > -9$.

17. Yukari said to Romare, "Think of a number between 1 and 20. Multiply it by 6. Subtract 19." If Romare's result was 59, what was his original number? Show your work.

18. Alonzo graphed the numbers shown in the box on a number line. Write an inequality that describes where on the number line all these numbers are located.

1.11	1.56	2.40	2.85	3.49
3.83	2.39	2.73	4.00	2.18
3.05	3.91	1.84	2.09	1.55
1.00	2.95	1.79	2.36	3.20

19. a. Use the distributive property to multiply $5(x + 9)$.
   b. Using $x = 7$, verify that both expressions in part a are equivalent.
   c. Solve $5(x + 9) = 110$. Show your work.

20. Carla needs to buy some gas because her car's tank is empty. Where she lives, gasoline is selling for between $1.60 and $1.75 per gallon. If Carla spends $14.00 on gas, how many gallons will she get? Explain how you solved the problem.

Chapter 5 Cumulative Review

Multiple-Choice Questions

1. What is the standard form of thirty-six million, twelve thousand, nine hundred eight?

   A. 3,612,908
   B. 360,012,980
   C. 36,120,098
   D. 36,012,908

2. Estimate 768×319.

   F. 210,000
   G. 240,000
   H. 270,000
   J. 320,000

3. $449,183 \div 547 =$

   A. 883 r 182
   B. 821 r 96
   C. 804 r 395
   D. 792 r 256

4. What is the prime factorization of 80?

   F. $4 \times 4 \times 5$
   G. $2^2 \times 4 \times 5$
   H. $2^4 \times 5$
   J. $2^5 \times 5$

5. Baldwin had $40 to spend at the mall. He bought a CD for $11.99 and bought lunch for $4.68. How much money did he have left?

   A. $23.33
   B. $24.43
   C. $22.15
   D. not enough information

6. $20.1 \div 0.25 =$

   F. 8.04   G. 8.4
   H. 80.4   J. 804

7. What part of the figure is shaded?

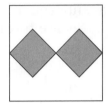

   A. $\frac{1}{6}$ B. $\frac{1}{4}$

   C. $\frac{1}{2}$ D. $\frac{2}{3}$

8. $9\frac{1}{4} - 4\frac{2}{3} =$

   F. $5\frac{7}{12}$ G. $4\frac{7}{12}$

   H. $5\frac{5}{12}$ J. $4\frac{1}{6}$

9. Multiply: $\frac{-3}{8} \times \frac{1}{5} \times (-20)$.

   A. $\frac{-4}{5}$

   B. $\frac{-3}{10}$

   C. 6

   D. $1\frac{1}{2}$

10. Solve: $0.2x + 2 = 14$.

   F. 2.4
   G. 24
   H. 60
   J. 120

Short-Response Questions

11. If $3x + 10 = 1$, find the value of each expression.

   a. $6x + 20 =$
   b. $-5x + 4 =$

12. Martina asked her boss for a salary of $500 per week. She was offered a salary of $19,500 per year. How much less was that than what she had requested?

13. The areas of some large bodies of water are shown in the table.

   a. Round each area to the nearest thousand square miles.
   b. Write each rounded area in scientific notation.

Name	Area (sq mi)	Rounded Area	Scientific Notation
Pacific Ocean	63,801,668		
Caribbean Sea	1,063,340		
Gulf of Mexico	595,760		

Extended-Response Questions

14. Buck said, "Gee whiz, there must be between 10 and 20 animals in the yard!" "Yes," said Chuck. "There is one more cat than there are dogs, and there is one more squirrel than there are cats." What is the greatest possible number of dogs in the yard? Explain.

15. A number is a common multiple of 7 and 11. It is the least such number with a sum of its digits equal to 11. Find the number.

16. Patty is a finance executive who works in a skyscraper. On Thursday, she rode up 17 floors from the floor where her office is located to go to the Accounting Department. Then she rode down 24 floors to Human Resources. Finally, she rode down 19 floors to the Employee Café on the 20th floor. On what number floor is Patty's office?

17. a. Find $5 + 4 \times 5 - 1$.
   b. Show three different ways to change the value of this expression using parentheses. Find the value of each.

18. a. Find the value of each expression: $(0.5)^3$, $(0.4)^2$, $(0.3)^4$.

   b. Write the expressions in order from least to greatest value.

19. Henry bought four T-shirts, each the same price. He gave the cashier $50.00 and received less than $3.00 change.

   a. What is the least that one shirt could cost? Show your work.
   b. What is the most that one shirt could cost? Explain.

20. Yvonne sewed a large square quilt made up of 121 same-sized squares. The squares are alternately turquoise and magenta. How many turquoise and how many magenta squares are there if the corner squares are turquoise?

Ratio, Proportion, and Percent

Chapter Vocabulary

ratio	rate	proportion
cross products	means	extremes
unit price	scale drawing	percent
selling price	discount	rate of discount
cost	markup	amount of markup
commission	rate of commission	interest
principal	interest rate	simple interest

6.1 Ratio

A **ratio** is a pair of numbers that compares two quantities or describes a rate. A ratio can be written in three ways:

Using the word *to*	3 to 4	
Using a colon	3:4	These are all read *3 to 4*.
Writing a fraction	$\frac{3}{4}$	

When a ratio is used to compare two different kinds of quantities, such as miles to gallons, it is called a **rate**.

Equal ratios make the same comparison. To find equal ratios, multiply or divide each term of the given ratio by the same number.

Model Problems

1. In a basket, there are 7 apples and 12 oranges. In three different ways, write the ratio of:

   a. apples to oranges   b. oranges to apples

   c. apples to pieces of fruit

   Solution

   a. apples to oranges   7 to 12   7:12 $\frac{7}{12}$

   b. oranges to apples   12 to 7   12:7 $\frac{12}{7}$

   c. apples to pieces of fruit   7 to 19   7:19 $\frac{7}{19}$

2. Find three ratios equal to $\frac{10}{12}$.

   Solution

   $$\frac{10}{12} = \frac{10 \div 2}{12 \div 2} = \frac{5}{6} \qquad \frac{5}{6} \text{ is the lowest terms fraction for this ratio.}$$

   $$\frac{10}{12} = \frac{10 \times 2}{12 \times 2} = \frac{20}{24}$$

   $$\frac{10}{12} = \frac{10 \times 5}{12 \times 5} = \frac{50}{60}$$

Practice

Multiple-Choice Questions

1. Which ratio is different from the others?

   A. 8 to 15
   B. 15:8
   C. 8:15
   D. $\frac{8}{15}$

2. In a room, there are 9 boys and 12 girls. The ratio of girls to boys is

   F. 9 to 12   G. 12 to 21
   H. 12:9   J. 21:9

3. In the word **BALLOONS**, the ratio of vowels to consonants is

   A. $\frac{3}{5}$ B. 3 to 8

   C. 5:3   D. $\frac{8}{5}$

4. A pet store has 8 cats, 12 dogs, and 3 rabbits. The ratio 8:23 compares

    F. dogs to cats
    G. cats to dogs
    H. rabbits to cats
    J. cats to all animals

5. Which ratio is equal to 15:20?

    A. 5 to 10
    B. 18:25
    C. 21 to 28
    D. 24:30

6. Which ratio is equal to $1\frac{1}{3}$:2?

    F. $\frac{2}{3}$ to $\frac{1}{2}$

    G. 4:6

    H. $3\frac{2}{3}$ to 5

    J. 8 to 11

7. If Joey earns $46.80 for 6 hours of work, what is Joey's rate of pay?

    A. $6.80 per hour
    B. $46.80 per week
    C. $62.00 per day
    D. $7.80 per hour

8. Which pair of ratios is NOT equal?

    F. 72 to 64, 27 to 24
    G. 25:10, 10 to 4
    H. 6:7, 30:35
    J. 7.5 to 18, 5 to 1

Short-Response Questions

9. The Blue Devils won 13 games and lost 7. Find each ratio.

    a. games won to games lost
    b. games won to games played
    c. games played to games lost

10. At a meeting, there were 30 Democrats and 20 Republicans. Write each ratio as a fraction in lowest terms.

    a. Democrats to Republicans
    b. Republicans to all meeting participants
    c. All meeting participants to Democrats

11. A vase contains 16 roses, 10 carnations, and 14 daisies. Write each ratio in lowest terms using *to*.

    a. roses to daisies
    b. carnations to all flowers
    c. daisies to roses and carnations

12. Mr. Nolan drove 228 miles using 12 gallons of gasoline. What was his mileage per gallon?

Find three equal ratios for each.

13. 5 to 8

14. $\frac{60}{40}$

15. 9:21

16. 25 to $33\frac{1}{3}$

Extended-Response Questions

17. Irene made 11 baskets out of 15 free throws.

    a. Write her ratio of baskets to free throws and three ratios equal to it.
    b. If she continues this way, how many baskets will she have after 90 free throws?

18. The ratios in the box are all equal. Find the value of each variable.

$$\frac{6}{x+2} \quad \frac{12}{10} \quad \frac{y-1}{15} \quad \frac{30}{z^2}$$

6.2 Proportion

A **proportion** is a statement that two ratios are equal. A proportion shows that the numbers in two different ratios compare to each other in the same way.

The proportion $\frac{2}{3} = \frac{10}{15}$ is read *2 is to 3 as 10 is to 15*.

In a proportion, the **cross products** are equal.

3 and 10 are the **means**, or the terms in the middle of the proportion.

2 and 15 are the **extremes**, or the terms at the beginning and end of the proportion.

$$2 \times 15 = 3 \times 10$$

$$30 = 30 \qquad \text{If the product of the means equals the product of the extremes, then you have a proportion.}$$

To solve a proportion with an unknown term represented by a variable, set the cross products equal to each other. Then solve the resulting equation using division.

 Model Problems

1. Tell whether the statement is a proportion: $\frac{6}{9} ? \frac{8}{12}$.

   Solution Determine if the cross products are equal.

   $$\frac{6}{9} ? \frac{8}{12}$$

   $9 \times 8 = 72 \qquad$ product of the means
   $6 \times 12 = 72 \qquad$ product of the extremes
   $72 = 72 \qquad$ The product of the means = the product of the extremes, so $\frac{6}{9} = \frac{8}{12}$.

   Answer The statement is a proportion.

2. Tell whether the statement is a proportion: $\frac{12}{7} ? \frac{9}{2}$.

   Solution Determine if the cross products are equal.

   $$\frac{12}{7} ? \frac{9}{2}$$

   $7 \times 9 = 63 \qquad$ product of the means
   $12 \times 2 = 24 \qquad$ product of the extremes
   $63 \neq 24 \qquad$ The product of the means does not equal the product of the extremes, so $\frac{12}{7} \neq \frac{9}{2}$.

   Answer The statement is not a proportion.

3. Solve the proportion $\frac{n}{18} = \frac{20}{45}$.

Solution   Set the cross products equal and solve for n.

$$18 \times 20 = n \times 45$$
$$360 = 45n$$
$$\frac{360}{45} = \frac{45n}{45} \quad \text{Divide both sides by 45 to solve the equation.}$$
$$8 = n$$

Answer $\frac{8}{18} = \frac{20}{45}$

4. Gleamo toothpaste commercials claim that 9 out of 10 people prefer it. If this is true, how many people out of 250 should prefer Gleamo?

Solution   Let n = number of people out of 250 who prefer Gleamo. Set up a proportion and solve for n.

$$\frac{9}{10} = \frac{n}{250}$$
$$10 \cdot n = 9 \cdot 250$$
$$10n = 2{,}250 \qquad \text{Divide both sides by 10.}$$
$$\frac{10n}{10} = \frac{2{,}250}{10}$$
$$n = 225$$

Answer   225 out of 250 people should prefer Gleamo.

Unit Pricing

A useful application of proportions is unit pricing. The **unit price** of an item is the price per unit of measure. The unit could be an ounce, quart, pound, or some other unit.

To find the unit price of an item, set up and solve the proportion:

$$\frac{\text{price paid}}{\text{quantity bought in units}} = \frac{\text{unit price}}{1 \text{ unit}}.$$

Since the denominator of the second ratio is 1, the proportion becomes: $\frac{\text{price paid}}{\text{quantity}} = \text{unit price}.$

Model Problems

5. A 12-ounce bottle of shampoo sells for $2.79. Find the unit price.

Solution   Let the unit be an ounce, and s = unit price.

$$s = \frac{\$2.79}{12 \text{ oz}} = \$0.2325 \text{ per ounce or } \$0.23 \text{ rounded to the nearest cent.}$$

Answer   The unit price is $0.23 per ounce.

6. Which is the better buy: 5 bars of soap for $2.29 or 4 bars for $1.89?

Solution The unit is 1 bar of soap.

$$\frac{\$2.29}{5} = \$0.458 \text{ or } \$0.46 \text{ per bar}$$

$$\frac{\$1.89}{4} = \$0.4725 \text{ or } \$0.47 \text{ per bar}$$

Answer The 5 bars are the better buy.

Practice

Multiple-Choice Questions

1. Which fraction could be used to form a proportion with $\frac{7}{8}$?

   A. $\frac{13}{16}$ B. $\frac{14}{15}$

   C. $\frac{21}{24}$ D. $\frac{35}{25}$

2. Which statement is a proportion?

   F. $\frac{2}{7} = \frac{10}{42}$

   G. $\frac{6}{24} = \frac{4}{16}$

   H. $\frac{5}{9} = \frac{15}{36}$

   J. $\frac{12}{30} = \frac{3}{5}$

3. Which statement is NOT a proportion?

   A. 4 is to 32 as 1 is to 8
   B. 51 is to 6 as 34 is to 4
   C. 9 is to 27 as 2 is to 6
   D. 60 is to 48 as 4 is to 5

4. Find the value of n in the proportion $\frac{32}{48} = \frac{38}{n}$.

   F. 56   G. 57
   H. 60   J. 64

5. Solve: $\frac{10.5}{x} = \frac{3}{5}$.

   A. 3.5
   B. 15.5
   C. 17.5
   D. 6.3

6. Solve: $\frac{35}{21} = \frac{x}{27}$.

   F. 41
   G. 45
   H. 36
   J. 49

7. A truck can travel 232 miles on 14.5 gallons of gasoline. How many gallons would the truck need to travel 400 miles?

   A. 25 gal
   B. 27.5 gal
   C. 16 gal
   D. 64 gal

8. Which bottle of hair conditioner is the best buy?

   F. 8 oz for $1.49
   G. 12 oz for $2.29
   H. 15 oz for $2.75
   J. 20 oz for $3.89

Short-Response Questions

9. Compare. Write $\neq$ or $=$ for each [ ].

   a. $\frac{5}{8}$[ ]$\frac{15}{24}$ b. $\frac{8}{10}$[ ]$\frac{36}{50}$

   c. $\frac{14}{18}$[ ]$\frac{70}{80}$ d. $\frac{48}{69}$[ ]$\frac{32}{46}$

10. Solve each proportion.

    a. $\frac{12}{n} = \frac{27}{36}$ b. $\frac{n}{26} = \frac{9}{39}$

    c. $\frac{21}{27} = \frac{n}{63}$ d. $\frac{n}{16} = \frac{25}{40}$

11. One of the ratios in a proportion is $\frac{4}{5}$. Each of the cross products is equal to 120. What is the other ratio?

12. If 4 apple pies require 22 apples, how many pies can be made with 77 apples?

13. Find the unit price for each. Round to the nearest cent.

    a. 3 cans of cat food for $0.85
    b. 14 ounces of olive oil for $3.59
    c. 8.5 pounds of ground beef for $19.00
    d. 4 pairs of socks for $11.00

14. Choose the better buy for each comparison.

    a. 3 cans of tomato sauce for $1.49 or 5 cans of tomato sauce for $2.39
    b. 24 ounces of juice for $2.25 or 40 ounces of juice for $3.79
    c. 4.4 pounds of sugar for $2.35 or 10 pounds of sugar for $4.99
    d. a bag of 8 rolls for $1.89 or a bag of 18 rolls for $3.79

Extended-Response Questions

15. The ratio of 12th graders to 11th graders at a school is 5 to 8. If there are 440 11th graders, how many 12th graders are there? Show your work.

16. A company sent out 150 questionnaires and 54 were returned with answers. At this rate, how many questionnaires must the company send out in order to get at least 135 returned? Show your work.

Use the figure below for problems 17 and 18. The small gear makes 8 turns for every 5 turns of the large gear.

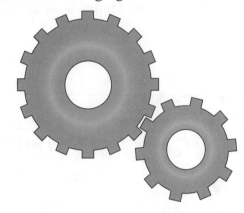

17. How many turns will the large gear make if the small gear makes 80 turns?

18. How many turns will the small gear make if the large gear makes 62.5 turns?

19. The ratio of female to male shoppers at a department store has been found to be 10 to 9. If 1,188 male shoppers were at the store one Saturday, how many shoppers were there in all that day? Show your work.

20. a. A box of national brand cornflakes is $2.09 for 12 ounces. A box of store label cornflakes is $3.09 for 20 ounces. Assuming both brands taste equally good, which is the better buy?
    b. If you have a 30¢-off coupon for the national brand, which is the better buy?

6.3 Scale Drawing

A **scale drawing** can be a reduction (such as a map or floor plan) or an enlargement (such as a drawing of a blood cell) of an actual object. The scale of the drawing is a ratio of the size of the drawing to the size of actual object, or:

$$\text{scale} = \frac{\text{size of drawing}}{\text{size of actual object}}$$

To find an unknown length for a scale drawing, write and solve a proportion.

 **Model Problems**

1. The scale of a New York State map is 1 in. = 50 mi. Find the actual distance from Albany to New York City if the distance on the map is 2.75 in.

   Solution   Let d represent the actual distance in miles.

   $$\frac{\text{scale length}}{\text{actual distance}} = \frac{\text{map length}}{\text{unknown actual distance}}$$

   $$\frac{1 \text{ in.}}{50 \text{ mi}} = \frac{2.75 \text{ in.}}{d}$$

   $1 \times d = 50 \times 2.75$ The product of the extremes = the product of the means.

   $d = 137.5$ miles

   Answer   The actual distance is 137.5 miles.

2. Roy made a scale drawing of the schoolyard using a scale of 2 cm:3 m. He measured the gym and found that its length was 60 m. What was the length of his drawing?

   Solution   Let d represent the length of the drawing in centimeters.

   $$\frac{\text{scale length}}{\text{actual distance}} = \frac{\text{drawing length}}{\text{schoolyard length}}$$

   $$\frac{2}{3} = \frac{d}{60}$$

   $3 \times d = 2 \times 60$ The product of the means = the product of the extremes.

   $3d = 120$

   $$\frac{3d}{3} = \frac{120}{3}$$ Divide both sides of the equation by 3 to solve.

   $d = 40$ cm

   Answer   The length of the drawing is 40 cm.

3. On a poster, a calculator that is actually 7 cm wide is shown as 0.5 m wide. If the calculator is actually 15.4 cm long, how long is it on the poster?

Solution Remember that the actual dimensions are enlarged proportionately.

Let l represent the poster length in meters.

$$\frac{\text{poster width}}{\text{poster length}} = \frac{\text{actual width}}{\text{actual length}}$$

$$\frac{0.5}{l} = \frac{7}{15.4}$$

$7 \times l = 0.5 \times 15.4$ The product of the means = the product of the extremes.

$7l = 7.7$

$\dfrac{7l}{7} = \dfrac{7.7}{7}$ Divide both sides of the equation by 7 to solve.

$l = 1.1$

Answer The length on the poster is 1.1 m.

Practice

Multiple-Choice Questions

1. The figure shows a scale drawing of a swimming pool. Use a centimeter ruler to measure the drawing. What are the actual dimensions of the pool?

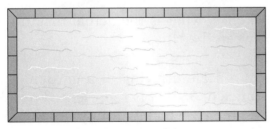

Scale: 2 cm = 5 m

A. 6 m by 14 m
B. 7.5 m by 17.5 m
C. 10.5 m by 19.5 m
D. 15 m by 21 m

2. A length of 20 miles is represented on a scale drawing by a line segment 4 centimeters long. What length is represented by a line segment 5.5 cm long?

F. 22.5 miles   G. 25 miles
H. 27.5 miles   J. 30 miles

3. On a map, a distance of 40 kilometers is represented by a line segment 3 centimeters long. What length segment would you use to represent a distance of 140 kilometers?

A. 3.5 cm   B. 7 cm
C. 10.5 cm   D. 12 cm

4. A scale drawing shows all dimensions $\frac{1}{16}$ actual size. What is the length of a computer screen that is represented by a line segment $1\frac{3}{4}$ inches long?

F. 28 inches   G. 36 inches
H. 20 inches   J. 9 inches

5. A New York State map uses a scale of 1.5 cm = 30 miles. On the map, the distance between Binghamton and Schenectady is approximately 6.5 centimeters. What is the approximate actual distance between the cities?

A. 97.5 miles   B. 110 miles
C. 130 miles   D. 195 miles

6. On a poster, a stamp that is actually 20 millimeters wide is shown 1 foot wide. If the stamp is actually 30 millimeters long, how long is it on the poster?

F. $\frac{2}{3}$ ft   G. $\frac{3}{4}$ ft

H. $1\frac{1}{3}$ ft   J. $1\frac{1}{2}$ ft

7. A scale drawing of a room uses a 9-centimeter line segment to represent a length of 6 meters. Which of these could be the scale of the drawing?

A. 2 cm:3 m   B. 3 cm:2 m
C. 3 cm:4 m   D. 4.5 cm:5 m

8. A model of a building uses a scale of 2 cm:1.5 m. The height of the building is 18 meters. What is the height of the model?

F. 12 cm   G. 24 cm
H. 27 cm   J. 36 cm

Short-Response Questions

 For problems 9 and 10, make scale drawings using a scale of 0.5 in. = 2 ft.

9. a rectangular kitchen 12 feet by 16 feet

10. a rectangular cooking island in the kitchen that is 4 feet by 5 feet

 The drawing below represents a hiking trail through a forest. For problems 11–14, use the drawing and a centimeter ruler to find the actual distances of the following.

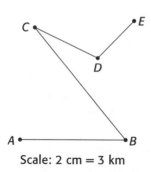

Scale: 2 cm = 3 km

11. A to B

12. B to C

13. C to D

14. the total length of the trail

Extended-Response Questions

15. A drawing of a city's downtown area uses a scale of 4 cm = 5 km. On the drawing, the length of a park is 1.8 cm. What is the actual length of the park?

16. A giant model of an insect is being built for a museum. The actual length of the insect is 0.6 inch and its width is 0.2 inch. The plans call for the model to be 30 inches long. How wide will the model be?

17. Wanda is 5 feet tall and her brother William is 6 feet tall. In a photograph of them standing side by side, William is 4.8 inches tall. How tall is Wanda in the photograph?

18. A map of the United States uses a scale of $\frac{1}{4}$ inch = 80 miles. If the map distance between two cities in Texas is $1\frac{5}{8}$ inches, what is the actual distance between the cities?

19. In a scale drawing of a garden, a distance of 35 feet is represented by a line segment 4 inches long. On the same drawing, what distance is represented by a line segment 14 inches long?

20. A drawing uses a scale of 3 in. to 50 ft. The actual area of a rectangular parking lot is 7,500 square feet. What are the possible dimensions of a rectangle that could represent this lot on the drawing? Show your work.

6.4 Percent

The word *percent* means *for each hundred*. A **percent** is a ratio whose second term is 100. The symbol % is used for percent.

$$100\% = \frac{100}{100} = 1$$

Percents less than 100% are numbers less than 1. Percents greater than 100% are numbers greater than 1.

$$75\% = \frac{75}{100} = \frac{3}{4} \qquad 160\% = \frac{160}{100} = 1\frac{3}{5}$$

Decimals, fractions, and percents can be converted from one form to another.

- To write a decimal as a percent, multiply the decimal by 100 and write a percent sign after it.
- To write a percent as a decimal, divide the percent by 100 and omit the percent sign.
- To write a fraction as a percent, first change the fraction to a decimal. Then use the method above. (Another way is to find an equal ratio with a denominator of 100.)
- To write a percent as a fraction, first write the percent as the numerator without the percent sign. Use 100 as the denominator. Then write the fraction in lowest terms.

 **Model Problems**

1. Write 0.625 as a percent.

   Solution $0.625 \times 100 = 62.5$

   To multiply by 100, move the decimal point 2 places to the right, then write a % sign.

   Answer   62.5%

2. Write 7% as a decimal.

   Solution $7\% \div 100 = 0.07$

   To divide by 100, move the decimal point 2 places to the left. Add zeros as placeholders if needed. Omit the % sign.

   Answer   0.07

3. Larry attended 4 out of 5 of the band rehearsals. What percent of the rehearsals did Larry attend?

Solution *Method I*   *Method II*

$$5 \overline{)\,4.0}$$
$$0.8$$
$$\underline{-4.0}$$
$$0$$

$$\frac{4}{5} = \frac{4 \times 20}{5 \times 20} = \frac{80}{100} = 80\%$$

Answer $\frac{4}{5} = 0.80 = 80\%$

4. Write each percent as a fraction in lowest terms.
   a. 46%   b. 13.5%   c. 128%

Solution

a. $46\% = \frac{46}{100} = \frac{23}{50}$

b. $13.5\% = \frac{13.5}{100} = \frac{13.5 \times 2}{100 \times 2} = \frac{27}{200}$

c. $128\% = \frac{128}{100} = \frac{32}{25} = 1\frac{7}{25}$

Percent Shortcuts

Some percents, like those in the table below, are easy to calculate mentally.

Percent	Fractional Equivalent	Shortcut for Calculating
1%	$\frac{1}{100}$	Divide by 100 or move the decimal point 2 places left.
10%	$\frac{10}{100} = \frac{1}{10}$	Divide by 10 or move the decimal point 1 place left.
25%	$\frac{25}{100} = \frac{1}{4}$	Divide by 4.
50%	$\frac{50}{100} = \frac{1}{2}$	Divide by 2.
100%	$\frac{100}{100} = 1$	100% equals the whole amount.
200%	$\frac{200}{100} = 2$	Double or multiply by 2.

Once you have learned the shortcuts in the table, you can combine them to calculate other percents mentally.

Examples

1. Find 75% of 24. 75% = 25% + 50%. Calculate 25% and 50% of 24 mentally and then add to find 75%.

    25% of 24 = 6   50% of 24 = 12   6 + 12 = 18   So, 75% of 24 = 18.

2. Find 3% of 146. 3% = 1% × 3. Calculate 1% of 146 mentally and then multiply by 3.

    1% of 146 = 1.46   1.46 × 3 = 4.38   So, 3% of 146 = 4.38.

Model Problems

5. Calculate each percent mentally.
    a. 1% of 78
    b. 10% of 300
    c. 25% of 40
    d. 50% of 62
    e. 100% of 1,395
    f. 200% of 9

    Solution
    a. 1% of 78 = 0.78   Move the decimal point 2 places left.
    b. 10% of 300 = 30   Move the decimal point 1 place left.
    c. 25% of 40 = 10   Divide by 4.
    d. 50% of 62 = 31   Divide by 2.
    e. 100% of 1,395 = 1,395   100% equals the whole amount.
    f. 200% of 9 = 18   Multiply by 2.

6. Calculate each percent mentally.
    a. 20% of 982
    b. 500% of 35
    c. 35% of 60

    Solution
    a. 20% of 982
        20% = 2 × 10%   Find 10% of 982 and multiply by 2.
        98.2 × 2 = 196.4
    b. 500% of 45
        500% = 100% × 5   Multiply the whole amount by 5.
        45 × 5 = 225
    c. 35% of 60
        35% = 25% + 10%   Find 25% and 10% of 60 and add.
        25% of 60 = 15
        10% of 60 = 6
        15 + 6 = 21

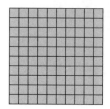 

Practice

Multiple-Choice Questions

1. What percent is shaded?

   A. 1.23%   B. 23%
   C. 123%   D. 177%

2. Write 0.938 as a percent.

   F. 9.38%   G. 93.8%
   H. 90.38%   J. 938%

3. Which number is greater than 1?

   A. 103%   B. 59.8%
   C. 87%   D. 95.08%

4. Jim read 296 pages of a 400-page book. What percent of the book did he read?

   F. 74%   G. 29.6%
   H. 148%   J. 296%

5. Which number is NOT equivalent to the others?

   A. 56%   B. $\frac{40}{75}$

   C. 0.56   D. $\frac{14}{25}$

6. Which is equal to 350%?

   F. 350   G. 0.35
   H. 3.5   J. 35

7. A bank is paying $5\frac{1}{2}$% on savings accounts. What fraction is equivalent?

   A. $\frac{11}{20}$ B. $\frac{11}{50}$

   C. $\frac{11}{100}$ D. $\frac{11}{200}$

8. Seven–eighths of the students in Ms. Wilson's class passed the mathematics test. What percent of the class did NOT pass the test?

   F. 12.5%   G. 25%
   H. 62.5%   J. 87.5%

9. Laurie makes 85% of her free throws. What fractional part of her free throws does Laurie make?

   A. $\frac{17}{50}$ B. $\frac{8}{15}$

   C. $\frac{17}{20}$ D. $\frac{19}{25}$

10. Which is the greatest?

    F. 1% of 3,999   G. 10% of 421
    H. 25% of 160   J. 100% of 42

Short-Response Questions

11. Write each as a percent.

    a. 0.03   b. 0.79
    c. 0.214   d. 1.18

12. Write each as a decimal.

    a. 32%   b. 176%
    c. 44.8%   d. 0.1%

13. Write each as a percent.

    a. $\frac{2}{5}$ b. $\frac{13}{20}$

    c. $\frac{4}{25}$ d. $\frac{9}{16}$

14. Write each as a fraction in lowest terms.

    a. 45%   b. 8%

    c. 77%   d. $7\frac{1}{4}$%

15. Write these numbers in order from least to greatest.

$3\frac{3}{4}$, 390%, 3.86, $\frac{11}{3}$, 3.6, 369%

Extended-Response Questions

16. Calculate 75% of 400 mentally. Explain how you arrived at your answer.

17. In a survey of 300 teenagers, 156 said they earned some of their own money.

a. What percent of the teenagers surveyed earned some of their own money?

b. What percent of the teenagers surveyed did NOT earn money?

18. A 1-hour radio program is divided into the categories shown here.

Music	18 minutes
News	15 minutes
Sports	12 minutes
Commercials	9 minutes
Call-ins	6 minutes

a. Write each as a percent of the whole program.

b. What should be true of the percents you wrote?

19. In 1988, the N.Y. Mets won the pennant in the National League East with a record of 100 wins and 60 losses. What percent of their games did the Mets win? Show your work.

20. This circle graph shows the methods of transportation used by students at Sunny Shores School.

Student Transportation

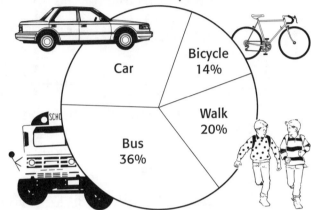

a. What percent of students come by car?

b. Express the part for each type of transportation as a fraction and a decimal.

6.5 Solving Percent Problems

There are three basic types of problems that involve percents: finding a percent of a number, finding what percent one number is of another, and finding a number when a percent of it is known. Each problem can be solved by using a proportion or by writing an equation that relates a number to the whole quantity to which it is being compared.

Model Problems

1. What number is 70% of 250?

    Solution One way to **find a percent of a number** is to first express the percent as a ratio, and then rewrite the question as a proportion, assigning a variable for the unknown number.

    $70\% = \dfrac{70}{100}$ You must find the number that has the same ratio to 250 that 70 has to 100, so the proportion is:

    $\dfrac{70}{100} = \dfrac{n}{250}$ Simplify $\dfrac{70}{100}$ to $\dfrac{7}{10}$ and then cross multiply.

    $10n = 7 \cdot 250$ The product of the means = the product of the extremes.

    $10n = 1{,}750$ Divide both sides of the equation by 10 to solve.

    $n = 175$

    Answer 70% of 250 is 175, which means that 70 has the same ratio to 100 as 175 has to 250.

2. The Martinez family spent 28% of its monthly income for housing. If the family's monthly income is $3,200, how much did they spend for housing?

    Solution Another way to find the percent of a number is to translate the question into an equation. Let n represent the unknown number. Write the percent as a decimal and use multiplication for "of."

    $0.28 \times \$3{,}200 = n$

    $\$896 = n$

    Answer The Martinez family spent $896 for housing.

3. What percent of 75 is 12?

    Solution The question asks **what percent one number is of another**. The unknown quantity is a percent. Write an equation, letting p represent the unknown percent.

    $p \times 75 = 12$

    $75p = 12$ Divide both sides of the equation by 75 to solve.

    $\dfrac{75p}{75} = \dfrac{12}{75}$

    $p = \dfrac{12}{75}$ Divide or use equivalent fractions to convert $\dfrac{12}{75}$ to a percent.

$$\begin{array}{r} 0.16 \\ 75\overline{)12.00} \\ \underline{7\ 5} \\ 4\ 50 \\ \underline{4\ 50} \\ 0 \end{array}$$
$\qquad$ or $\qquad \dfrac{12}{75} = \dfrac{4}{25} = \dfrac{4 \times 4}{25 \times 4} = \dfrac{16}{100}$

$p = 0.16$

Answer So, 12 is 16% of 75.

4. The Recommended Daily Allowance (RDA) of calcium for teenagers is 1200 milligrams. One day, Tina determined that she consumed 1500 milligrams of calcium. What percent of the RDA did Tina consume?

Solution What percent $\times$ 1200 = 1500?

$$p \times 1200 = 1500$$
$$1200p = 1500$$
$$\frac{1200p}{1200} = \frac{1500}{1200} \qquad \text{Divide both sides of the equation by 1200 to solve.}$$
$$p = \frac{5}{4} \text{ or } 1.25 \text{ or } 125\%$$

Answer 1500 milligrams is 125% of the RDA of 1200 milligrams.

5. How much money did Pete earn if $21 is 30% of the total?

Solution This problem asks you to **find a number when a percent of it is known**. You can rewrite the question as a proportion and solve for the unknown amount.

30% of what number is 21?

$$\frac{30}{100} = \frac{21}{n} \qquad \text{Cross multiply.}$$
$$2{,}100 = 30n \qquad \text{Divide both sides of the equation by 30 to solve.}$$
$$70 = n$$

Answer Pete earned $70.

6. A container of yogurt states that it provides 6 grams of protein, which is 12% of the RDA. How much protein is recommended daily?

Solution

12% of what number is 6?

$$0.12 \times n = 6$$
$$0.12n = 6$$
$$\frac{0.12n}{0.12} = \frac{6}{0.12} \qquad \text{Divide both sides of the equation by 0.12 to solve.}$$
$$n = 50$$

Answer The recommended number of grams of protein is 50.

Practice

Multiple-Choice Questions

1. What percent of 72 is 54?

   A. 62.5%
   B. 75%
   C. 87%
   D. $133\frac{1}{3}\%$

2. How much is 45% of 240?

   F. 53
   G. 84
   H. 108
   J. 118

3. 225% of what number is 81?

   A. 36
   B. 45
   C. 108
   D. 182.25

4. Brian saved 35% of the money he earned. If Brian earned $260, how much did he save?

   F. $46.50
   G. $74.28
   H. $91
   J. $117

5. Of 1,600 radios inspected, 112 were defective. What percent of the radios were defective?

   A. 7%
   B. 14.3%
   C. 18.6%
   D. 22%

6. The 15% tip on Craig's lunch bill came to $2.40. What was the amount of the bill before the tip?

   F. $3.60
   G. $16.00
   H. $22.00
   J. $36.00

7. A charity walkathon raised $1,200. This amount was 125% of the amount raised last year. How much was raised last year?

   A. $480
   B. $960
   C. $1,004
   D. $1,500

8. Diane's salary is $32,000 per year. Her car payments total $2,880 per year. What percent of her salary is spent on car payments?

   F. 9%
   G. 11%
   H. 13%
   J. 19%

9. A jewelry store is reducing the price of all items by 15%. The price of a silver bracelet will be $5.70 less than the regular price. Which equation can be used to find c, the regular price.

   A. $0.15 \times 5.70 = c$
   B. $5.70 - 0.15 = c$
   C. $0.15c = 5.70$
   D. $5.70c = 0.15$

10. 58% of what number is 580?

    F. 336.4
    G. 1,000
    H. 3,364
    J. 10,000

Short-Response Questions

11. Find the percent of each number.

    a. 3% of 280
    b. 18% of 550
    c. 135% of 160
    d. $\frac{1}{2}\%$ of 420

12. Find each percent.

   a. What percent of 14 is 56?
   b. What percent of 1,500 is 120?
   c. 65 is what percent of 200?
   d. 153 is what percent of 90?

13. The dinner bill for the Rao family was $58. Mr. Rao left a tip of 15% of the bill. What was the total cost of the family's dinner?

14. Four theater tickets cost $116. The sales tax on the tickets was $6.96. What was the sales tax rate?

Extended-Response Questions

Use the circle graph below for problems 15 and 16.

Favorite TV Sports

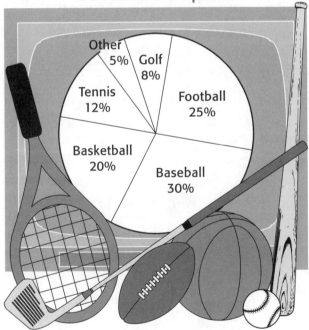

Other 5%
Golf 8%
Tennis 12%
Football 25%
Basketball 20%
Baseball 30%

15. In a survey, 500 teenagers were asked to name their favorite sport to watch on television. How many chose:

   a. basketball   b. golf
   c. football   d. baseball

16. What is the greatest number of teenagers in the survey who might have named bowling as their favorite? Explain.

17. The neighborhood office supply store gives a 15% discount to senior citizens and an 8% discount to students. Phyllis, a ninth-grade student, got $1.64 off her purchase and her grandfather got $2.40 off his purchase. How much was each purchase before the discount?

18. At a fast-food restaurant, a large serving of fries contains 560 calories. Of these calories, 240 come from fat. To the nearest tenth, what percent of the total calories comes from fat?

19. The table shows the sales of different beverages at the Quik Shop for one week. What percent of the total number of bottles sold does each beverage represent?

Beverage	Number of Bottles
Coffee Frappe	54
Orange Gold	110
Great Grape	46
Fruit Zinger	132
Tea Time	58

20. Last year, Sam earned $22,000 and saved 8% of his earnings. This year, Sam earned twice as much but his rate of savings was only half of last year.

   a. How does the amount of money Sam saved this year compare to the amount saved last year?
   b. What was the total amount Sam saved during the two years?
   c. What percent of Sam's total earnings does his total savings represent. (Round to the nearest tenth of a percent.)

6.6 Percent of Increase and Decrease

To find the percent of increase or decrease, first subtract to find the amount of change. Then compare the amount of change to the original amount.

$$\frac{\text{percent of change (increase or decrease)}}{} = \frac{\text{amount of change (increase or decrease)}}{\text{original amount}}$$

Model Problems

1. The membership of the Film Club increased from 24 to 30. Find the percent of increase.

   Solution

   $30 - 24 = 6$ Subtract to find the amount of increase.

   $\frac{6}{24} = 0.25$ Divide the amount of increase by the original amount.

   $0.25 = 25\%$ Write the decimal as a percent.

   Answer   The membership increased 25%.

2. During a thunderstorm, the outdoor temperature went from 88°F to 77°F. Find the percent of decrease.

   Solution

   $88 - 77 = 11$ Subtract.

   $\frac{11}{88} = 0.125$ Divide the amount of decrease by the original

   $0.125 = 12.5\%$ amount.

   Answer   The temperature decreased 12.5%.

Practice

Multiple-Choice Questions

1. The price of a share of stock went from $30 to $27. What was the percent of decrease?

   A. 3%   B. 10%
   C. 11%   D. 19%

2. The population of a village rose from 5,000 to 5,800. What was the percent of increase?

   F. 9.6%   G. 13.8%
   H. 16%   J. 24%

3. A loaf of bread that used to cost $1.50 now costs $1.70. What was the percent of increase in the price?

   A. 11.8%   B. $13\frac{1}{3}$%

   C. 20%   D. 88%

4. A house was originally listed for sale at $250,000. After 3 months, the seller dropped the price to $235,000. What was the percent of decrease in the price?

   F. 6%   G. 9.2%
   H. 15%   J. 94%

5. Which represents a percent increase of 20%?

   A. 20 to 30   B. 30 to 40
   C. 50 to 60   D. 90 to 100

6. Which represents a percent decrease of 12.5%?

   F. 40 to 32   G. 72 to 60
   H. 135 to 122.5 J. 168 to 147

7. Which percent increase in temperature is the greatest?

   A. 25°F to 31°F   B. 32°F to 40°F
   C. 50°F to 59°F   D. 75°F to 90°F

8. Which student had the smallest percent decrease in test scores from week 1 to week 2?

Test Scores		
Student	Week 1	Week 2
Jasmine	90	81
Randy	86	76
Maya	75	68
Steven	95	83

   F. Jasmine   G. Randy
   H. Maya   J. Steven

9. The percent increase in the price of a car was 4%. The new price is $15,600. What was the price before the increase?

   A. $15,000   B. $14,976
   C. $14,600   D. $14,000

10. The table shows Pam's hours of community service at the Senior Center. Which of the following is true?

Pam's Community Service	
Month	Hours
May	10
June	20
July	30

   F. The percent increase from May to June is greater than the percent increase from June to July.
   G. The percent increase from May to June is less than the percent increase from June to July.
   H. The percent increase from May to June is equal to the percent increase from June to July.
   J. The percent increase from month to month cannot be determined.

11. Find the percent of increase.

    a. $5 to $7   **b.** 400 to 1,000
    c. 65 to 195   **d.** 1,256 to 1,413

12. Find the percent of decrease.

    a. $40 to $34   **b.** 80°F to 75°F
    c. 200 to 148   **d.** $17.50 to 0

13. Estimate each percent of change. Show your work.

    a. 97 to 74   **b.** 253 to 201
    c. 789 to 1,102   **d.** 1,006 to 1,160

14. **a.** What is the percent change from 20 to 25?
    b. What is the percent change from 25 to 20?
    c. Explain why the answers are different.

Extended-Response Questions

 Use a calculator for problems 15 and 16.

15. In 1990, the population of Wyoming was 453,588. In 2000, the population was 493,782. To the nearest tenth of a percent, what was the percent of increase in the population?

16. In 1990, the population of Washington, D.C., was 606,900. In 2000, the population was 572,059. To the nearest tenth of a percent, what was the percent of decrease in the population?

17. The numbers of scooters sold by a manufacturer are shown in the table.

Monthly Scooter Sales	
September	4,000
October	4,400
November	4,840
December	?

    a. Explain the pattern.
    b. Assuming the pattern continues, find the sales for December.

18. A computer stock opened at $150 a share.

    a. There was a 20% increase in the share's price. What was the new price?
    b. Next, there was a 20% decrease in the share's price. What was the final price?
    c. What was the percent change from the opening price to the final price? Indicate if the change was an increase or decrease.

6.7 Applications of Percent

There are many everyday situations that involve the use of percent. Determining the **selling price** of an item is a common problem for retailers and consumers. For the buyer, the amount a regular or list price is reduced is called a **discount**.

The **rate of discount** is the percent that the regular price is reduced.

- discount = regular price × rate of discount

- sale price = regular price − discount

For the retailer, the **cost** is the original amount paid for an item.

The **markup** is the percent of increase. The **amount of markup** is the increase.

- amount of markup = cost × markup

- selling price = cost + amount of markup

Model Problems

1. A television set is on sale at 15% off the regular price of $459. What is the sale price of the television?

   Solution   15% = 0.15
   discount = $459 × 0.15 = $68.85
   sale price = $459 − $68.85 = $390.15

   Answer   The sale price of the television is $390.15.

2. A department store buyer pays $20 for each sweater that the store will sell. The store sells the sweaters at a 75% markup. Find the selling price of each sweater.

   Solution   75% = 0.75
   amount of markup = $20 × 0.75 = $15
   selling price = $20 + $15 = $35

   Answer   The selling price of each sweater is $35.

A **commission** is an amount of money that a salesperson earns on a sale. A percent is used to show the **rate of commission**.

- amount of commission = amount of sales × rate of commission

3. Vanessa is a salesperson at a jewelry store. She earns an 8% commission on each sale. How much does she earn on the sale of a $2,400 diamond ring?

   Solution   8% = 0.08
   amount of commission = $2,400 × 0.08 = $192

   Answer   Vanessa earns a commission of $192.

Calculating Interest

Interest is a charge for money that is borrowed. When you take a loan from a bank, you pay the bank interest. When you deposit money in a bank, the bank pays you interest because it is using your money to carry on its business.

The **principal** (s) is the amount of money that is invested or borrowed.

The **interest rate** (r) is the percent charged or earned. The interest rate is usually based on 1 year.

- Interest (I) = principal × interest rate × time (in years) $I = p \times r \times t$

- Total amount owed or earned = principal + interest

For the calculation of **simple interest**, the principal and amount of interest earned each year stay the same.

---

4. Roberto has $700 in his savings account. If the bank pays 5% interest, how much money will Roberto have in his account:

   a. 1 year from now?   b. 2 years from now?

   Solution

   a. $p = \$700, r = 5\% = 0.05, t = 1$
   $I = p \times r \times t$
   $\quad = \$700 \times 0.05 \times 1$
   $\quad = \$35$
   Total amount =
   $\quad \$700 + \$35 = \$735$

   b. $p = \$700, r = 5\% = 0.05, t = 2$
   $I = p \times r \times t$
   $\quad = \$700 \times 0.05 \times 2$
   $\quad = \$70$
   Total amount =
   $\quad \$700 + \$70 = \$770$

   Answer Roberto will have $735 after 1 year and $770 after 2 years if he makes no additional deposits.

5. Mr. Ito took out a 6-month loan of $1,500 at an 8% yearly interest rate. How much must Mr. Ito repay at the end of the 6 months?

   Solution *Think:* 6 months = $\frac{1}{2}$ year, so $t = \frac{1}{2}$ or 0.5

   $p = \$1,500, r = 0.08, t = 0.5$
   $I = p \times r \times t$
   $\quad = \$1,500 \times 0.08 \times 0.5$
   $\quad = \$60$
   Total = $\$1,500 + \$60 = \$1,560$

   Answer Mr. Ito must repay $1,560.

Practice

Multiple-Choice Questions

1. A tie that has a regular price of $16 is on sale at 20% off. What is the sale price?

   A. $3.20   B. $12.80
   C. $14.00   D. $15.68

2. A stereo that regularly sells for $330 is on sale at 15% off. How much will a customer save on the stereo during the sale?

   F. $30   G. $49.50
   H. $82.50   J. $280.50

3. Lauren bought a winter coat that was on sale for 40% off the regular price of $199. How much did Lauren pay for the coat?

   A. $119.40   B. $159.00
   C. $179.60   D. $191.04

4. Mrs. Santiago bought a set of dishes at a factory outlet for $120. The same set of dishes is sold at a 60% markup in a department store. How much did she save?

   F. $48   G. $60
   H. $72   J. $96

5. A restaurant manager figured that the food for the nightly steak special actually cost $8.00. The restaurant will sell the special at an 80% markup. How much will a customer pay for the steak special?

   A. $10.90   B. $12.60
   C. $14.40   D. $16.00

6. Martin is paid a 12% commission on his shoe sales. He sold $780 worth of shoes. Find his commission.

   F. $56.60   G. $93.60
   H. $124.80   J. $156.00

7. Andrea is paid an 18% commission on her cosmetic sales. She sold $550 worth of cosmetics. Find her commission.

   A. $220   B. $199
   C. $118   D. $99

8. Ian deposits $900 in a savings certificate that pays $6\frac{1}{2}$% annually. How much money will Ian have at the end of one year?

   F. $958.50   G. $965.80
   H. $1,048.50   J. $1,485.00

9. Doreen invested $1,200 at 7% for 3 years. What will Doreen's investment be worth at the end of the 3 years?

   A. $1,225.20
   B. $1,452.00
   C. $1,704.00
   D. $2,520.00

10. Mr. Jaffrey borrowed $3,500 at 9% interest. He paid back the loan after 9 months. How much did he repay in all?

    F. $6,335.00
    G. $5,862.50
    H. $3,736.25
    J. $3,657.50

Short-Response Questions

11. Karyn bought a $49.95 leather tote bag on sale for 20% off. The sales tax in the area is 7%. To the nearest cent, what was the total cost of the tote bag?

12. A clothes boutique pays $32 for pairs of designer jeans, which it then marks up 65%. During March, the boutique has a 25% off sale. What is the sale price of the jeans?

Extended-Response Questions

13. Jared has priced a laptop computer at two stores. At Royal Electronics, the computer sells for $1,150, and there is a 15% off sale this month. At Crown Computers, the same laptop sells for $1,240, and there is a 20% off sale this month. Where will the computer cost less? How much will it cost?

14. A sweatshirt with a regular price of $34 is first discounted by 20% and then by an additional 10%.

    a. What is the final sale price?
    b. How do these two discounts compare to a single discount of 30%? Explain.

15. Beth receives a base salary of $215 per week plus a commission of 8% on all sales. How much does Beth earn in a week in which her sales total $2,200?

16. Nahum receives a commission of 18% on appliance sales in excess of $2,000 per week. One week, his sales were $6,850. How much was his commission that week?

17. Ms. O'Rourke purchased a certificate of deposit for $2,800. The certificate paid $5\frac{1}{2}$% interest per year. How much money will Ms. O'Rourke have after 2 years?

18. Mr. Sipowitz borrowed $25,000 to expand his cheese store. The interest rate was $10\frac{1}{2}$% per year. How much will he have to repay at the end of 2 years?

19. Kyra loaned $1,400 to her brother Ken. Ken agreed to repay the loan in 6 months at a rate of 11% per year. How much interest will Kyra receive?

20. A bank pays $7\frac{1}{2}$% interest on its savings certificates, and the customer can choose any whole number of years for the term. Joseph has $2,000 to invest. If he wants to double his money, what is the shortest term he can request for his certificate? Show your work.

Chapter 6 Review

Multiple-Choice Questions

1. Which ratio makes the same comparison as 8 dogs to 6 cats?

    A. 6 cats to 8 dogs
    B. 12 dogs to 9 cats
    C. 12 cats to 9 dogs
    D. 10 dogs to 8 cats

2. Solve: $\frac{112}{3.2} = \frac{0.7}{m}$.

    F. 0.02   G. 2.24
    H. 20.0   J. 24.5

3. A soup recipe calls for $1\frac{1}{2}$ quarts of chicken broth. This recipe serves 8 people. How much broth would be needed to make soup for 28 people?

    A. 42 quarts
    B. 30 quarts
    C. $5\frac{1}{4}$ quarts
    D. $3\frac{3}{4}$ quarts

4. Yani answered correctly 34 of the 40 questions on a test. What percent of the questions did she answer correctly?

F. 68%   G. 74%
H. 78%   J. 85%

5. Write 2.065 as a percent.

A. 2.65%   B. 26.5%
C. 206.5%   D. 265%

6. Walter spent 60% of the $37.50 he had for a video game. What was the cost of the game?

F. $24.00   G. $22.50
H. $15.00   J. $6.25

7. Sun Bright liquid detergent comes in four sizes. Which size is the best buy?

A. 16 ounces for $1.29
B. 28 ounces for $2.19
C. 40 ounces for $2.89
D. 96 ounces for $7.25

8. A blueprint of an apartment building uses a scale of $\frac{1}{2}$ inch = 4 feet. What is the actual length of a hallway that measures $1\frac{3}{4}$ inches on the blueprint?

F. 11 feet   G. 14 feet
H. 20 feet   J. 24 feet

9. What percent of 2 is 20?

A. 10%   B. 40%
C. 200%   D. 1,000%

10. A $64 pair of boots is on sale for 15% off. What is the sale price of the boots?

F. $29.60   G. $48.20
H. $54.40   J. $56.60

Short-Response Questions

11. The ratio of cups to cones sold at an ice cream shop is 8 to 9. If the store sold 1,098 cones in one week, how many cups and cones did it sell in all that week?

12. Express $\frac{97}{20}$

a. as a decimal   b. as a percent.

13. Ms. Sanchez earns $1,100 each week. Her employer withholds $308 for federal taxes. What percent of Ms. Sanchez's earnings is withheld for taxes?

Extended-Response Questions

14. A wholesaler of kitchen appliances offers a discount of 12% on the first $10,000 worth of merchandise and 17% on additional merchandise. What would be the final cost of $18,000 worth of appliances? Show your work.

15. Troy bought a $1,200 savings certificate that paid simple interest. After 2 years, his money had earned $108. What was the annual interest rate?

16. A newspaper's average daily circulation increased from 510,000 to 545,700. What was the percent increase in circulation?

17. In a survey, 800 tourists in New York City were asked to name their favorite type of international restaurant. The results are shown in the table.

Favorite Type of Restaurant	
Italian	39%
Chinese	25%
French	23%
Mexican	8%
Indian	5%

a. How many of those surveyed chose Italian?
b. How many more people chose French than Indian?

18. Use a scale of 2 cm = 3 m.

   a. Make a scale drawing of a rectangular garden that is 13.5 meters long and 21 meters wide.

   b. On the drawing, show a 1.5-cm square and label it *Lettuce*. What is the actual length of a side of the square bed for lettuce?

19. Find the next number in the pattern below:

1500, 2100, 2940, . . .

20. On a 75-question test, Alec answered 9 questions incorrectly. If each question was worth the same amount, how many points did he earn out of 100 points?

Chapter 6 Cumulative Review

Multiple-Choice Questions

1. Find the sum of 684.3 + 2,816.35 + 82.35 + 8.

   A. 6,583.08
   B. 3,591
   C. 3,583.8
   D. 2,906.7

2. How much greater is 64.7 than 6.9398?

   F. 30.76
   G. 56.872
   H. 57.7602
   J. 71.6398

3. Find the product of 8.75 and 0.734.

   A. 6.4225
   B. 64.2254
   C. 64.226
   D. 642.25

4. Divide 7.905 by 0.85.

   F. 8.36
   G. 9.03
   H. 9.3
   J. 93.6

5. Evaluate $16 - 3^2 + 2(4 - 2)$.

   A. 3
   B. 11
   C. 14
   D. 15

6. Evaluate $70 - 30 \div 5 + 3 \times 6$.

   F. 26
   G. $60\frac{3}{8}$
   H. 66
   J. 82

7. Which of these is the same as 2^5?

   A. 2×5
   B. $2 \times 2 \times 2 \times 2 \times 2$
   C. 5×5
   D. 50

8. Keith buys an entertainment system for $100 down and $25 per month for 3 years. Which of these expressions represents the total amount he will pay for the system?

   F. $\$100 + 3 \cdot \$25$
   G. $\$100 \cdot 3 \cdot \$25$
   H. $\$100 + \$25 + (3 \times 12)$
   J. $\$100 + \$25 \cdot 3 \cdot 12$

9. Coffee beans cost $7.39 a pound. Mrs. Romano buys a package for $6.18. Which of these expressions could be used to find the weight, in pounds, of the package?

A. $\dfrac{1}{\$7.39} = \dfrac{x}{\$6.18}$

B. $\dfrac{1}{\$6.18} = \dfrac{x}{\$7.39}$

C. $\dfrac{\$7.39}{\$6.18} = \dfrac{x}{1}$

D. $\dfrac{\$7.39}{x} = \dfrac{1}{\$6.18}$

10. What percent of 6 is 15?

F. 40%

G. $66\dfrac{2}{3}$

H. 150%

J. 250%

Short-Response Questions

11. The scale of a map is 0.5 cm = 8.3 km. Use a centimeter ruler to draw the line segment that would represent a distance of 41.5 km on the map. Show how you found the length.

12. A taxi service charges $1.35 for the first one tenth of a mile and $0.40 for each additional one tenth of a mile. Before the tip, Wendy's ride cost $6.15. What distance did she travel?

13. Insert parentheses so that the expression below simplifies to 6. Show the simplification steps.

$$\dfrac{15 \times 3 + 5^2 - 6}{44 - 43 \times 69}$$

14. The sale price of a sofa is $501.50 after a 15% discount has been given. What was the original price of the sofa?

Extended-Response Questions

15. Star Disposal picks up trash every Monday from the following companies:

Company	Tons of Trash
All-Time Toys	$2\dfrac{1}{2}$
Brite Products	$2\dfrac{4}{5}$
Cruz Inc.	3
Domino Partners	$3\dfrac{3}{10}$

If Star charges $85.00 per ton, find the total amount Star will earn for Monday's pickup.

16. Does subtraction of signed numbers follow the commutative property? Give examples to support your answer.

17. Last year, Mr. Prilik saved 8% of his $36,200 income. At the end of the year, he invested his savings at 5% interest. How much interest will his savings earn in one year? Show your work.

18. If 3 is subtracted from 4 times a number, the result is greater than 5.

a. Represent the statement above algebraically.
b. Find and graph the solution.

19. Four children chipped in to buy a Mother's Day gift. Brittany gave half as much as Andrew. Clinton gave half as much as Brittany. Darlene gave $7, which was the same amount as her brother Clinton. How much money was there for the gift?

20. Mr. Lawrence has 24 coins worth $2.20. The coins are all nickels and dimes. How many of each coin does he have?

CHAPTER 7

Measurement

Chapter Vocabulary

unit of measure
greatest possible error
  (GPE)

elapsed time
precision
relative error

time zones
accuracy
significant digits

7.1 Customary Units of Measure

The process of measurement allows you to compare a property of an object, such as length, width, weight, capacity, or temperature, to a standard **unit of measure**. Determining the number of units may require the use of a measurement device such as ruler for length or width, a scale for weight, or a thermometer for temperature.

The following table shows the relationships among customary units of measurement.

Length	Weight	Capacity
12 inches (in.) = 1 foot (ft)	16 ounces (oz) = 1 pound (lb)	8 fluid ounces (fl oz) = 1 cup (c)
3 feet = 1 yard (yd)	2,000 pounds = 1 ton (T)	2 cups = 1 pint (pt)
5,280 feet = 1 mile (mi)		2 pints = 1 quart (qt)
1,760 yards = 1 mile		4 quarts = 1 gallon (gal)

Converting Units

To convert from a larger unit to a smaller unit, multiply.
To convert from a smaller unit to a larger unit, divide.

Model Problems

1. Change 75 inches to feet.

   Solution   Inches are smaller units than feet. Divide the number of inches by 12.

   $75 \text{ in.} = (75 \div 12) \text{ ft} = 6\frac{1}{4} \text{ ft}$ There are fewer feet than inches because feet are larger units.

   Answer $75 \text{ in.} = 6\frac{1}{4} \text{ ft}$

2. Change 3 pounds 14 ounces to ounces.

   Solution   Ounces are smaller units than pounds. Multiply the number of pounds by 16.

   $3 \text{ lb} = (3 \times 16)\text{oz} = 48 \text{ oz}$ There are more ounces than pounds because ounces are smaller units.

   Add the ounces. 48 oz + 14 oz = 62 oz

   Answer   3 lb 14 oz = 62 oz

3. 35 qt = _____ gal _____ qt

   Solution   Since 4 quarts equal 1 gallon, divide the number of quarts by 4.

   $$\begin{array}{r} 8 \\ 4\overline{)35} \\ -32 \\ \hline 3 \end{array}$$

   35 quarts can be divided into 8 gallons, with a remainder of 3 quarts.

   Answer   35 qt = 8 gal 3 qt

4. $11\frac{1}{2}$ ft = _____ yd _____ in.

> **Solution** One way to find the answer is to convert feet to inches, then convert inches to yards.
>
> Multiply by 12 to convert feet to inches.
>
> $$11\frac{1}{2} \text{ ft} = \left(11\frac{1}{2} \times 12\right) \text{in.} = \left(\frac{23}{2} \times 12\right) \text{in.} = 138 \text{ in.}$$
>
> 1 yd = 36 in., so divide 138 by 36 to find out how many yards are in 138 inches.
>
> $$\begin{array}{r} 3 \\ 36\overline{)138} \\ -108 \\ \hline 30 \end{array}$$ 138 inches equal 3 yards, with a remainder of 30 inches.
>
> **Answer** $11\frac{1}{2}$ ft = 3 yd 30 in.
>
> *Note:* If you are asked to give the answer just in yards, then write 3 yd 30 in. as $3\frac{30}{36}$ yd. Simplify to get $3\frac{5}{6}$ yd.

Practice

Multiple-Choice Questions

1. The most reasonable estimate for the length of a family car is

 A. 80 in.
 B. 8 ft
 C. 80 ft
 D. 8 yd

2. Which is the heaviest weight?

 F. 4 lb 14 oz
 G. 68 oz
 H. $4\frac{3}{4}$ lb
 J. 3 lb 28 oz

3. The capacity of a large soup pot is most likely

 A. 4 c
 B. 4 pt
 C. 4 qt
 D. 4 gal

4. If the heights of students are being rounded to the nearest inch, a student listed as 5 ft 7 in. could be as short as

 F. 5 ft 5 in.   G. 5 ft $6\frac{1}{4}$ in.
 H. 5 ft $6\frac{1}{2}$ in.   J. $5\frac{1}{2}$ ft

5. Which distance is the same as $\frac{1}{4}$ mile?

   A. 1,220 ft
   B. 1,640 ft
   C. 440 yd
   D. 1,320 yd

6. Which length is the same as 351 inches?

   F. $27\frac{1}{4}$ ft
   G. 8 yd 21 in.
   H. $9\frac{3}{4}$ yd
   J. $10\frac{1}{3}$ yd

7. Akiko needs $4\frac{1}{2}$ yards of fabric for a dress she is designing. How many inches of fabric does she need?

   A. 54 in.
   B. 72 in.
   C. 148 in.
   D. 162 in.

8. How many pints are there in $7\frac{1}{2}$ quarts?

   F. 11 pt
   G. 15 pt
   H. 21 pt
   J. 30 pt

9. A contractor used $1\frac{3}{4}$ tons of gravel for a driveway. How many pounds of gravel was this?

   A. 2,750 lb
   B. 3,500 lb
   C. 3,750 lb
   D. 7,000 lb

10. How many fluid ounces are there in $2\frac{1}{2}$ gallons?

   F. 160 fl oz
   G. 248 fl oz
   H. 320 fl oz
   J. 512 fl oz

Short-Response Questions

11. Convert to feet.

   a. 108 in.
   b. $4\frac{2}{3}$ yd
   c. 5 yd 8 in.

12. Convert to pounds.

   a. 176 oz   b. 58 oz
   c. $6\frac{3}{4}$ T   d. 3 T 600 lb

13. Convert to pints.

   a. 96 fl oz
   b. $5\frac{1}{2}$ qt
   c. 17 c
   d. 3 gal 3 qt

14. Write four other measures that are equivalent to 104 cups.

15. Write the following lengths in order from least to greatest:

   49 ft $15\frac{3}{4}$ yd   620 in. $\frac{1}{100}$ mi

Extended-Response Questions

16. Each guest at a party drank 3 cups of punch. There were 24 guests. How many gallons of punch did the guests drink?

17. A motor scooter is traveling at 40 miles per hour. How many yards does it travel in $\frac{1}{4}$ hour?

18. The Leaning Tower of Pizza uses 20 quarts of pizza sauce every day. How many gallons of sauce does the restaurant use per week?

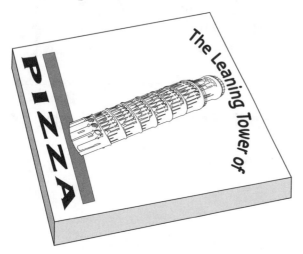

19. Marlena needs $4\frac{1}{2}$ yards of checked cotton to make curtains. At the fabric shop, some leftover pieces of the cotton she wants are on sale. The lengths are 156 inches, 166 inches, and 178 inches. Which is the smallest piece that Marlena can purchase?

20. Jack and Jill kept a record of the amount of water they drank each day for five days. Their records are shown on the table. Who drank more water during the five days? How much more? Give the answer in fluid ounces and in pints.

	Jack	Jill
Day	Fluid Ounces	Pints
Monday	56	4
Tuesday	60	3.5
Wednesday	72	4.5
Thursday	88	3.75
Friday	48	5

7.2 Computing with Customary Units

Measures must be expressed in the same units in order to be added or subtracted. To add or subtract measures that are given in different units (for example, one in feet and the other in yards):

- Write one of the given measures with the same unit as the other.
- Add or subtract the two measures that are now expressed in terms of the same unit and label the result with this unit.
- Rename the result if necessary. For example, 15 in. can be renamed 1 ft 3 in.

Note: Some subtraction problems will require regrouping.

Model Problems

1. Add 8 ft 9 in. to 5 ft 10 in.

Solution

    8 ft 9 in.
 + 5 ft 10 in.   Line up the units. Add feet and inches separately.
  13 ft 19 in.

Rename 19 in. as 1 ft 7 in. Add 1 ft to 13 ft.

Answer 14 ft 7 in.

2. Subtract 7 lb 8 oz from 12 lb 5 oz.

Solution

12 lb 5 oz	5 oz < 8 oz, so regroup to subtract.	11 lb 21 oz
− 7 lb 8 oz	1 lb = 16 oz	− 7 lb 8 oz
		4 lb 13 oz

No renaming is necessary.

Answer 4 lb 13 oz.

3. Subtract 2 gal 3qt from 7 gal.

Solution

7 gal		6 gal 4 qt
− 2 gal 3 qt	Regroup. 1 gal = 4 qt	− 2 gal 3 qt
		4 gal 1 qt

Answer 4 gal 1 qt

4. A math book weighs 2 lb 6 oz. How much do 6 of these books weigh?

Solution

  2 lb 6 oz
     × 6   Multiply pounds and ounces separately.
 12 lb 36 oz

Rename using both units or just one.

    1 lb = 16 oz, so 36 oz = 2 lb 4 oz.
    Add 2 lb 4 oz + 12 lb to get 14 lb 4 oz.

$$14 \text{ lb } 4 \text{ oz} = 14\frac{4 \text{ oz}}{16 \text{ oz}} = 14\frac{1}{4} \text{ lb}$$

Answer 14 lb 4 oz or $14\frac{1}{4}$ lb

5. A piece of rope 10 ft 8 in. long is cut into 4 equal pieces. How long is each piece?

Solution   Divide 10 ft 8 in. by 4.

Method I   Express as feet, then divide.

$$10 \text{ ft } 8 \text{ in.} = 10\tfrac{2}{3} \text{ ft}$$

$$10\tfrac{2}{3} \div 4 = \frac{32}{3} \div 4 = \frac{32^8}{3} \times \frac{1}{4^1} = \frac{8}{3}$$ Rewrite the improper fraction as a mixed number.

Answer $2\tfrac{2}{3}$ ft = 2 ft 8 in.

Method II   Express as inches, then divide.

10 ft 8 in. = 120 in. + 8 in. = 128 in.

$$
\begin{array}{r}
32 \text{ in.} \\
4\overline{)\ 128 \text{ in.}} \\
\underline{-12} \\
08 \\
\underline{-8} \\
0
\end{array}
$$

Answer   32 in. = 2 ft 8 in.

Practice

Multiple-Choice Questions

1. Add 8 ft 11 in. and 7 ft 9 in.

  A. 15 ft 8 in.   B. 16 ft 4 in.
  C. 16 ft 8 in.   D. 17 ft 2 in.

2. Add 16 lb 14 oz and 9 lb 10 oz

  F. 25 lb 4 oz
  G. 26 lb 6 oz
  H. 26 lb 8 oz
  J. 27 lb

3. A chef had 6 gal 2 qt of salad dressing. During the day, 2 gal 3 qt were used. How much salad dressing is left?

  A. 4 gal 1 qt
  B. 4 gal 3 qt
  C. 3 gal 3 qt
  D. 5 gal 1 qt

4. Jodi had $2\tfrac{1}{2}$ yd of ribbon. She cut off a piece 22 in. long. How much ribbon is left?

  F. 8 in.   G. 46 in.
  H. 52 in.   J. 68 in.

5. The length of a room is 20 ft 3 in. and its width is 12 ft 9 in. How much longer is the room than it is wide?

  A. $7\tfrac{1}{2}$ ft   B. $7\tfrac{5}{6}$ ft
  C. $8\tfrac{1}{4}$ ft   D. $8\tfrac{2}{3}$ ft

6. A vase weighs 3 lb 7 oz. How much will 7 of the vases weigh?

  F. 23 lb 3 oz   G. 24 lb 1 oz
  H. 25 lb 1 oz   J. 25 lb 5 oz

7. A piece of wire 6 yd 2 ft long is cut into 5 equal pieces. Which measure is NOT equivalent to the length of each piece?

   A. 1 yd 1 ft

   B. 4 ft

   C. $1\frac{1}{3}$ yd

   D. 64 in.

8. Add 8 yd 1 ft 9 in. and 4 yd 2 ft 5 in.

   F. 12 yd 2 ft 2 in.

   G. 13 yd 1 ft 2 in.

   H. 12 yd 2 ft 4 in.

   J. 14 yd 1 in.

9. 25 oz + 33 oz + 42 oz is equal to

   A. $5\frac{5}{8}$ lb

   B. $6\frac{1}{4}$ lb

   C. $7\frac{1}{2}$ lb

   D. $8\frac{1}{3}$ lb

10. At the Icy Delight frozen yogurt shop, a small scoop contains $4\frac{1}{2}$ fl oz of yogurt. How many quarts of yogurt are used for 160 scoops?

   F. $11\frac{1}{4}$ qt   G. 16 qt

   H. $22\frac{1}{2}$ qt   J. 45 qt

Short-Response Questions

11. Add or subtract.

   a.   7 ft 6 in.   b.   9 lb 3 oz
         + 4 ft 10 in.   − 2 lb 7 oz

   c.   5 gal 3 qt   d.   9 yd
        + 3 gal 1 qt   − 1 yd 25 in.

12. Multiply or divide.

   a. 5 lb 6 oz   b. 2 ft 9 in.
       ×   8   ×   5

   c. 8)19 yd 12 in.   d. 7)10 gal 2 qt

13. The maximum weight for a vehicle crossing a bridge is 12,500 pounds. An empty delivery truck weighs $2\frac{1}{2}$ tons. How many pounds can the truck safely carry? How many tons?

14. Arianna had 10 yards of fabric. She used 3 yards 10 inches to make a bedspread and 2 yards 4 inches to make pillow covers. How much fabric does she have left?

15. Adam's dog weighs 34 lb 13oz. Betty's dog weighs 51 lb 5 oz How much more does Betty's dog weigh?

16. The length of one side of a square garden is 14 ft 10 in. How many feet of fencing are needed to enclose the garden?

17. To make 6 qt of fruit punch, Fernando used 2 qt 1 pt of orange juice and 1 qt 1 pt of cranberry juice. The rest was sparkling water. How much sparkling water did he use?

Extended-Response Questions

18. Tina's distances for the triple jump were 4 ft 7 in., 5 ft 3 in., and 4 ft 11 in.

   a. What was the combined distance she jumped?

   b. Kate's combined distance was $15\frac{1}{2}$ ft. How much farther did she jump than Tina?

19. A box held 7 lb 3 oz of jelly beans. Elliot repacked the candy into several plastic bags that each weighed 1 lb 7 oz when full. How many plastic bags did he fill?

20. To make decorations for a July 4th barbecue, Ross used $9\frac{1}{2}$ yd of red ribbon, 6 ft 9 in. of white ribbon, and 2 yd 1 ft of blue ribbon. How many yards of ribbon did he use in all?

7.3 Metric Units of Measure

The names of metric units are made up of a prefix combined with a basic unit. The basic units are meter (length), gram (mass), and liter (capacity). All metric prefixes name powers of 10.

For example,

1 hectometer is a unit of length = 100 meters
1 centigram is a unit of mass = 0.01 gram
1 milliliter is a unit of capacity = 0.001 liter

Prefix	Symbol	Meaning
kilo-	k	1000 or 10^3
hecto-	h	100 or 10^2
deka-	da	10 or 10^1
deci-	d	0.1 or 10^{-1}
centi-	c	0.01 or 10^{-2}
milli-	m	0.001 or 10^{-3}

The metric relationships you will use most often are shown below.

Length	Mass	Capacity
10 millimeters (mm) = 1 centimeter (cm)	1000 milligrams (mg) = 1 gram (g)	1000 milliliters (mL) = 1 liter (L)
1000 millimeters = 1 meter (m)	1000 grams = 1 kilogram (kg)	1000 liters = 1 kiloliter (kL)
100 centimeters = 1 meter	1000 kilogram = 1 metric ton (t)	
1000 meters = 1 kilometer (km)		

This metric diagram can be used to change metric units.

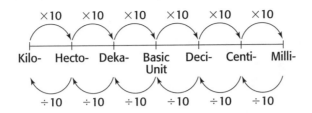

- To change from a larger unit to a smaller unit, multiply. Move to the right on the diagram. Move the decimal point the same number of places to the right.
- To change from a smaller unit to a larger unit, divide. Move to the left on the diagram. Move the decimal point the same number of places to the left.

 **Model Problems**

1. 5 g = ? mg

Solution
Method I
  1 g = 1000 mg
  5 g = (5 × 1000) mg
  5 g = 5000 mg

Method II
On the diagram, milligrams are 3 steps to the right of grams. Move the decimal point 3 places to the right to multiply by 1000.
  5 g = 5000 mg

Answer 5000 mg
Note: Since milligrams are smaller units than grams, there are more of them.

2. 38 m = ? km

Solution
Method I
  1000 m = 1 km
    38 m = (38 ÷ 1000) km
         = 0.038 km

Method II
Changing meters to kilometers is 3 moves left. Move the decimal point 3 places left to divide by 1000.

> 38 m = 0.038 km

Answer 0.038 km

Note: Since kilometers are larger units than meters, there are fewer of them.

3. There are 4.5 meters of tape on a brand-new roll. How many centimeters of tape are on the roll?

Solution Changing meters to centimeters is 2 steps right. Move the decimal point 2 places right to multiply by 100.

> 4.5 m = 450 cm

Answer The roll holds 450 cm of tape.

Commonsense Unit Comparisons

The following comparisons can help you estimate conversions between metric and customary units.

Length

- A meter is a little longer than a yard.
- An inch is about 2.5 centimeters.
- A centimeter is a little less than $\frac{1}{2}$ inch.
- A kilometer is a little longer than $\frac{1}{2}$ mile.

Mass/Weight

- A kilogram is a little more than 2 pounds.
- There are about 28 grams in an ounce.
- A metric ton is a little more than a customary ton.

Capacity

- A liter is a little more than a quart.

4. A greeting card is 8 inches long. About how many centimeters long is it?

Solution *Think:* 1 in. is about 2.5 cm
8 in. is about (8 × 2.5) cm = 20 cm

Answer The card is about 20 cm long.

Practice

Multiple-Choice Questions

1. The height of a bicycle seat from the ground is about

   A. 10 centimeters
   B. 1 meter
   C. 10 meters
   D. 1 kilometer

2. Which of these objects is most likely to have a mass of 10 kilograms?

   F. a motorcycle
   G. a basketball
   H. a math textbook
   J. a portable TV

3. How many grams are equivalent to 3.04 kilograms?

   A. 0.0304   B. 30.4
   C. 304   D. 3040

4. How many meters are equivalent to 762 centimeters?

   F. 0.0762   G. 0.762
   H. 7.62   J. 76.2

5. Which measure is NOT equivalent to the others?

   A. 0.0048 km   B. 4800 mm
   C. 48 cm   D. 4.8 m

6. Joan drank one fourth of a liter container of juice. How many milliliters did she drink?

   F. 2.5   G. 25
   H. 250   J. 2500

7. A cook uses 200 grams of ground beef for each hamburger patty. How many hamburgers can be made from a 5-kilogram package of ground beef?

   A. 25   B. 40
   C. 250   D. 400

8. Marc ran a 10-kilometer race. About how many miles did he run?

   F. 6 miles   G. 11 miles
   H. 21 miles   J. 26 miles

9. Melanie bought 6 bottles of soda for a party. Each bottle contained 2 liters. About how many quarts of soda did she buy?

   A. 7 quarts   B. 13 quarts
   C. 25 quarts   D. 49 quarts

10. The length of a hallway was measured to be 1786 centimeters. What is the length of the hallway to the nearest meter?

    F. 2 meters   G. 18 meters
    H. 179 meters J. 1786 meters

Short-Response Questions

11. Complete.

    a. 950 cm = ____ m
    b. ____ mg = 0.3 kg
    c. 4.6 L = ____ mL

12. Change each measurement to meters.

    a. 1360 cm
    b. 425 mm
    c. 0.94 km

13. Change each measurement to grams.

    a. 0.08 kg
    b. 53,400 mg
    c. 29 t

14. Change each measurement to liters.

    a. 630 mL
    b. 0.28 kL
    c. 45,000 mL

15. Write three measurements equivalent to 1670 cm.

Extended-Response Questions

16. Annette drank half a liter of water while she worked in the garden. She drank another 750 milliliters of water after sweeping the patio. How many liters of water did Annette drink in all?

17. A potato salad recipe calls for 600 grams of boiled potato. About how many ounces of potato is this? Show how you estimate.

18. A swimming pool is 50 meters long. An athlete practicing for a triathlon competition swims 60 lengths of the pool each day. How many kilometers does the athlete swim in a week of practice? Show your work.

19. Ms. Clancy has been advised by her doctor to limit her sodium intake to 1200 mg per day. On Monday, she determined that her breakfast contained 280 mg of sodium. For lunch, she had 4 ounces of chicken, 2 slices of whole wheat bread, 1 tablespoon of mayonnaise, and 1 cup of spinach. Later, she snacked on an apple. How many milligrams of sodium can Ms. Clancy have for dinner if she wants to stay within the limit?

Food	Amount of Sodium
Chicken (4 oz)	75 mg
Whole wheat bread (1 slice)	120 mg
Mayonnaise (1 tbsp)	90 mg
Spinach ($\frac{1}{2}$ cup)	44 mg
Apple (1)	2 mg

20. A factory produced 150 liters of Ambiance perfume. First, 400 bottles were filled with 325 milliliters each. Then, the remaining perfume was used to fill smaller bottles with 125 milliliters each. How many of the smaller bottles were filled?

7.4 Temperature

In the metric system, temperatures are measured using the Celsius scale. The degree Celsius (°C) is the unit of temperature for this scale.

In the customary system, temperatures are measured using the Fahrenheit scale. The degree Fahrenheit (°F) is the unit of temperature for this scale.

The figure shows a comparison of some familiar temperatures. Celsius degrees are larger units than Fahrenheit degrees, so fewer Celsius degrees are needed to measure a given amount of heat. As you can see by looking at the two thermometers, a Celsius temperature is always a smaller number of units than the corresponding Fahrenheit temperature.

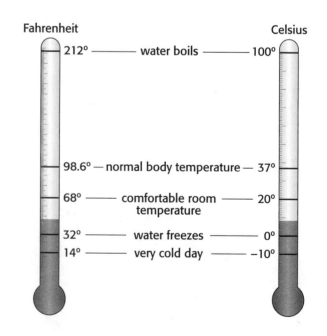

Recall that in Section 5.8 you worked with the following formulas to convert between Celsius and Fahrenheit temperatures.

$$°C = \frac{5}{9}(°F - 32) \qquad °F = \frac{9}{5}°C + 32$$

Model Problems

1. Which temperature is more reasonable for a bowl of very hot soup, 95°C or 95°F?

Solution 95°C is a little less than the temperature of boiling water. 95°F is close to normal body temperature, and the soup is much hotter than that.

Answer 95°C is the more reasonable temperature.

2. On the Fahrenheit scale, what temperature is 28° above the freezing point of water?

Solution The freezing point of water is 32°F.
    32 + 28 = 60

Answer 60°F

3. Find the change in temperature from 43°C to 17°C. Was the change an increase or a decrease?

Solution To determine the change in temperature, subtract the first value from the second.
    17 − 43 = −26

Answer The change is negative, which indicates that the temperature decreased by 26°.

4. Find the Fahrenheit temperature that corresponds to a Celsius temperature of 70°C.

Solution The formula that converts °C to °F is $°F = \frac{9}{5}°C + 32$. Substitute 70° for °C in the formula and solve.

$$°F = \frac{9}{5}(70) + 32$$

$$°F = \frac{9}{5^1} \cdot \frac{\cancel{70}^{14}}{1} + 32$$

$$°F = 126 + 32$$

Answer 70°C = 158°F
The answer makes sense since the number of °F is much greater than the number of °C.

Multiple-Choice Questions

1. The most reasonable temperature for a good beach day is

   A. 91°C   B. 86°F
   C. 50°F   D. 10°C

2. The most reasonable temperature for a glass of cold milk is

   F. 35°C   G. 20°C
   H. 5°C   J. −5°C

3. If it is 40°C in Chicago, which month of the year is it most likely to be?

   A. October   B. December
   C. March   D. July

4. Marianne is planning to roast a chicken. The oven temperature she should use is

   F. 375°F   G. 250°F
   H. 195°F   J. 150°F

5. Which thermometer shows −4°C?

   A.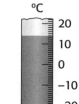
   B.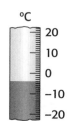

   C.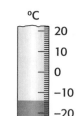
   D.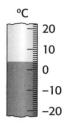

6. On the Celsius scale, what temperature is 18° below the freezing point of water?

   F. −18°C   G. 2°C
   H. 14°C   J. 82°C

7. On the Fahrenheit scale, what temperature is about 20° below normal body temperature?

   A. 108°F
   B. 78°F
   C. 52°F
   D. 17°F

8. Find the change in temperature from −6°C to 17°C.

   F. −23°C
   G. −11°C
   H. 11°C
   J. 23°C

9. Find the change in temperature from −8°F to −21°F.

   A. −29°F
   B. −13°F
   C. 13°F
   D. 29°F

10. Use the formula $°C = \frac{5}{9}(°F - 32)$ to find the Celsius temperature that corresponds to 122°F.

    F. 154°C
    G. 85.5°C
    H. 50°C
    J. 36°C

Short-Response Questions

11. State whether each temperature is most likely Fahrenheit or Celsius.

    a. a cup of hot tea: 110°F or 110°C
    b. a cool autumn day: 15°F or 15°C
    c. swimming pool water: 68°F or 68°C

12. Find each change in temperature.

    a. from 24°C to 52°C
    b. from −7°F to 25°F
    c. from −2°C to −16°C
    d. from 19°F to −11°F

13. Find the change in temperature from the first thermometer reading to the second.

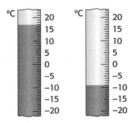

Extended-Response Questions

14. On the way into her house, Mrs. Green did not realize she had dropped a container of ice cream from the grocery bags. Overnight, the temperature was 12°C. What was the condition of the ice cream when she found it the next morning? Explain.

15. For a science experiment, Lisa and Ronald each began with a beaker of the same tap water. Lisa heated her water until its temperature rose 20 Fahrenheit degrees and Ronald heated his water until its temperature rose 20 Celsius degrees. Would the water in one beaker feel as hot as the other? If not, which is hotter and why?

16. A board of health inspector told a grocery store owner that the temperature of the diary cooler must not exceed 40°. Is it more likely that the inspector meant 40°C or 40°F? Explain.

17. Use the formulas on page 211. What Fahrenheit temperature corresponds to −40°C?

18. There is a 63-degree difference between 37°C and 100°C. What is the difference between the equivalent temperatures on the Fahrenheit scale?

7.5 Time and Time Zones

The same units of time are used in both the metric and customary systems.

60 seconds (s) = 1 minute (min)	52 weeks = 1 year (yr)
60 minutes = 1 hour (h)	12 months (mo) = 1 year
24 hours = 1 day (d)	100 years = 1 century
7 days = 1 week (wk)	1,000 years = 1 millennium

The hours between midnight and noon are A.M. hours.
The hours between noon and midnight are P.M. hours.
To compute the **elapsed time** (time gone by) between two given times, you may use different methods. If the times are both A.M. or both P.M. in the same day, just subtract the earlier time from the later time. If one time is A.M. and one time is P.M., you can think about a clock and count forward from the first time.

Model Problems

1. The accounting department used 696 minutes of computer time. How many hours did it use?

   Solution   To change to a larger unit, divide.

   60 min = 1 h

   $$\begin{array}{r} 11.6 \\ 60\overline{)\,696.0} \\ -60 \\ \hline 96 \\ -60 \\ \hline 36\,0 \\ -36\,0 \\ \hline 0 \end{array}$$

   Or, using a remainder

   $$\begin{array}{r} 11 \text{ r } 36 \\ 60\overline{)696} \\ -660 \\ \hline 36 \end{array}$$

   Answer   11.6 h or 11 h 36 min
   Note: If the time were to be given in hours and minutes, you would use the remainder.

2. How much time has elapsed between 2:25 P.M. and 8:40 P.M.?

   Solution   Since both times are P.M., subtract directly.

   $$\begin{array}{r} 8:40 \\ -2:25 \\ \hline 6:15 \end{array}$$

   Answer   6 h 15 min have elapsed.

3. School begins at 8:30 A.M. and ends at 2:50 P.M. How long is the school day?

   Solution
   First find the hours.
   There are 6 h between
   8:30 A.M. and 2:30 P.M.

   Then find the minutes.
   There are 20 min between
   2:30 P.M. and 2:50 P.M.

   Answer   The school day is 6 h 20 min long.

4. Dennis worked 5 hours and 10 minutes on Monday and 2 hours and 55 minutes on Tuesday. How much longer did he work on Monday than Tuesday?

Solution The problem asks you to find the difference between two amounts of time, so you must subtract. Regrouping may be needed when times are added or subtracted.

$$
\begin{array}{r}
5 \text{ h } 10 \text{ min} \\
-2 \text{ h } 55 \text{ min} \\
\hline
\end{array}
\qquad \text{Regroup 1 h as 60 min.} \qquad
\begin{array}{r}
4 \text{ h } 70 \text{ min} \\
-2 \text{ h } 55 \text{ min} \\
\hline
2 \text{ h } 15 \text{ min}
\end{array}
$$

Answer He worked 2 h 15 min longer on Tuesday.

Time Zones

When a problem involves times in different geographical locations in the United States, you must consider **time zones** when finding a solution. Most of the United States is divided into four time zones. From the East Coast to the West Coast the zones are: Eastern Time, Central Time, Mountain Time, and Pacific Time. Starting in the Eastern Time zone, the consecutive time zones are each one hour earlier.

The map below shows the time zones in the United States. The clocks show the time in each zone at the same moment. For example, when it is 4:00 P.M. in the Eastern Time zone, it is 3 hours earlier or 1:00 P.M. in the Pacific Time zone.

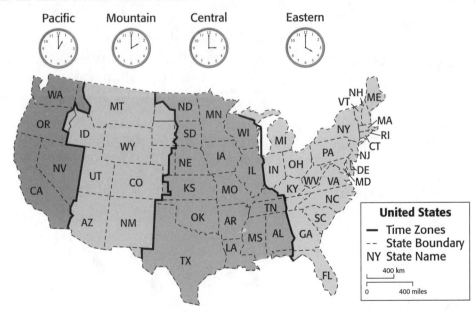

5. When it is 10:30 A.M. in Colorado, what time is it in

a. Ohio? b. Nevada?

Solution a. Ohio is two time zones east of Colorado, so the time is 2 hours later. It is 12:30 P.M. in Ohio.
b. Nevada is one time zone west of Colorado, so the time is 1 hour earlier. It is 9:30 A.M. in Nevada.

6. A plane left Atlanta, Georgia, at 8:00 A.M. and landed in San Francisco, California, $6\frac{1}{2}$ hours later. What was the time in San Francisco?

Solution When the plane landed in San Francisco, the time in Atlanta was 8:00 A.M. $+ 6\frac{1}{2}$ hours, or 2:30 P.M.

Traveling west from Atlanta to San Francisco, the plane crossed 3 time zones. The time in San Francisco was 3 hours earlier.

2:30 P.M. $-$ 3 hours = 11:30 A.M.

Answer 11:30 A.M.

Practice

Multiple-Choice Questions

1. 540 min = ____ h

   A. 6
   B. 7
   C. 8
   D. 9

2. How many seconds are equivalent to 5 hours?

   F. 300 s
   G. 3,000 s
   H. 18,000 s
   J. 108,000 s

3. Which is the greatest amount of time?

   A. 100 hours
   B. 4 days
   C. 4,320 minutes
   D. 90,000 seconds

4. Choose the missing unit. It takes about 3 ____ for a flower seed to sprout.

   F. minutes
   G. weeks
   H. months
   J. years

5. Vinelle must get to work by 9:20 A.M. If her trip from home takes 55 minutes, by what time must she leave home in order to be on time?

   A. 7:45 A.M.   B. 8:10 A.M.
   C. 8:35 A.M.   D. 8:25 A.M.

6. The Roosevelt Branch Library is open daily from 9:00 A.M. until 4:30 P.M., 255 days a year. How many hours in all is the library open to the public each year?

   F. 114.75 h   G. 191.25 h
   H. 1,147.5 h   J. 1,912.5 h

7. How much time has elapsed from the first clock to the second?

A.M.

P.M.

   A. 4 h 25 min   B. 5 h 15 min
   C. 7 h 35 min   D. 19 h 15 min

8. Oscar began reading a book at 11:20 A.M. When he finished the book he realized $4\frac{1}{4}$ hours had passed. What time was it when he finished the book?

   F. 3:35 P.M.
   G. 3:45 P.M.
   H. 4:15 P.M.
   J. 5:30 P.M.

9. When it is 10 A.M. in the Eastern Time zone, what time is it in the Pacific Time zone?

   A. 1:00 P.M.
   B. 12 noon
   C. 8:00 A.M.
   D. 7:00 A.M.

10. If it is 6:25 P.M. in the Mountain Time zone, what time is it in the Central Time zone?

   F. 4:25 P.M.
   G. 5:25 P.M.
   H. 7:25 P.M.
   J. 8:25 P.M.

Short-Response Questions

11. Complete.

   a. _____ yr = 48 mo
   b. 780 s = _____ min
   c. 204 h = _____ d

12. Write in order from least amount of time to greatest:

   95 minutes, $1\frac{1}{2}$ hours, 0.1 day, 2,700 seconds

13. How much time has elapsed between the first time and the second time?

   a. from 3:45 P.M. to 9:20 P.M.
   b. from 2:05 A.M. to 10:25 A.M.
   c. from 11:03 P.M. to 8:42 A.M.
   d. from 9:28 A.M. to 11:40 P.M.

14. Add or subtract.

   a.   4 h 48 min
       + 3 h 19 min

   b.   10 h 24 min
       − 3 h 39 min

   c.   5 d 13 h
       + 4 d 11 h

   d.   15 h 5 min
       − 2 h 53 min

Extended-Response Questions

15. The table shows the daily high tides at several locations along the East Coast.

High Tides		
Bridgeport	3:04 A.M.	3:44 P.M.
Northport	3:08 A.M.	3:47 P.M.
Port Washington	2:58 A.M.	3:38 P.M.
Tarrytown	1:43 A.M.	2:46 P.M.
Willets Point	2:51 A.M.	3:31 P.M.

   a. How much time elapses between the first and second high tides at Northport?
   b. At which location does more time elapse between the first and second high tides: Bridgeport or Tarrytown? How much longer?

16. On April 28, 2001, sunrise in the New York area was at 5:59 A.M. Sunset was 13 hours and 51 minutes later. What time did the sun set?

17. Use the Time Zone Map on page 215. A plane leaves New York City at 9:25 A.M. and lands in Chicago, Illinois, 2 hours and 40 minutes later. What time is it in Chicago when the plane lands?

18. Each day in the United States, approximately 60 million plastic bottles are thrown away. Approximately how many plastic bottles are thrown away in the United States per hour? Show your work.

19. David's weekly time card is shown on the right. David earns $6.30 an hour. Complete the hours column. Find David's total earnings before taxes. Show your work.

20. Ginelda finished her homework at 8:40 P.M. She had spent 25 minutes on English, 1 hour and 10 minutes on math, 40 minutes on science, and 35 minutes on history. If she worked without interruption, at what time had she started?

BVS Pharmacy			Employee Time Card		
Employee Name: David Ortega					
Date	In	Out	In	Out	Hours
5/1	8:30	12:30	1:00	5:00	
5/2	9:45	1:00	1:30	6:45	
5/3	3:30	7:00			
5/4	3:10	6:40			
5/5	4:00	9:00			
				Total Hours	_____

7.6 Precision in Measurement

All measurements are approximate. To the nearest unit, the length of the bracelet below is 6 inches.

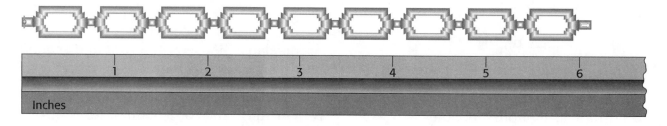

With a ruler marked in whole units, any length from $5\frac{1}{2}$ inches to $6\frac{1}{2}$ inches is reported as 6 inches. The **greatest possible error** (GPE) in measurements using such a ruler is $\frac{1}{2}$ inch. You could write the length of the bracelet as:

Measure to the Nearest Unit ± Greatest Possible Error, or $6 \pm \frac{1}{2}$

Because the closest marks on this ruler are 1 inch apart, the **precision** of measurements using this ruler is 1 inch.

If the same bracelet is measured using a ruler marked in half units, a length reported as 6 inches could actually be between $5\frac{3}{4}$ inches and $6\frac{1}{4}$ inches. In this case, the precision is $\frac{1}{2}$ inch and the greatest possible error is $\frac{1}{4}$ inch. The length of the bracelet could be written $6 \pm \frac{1}{4}$ inches.

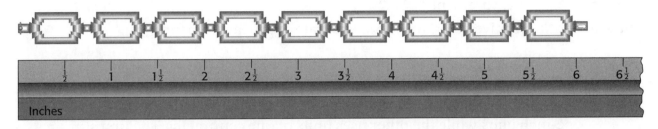

The greatest possible error of any measurement is one half the precision of the measurement. The smaller the GPE, the greater the precision of the measuring tool and the measurements. In other words, the smaller the unit of measure, the more precise the measurement.

Model Problems

1. What are the precision and GPE of any measurement made with the ruler below?

Solution   The closest marks on the ruler are 0.1 cm apart; therefore the precision of this ruler is 0.1 cm. (Since 0.1 cm = 1 mm, the precision can also be stated as 1 mm.)

Answer   Using this ruler, the precision is 0.1 cm, and the GPE is half of 0.1 cm or 0.05 cm.

2. The GPE of a measurement made using a certain ruler is $\frac{1}{16}$ inch. What is the precision of the ruler?

Solution   The GPE is half the precision, so the precision is double the GPE.

$$\text{precision} = 2 \times \frac{1}{16} \text{ in.} = \frac{1}{8} \text{ in.}$$

Answer   The precision of the ruler is $\frac{1}{8}$ in.

Accuracy

The **accuracy** of a measurement tells you how correct or true it is. To determine which of two measurements is more accurate, calculate the **relative error** for each. The relative error is expressed as a percent.

$$\text{relative error} = \frac{\text{GPE}}{\text{measurement}} \times 100$$

The smaller the relative error, the more accurate the measurement.

---

3. Two students measure the length of the chalkboard. One student uses one-inch units while the other uses units of one centimeter. The first student measures the length to be 100 inches and the second student reports the length as 254 centimeters. Which student had the more accurate measurement?

Solution

Measurement	Unit	GPE	Relative Error
100 in.	1 in.	± 0.5 in.	$\frac{0.5}{100} \cdot 100 = 0.5\%$
254 cm	1 cm	± 0.5 cm	$\frac{0.5}{254} \cdot 100 = 0.2\%$ (rounded)

Answer The measurement of 254 cm is more accurate because the relative error is smaller.

Significant Digits

Significant digits are the digits of a measurement that represent meaningful data. Significant digits show the accuracy of the measurement. A measurement such as 16 feet is accurate to 2 significant digits. It is understood that the final digit, the 6, could be wrong, so the last significant digit in any measurement is always the estimated digit.

The following rules are used to identify significant digits:

• All nonzero digits are significant.

Zero is significant in the following situations:

• When zero is used as a placeholder between two nonzero digits; for example, 105 or 1,004
• When final zeros appear after the decimal point; for example, 4.20 or 15.300
• When the final zero (or zeros) in whole numbers are marked to show they are significant; for example, 2<u>00</u> or 5,<u>000</u>
• Initial zeros are never significant; for example, 0.0056 has only two significant digits, 5 and 6.

4. Indicate the significant digits in each number.

                      Solution

    a. 584   3 significant digits 5, 8, 4

    b. 46.290   5 significant digits 4, 6, 2, 9, 0

    c. 0.400   3 significant digits 4,0, 0

    d. 1,500   2 significant digits 1, 5

    e. 1,5<u>00</u>   4 significant digits 1, 5, 0, 0

    f. 0.0871   3 significant digits 8, 7, 1

5. Give the possible range of the actual measurement for each.

    a. 140 cm   b. 14<u>0</u> cm

    Solution

    a. There are 2 significant digits, 1 and 4. Since 4 is the estimated digit and indicates tens, the actual measurement is 140 ± 5 cm or 135 cm ≤ actual measure ≤ 145 cm.

    b. There are 3 significant digits, 1, 4, and 0. Since 0 is the estimated digit and indicates ones, the actual measurement is 140 ± 0.5 cm or 139.5 ≤ actual measure ≤ 140.5 cm.

Practice

Multiple-Choice Questions

1. The precision of a measurement is 0.01 unit. What is the GPE for this measurement?

    A. 0.005   B. 0.02

    C. 0.5   D. 0.1

2. The precision of a measurement is 0.5 liter. What is the GPE for this measurement?

    F. 1.0 L   G. 0.5 L

    H. 0.25 L   J. 0.05 L

3. The GPE of a measurement is $\frac{1}{32}$ inch. What is the precision of this measurement?

    A. $\frac{1}{64}$ in.   B. $\frac{1}{16}$ in.

    C. $\frac{1}{8}$ in.   D. $\frac{1}{4}$ in.

4. What is the GPE of any measurement made with this ruler?

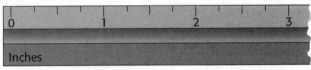

    F. $\frac{1}{64}$ in.

    G. $\frac{1}{8}$ in.

    H. $\frac{1}{4}$ in.

    J. $\frac{1}{2}$ in.

5. The length of a room was measured to be 25 meters. What is the relative error for this measurement?

    A. 50%

    B. 20%

    C. 4%

    D. 2%

6. How many significant digits are there in the measurement 1,708.30 units?

    F. 3   G. 4
    H. 5   J. 6

7. Which measurement is the most precise?

    A. 379 cm
    B. 58.6 km
    C. 1472 mm
    D. 265.4 m

8. The length of a machine part is required to be 3.0 ± 0.05 cm. Which of the following lengths would NOT be acceptable?

    F. 2.98 cm
    G. 3.04 cm
    H. 2.93 cm
    J. 3.0 cm

9. The mass of a piece of quartz is reported as 39.1 g. The actual measurement is between

    A. 39.0 g and 39.2 g
    B. 39.05 g and 39.15 g
    C. 38.6 g and 39.6 g
    D. 38.10 g and 40.10 g

10. Which number has exactly 4 significant digits?

    F. 2700
    G. 136.02
    H. 4,008.19
    J. 0.0065

Short-Response Questions

11. a. Give the precision and GPE of any measurement made with this ruler.

    b. What are the smallest and largest measures that would be reported as 2 in. using this ruler?

12. Complete the table below.

	Measurement	Precision	GPE	Actual measurement is between
a.	9 m			
b.	6.5 L			
c.	26.0 cm			
d.	300 g			

13. Indicate how many digits are significant in each number. Identify these digits.

    a. 410,600   b. 0.008
    c. 209.325   d. 15.200
    e. 5,<u>000</u>

14. For each measurement below, determine the GPE and relative error (accuracy) to the nearest tenth of a percent.

    a. 8 ft
    b. 125 cm
    c. 2<u>00</u> mi

15. Write these measurements in order from most precise to least precise:

    43.76 L, 43.760 kL, 4376 mL, 43.7 L

Extended-Response Questions

16. Simon and Stella each measured the length of a table. Simon reported the measure as 1.62 m and Stella reported it as 162 cm. Whose measurement was more precise? Explain.

17. a. Measure the given line segment using a metric or customary ruler.

    b. Indicate the precision and GPE of your measurement.
    c. Determine the relative error (accuracy) of your measurement.

18. a. Write a 5-digit number that has only 3 significant digits.

b. Show how the number you wrote could be changed to make all 5 digits significant.

19. a. An industrial designer specified that the width of an iron rod should be 2.6 ± 0.05 cm. Give the smallest and largest acceptable measures for this rod.

b. Which of the widths below could NOT be used?
2.64 cm, 2.53 cm, 2.585 cm, 2.608 cm, 2.672 cm

20. a. Explain the difference between these measures: 270 m and 27<u>0</u> m.

b. Give an actual measurement that would fall into the range for 270 m, but not for 27<u>0</u> m.

Chapter 7 Review

Multiple-Choice Questions

1. The most reasonable estimate of the length of a kitchen table is

A. 16 in. B. 6 ft
C. 8 yd D. $\frac{1}{2}$ mi

2. A store received 576 ounces of cheese from a dairy. How many pounds of cheese did the store receive?

F. 36 lb G. 48 lb
H. 72 lb J. 96 lb

3. Which measure is NOT equivalent to the others?

A. $5\frac{3}{4}$ gal B. 23 qt
C. 44 pt D. 92 c

4. Wesley is 6 ft 2 in. tall. His little sister Alicia is 4 ft 8 in. tall. How much taller is Wesley than his sister?

F. 2 ft 10 in.
G. 2 ft 6 in.
H. 1 ft 10 in.
J. 1 ft 6 in.

5. Which measure is equivalent to 470 cm?

A. 470 mm B. 47 m
C. 4.7 dam D. 0.0047 km

6. Which item has a mass of about 100 g?

F. a pair of ice skates
G. a lemon
H. a plastic sandwich bag
J. a hockey stick

7. A restaurant puts 75 g of turkey in each sandwich. How many sandwiches can the restaurant make from 3 kg of turkey?

A. 25
B. 40
C. 56
D. 60

8. Evan fell asleep at 11:33 P.M. He awoke the next morning at 8:12 A.M. How long did he sleep?

F. 3 h 21 min
G. 7 h 27 min
H. 8 h 39 min
J. 19 h 45 min

9. In the New York area, during which season is a temperature of 34°C most likely?

A. summer
B. fall
C. winter
D. spring

10. What is the precision of the measurement 0.018 km?

F. 0.0005 km
G. 0.005 km
H. 0.001 km
J. 0.0001 km

Short-Response Questions

11. Indicate whether the mass of each object is closest to 1 g, 10 g, 100 g, or 1 kg.

a. hair dryer
b. plum
c. rubber band
d. pencil

12. At birth, a baby weighed 6 lb 9 oz. After one month, the baby weighed 8 lb 1 oz. How much weight did the baby gain during the month?

13. A meat loaf recipe calls for 1.5 kilograms of ground beef. The chef has one package marked 0.95 kilogram. How many more grams of beef are needed?

14. The scale on a map shows that 2.5 centimeters represent 35 kilometers. About how many miles does 1 inch on the map represent? Explain how you estimated.

15. Complete the flight chart below. The time zone for each city is given in parentheses.

16. Bruce rode for a distance of 2,400 feet along the parkway until he reached the bicycle path. Then he rode for 4.5 miles on the path before stopping for lunch. How many yards did Bruce bicycle before lunch?

Extended-Response Questions

17. When he served the pizza, Mario told his customers to be careful because the pizza was about 150°. Do you think Mario meant Fahrenheit or Celsius degrees? Explain.

18. a. Sketch a ruler that has a precision of 0.2 inch.
   b. What is the GPE of a measurement made with this ruler?
   c. A nail measured with this ruler is reported to have a length of 3.4 in. In what range is the actual length?

19. On Thursday, Nikki worked at the Seaside Shop from 3:30 P.M. until 6:06 P.M. without a break. She calculated that she earned $19.50 on Thursday. On Saturday, Nikki worked from 8:40 A.M. until 1:00 P.M. If her rate of pay was the same on both days, how much did Nikki earn on Saturday? Show your work?

20. a. Identify the significant digits for this length: 50.0 in.
   b. Determine the percent error (accuracy) of the above measurement.

Departure City	Departure Time	Flight Time	Arrival City	Arrival Time (local time)
a. Boston (ET)	9:35 A.M.	?	Los Angeles (PT)	12:40 P.M.
b. Denver (MT)	6:10 P.M.	1 h 40 min	St. Louis (CT)	?
c. Chicago (CT)	?	3 h 45 min	Phoenix (MT)	11:05 A.M.

Multiple-Choice Questions

1. Which group of decimals is in order from the least to the greatest?

   A. 0.6, 0.07, 0.67, 0.655, 0.753
   B. 0.07, 0.6, 0.67, 0.655, 0.753
   C. 0.07, 0.6, 0.655, 0.67, 0.753
   D. 0.655, 0.6, 0.67, 0.07, 0.753

2. A gold bracelet that sells for $320 is on sale for 40% off. One way to find the amount by which the regular price is reduced is to multiply $320 by

   F. $\frac{3}{10}$ G. $\frac{2}{5}$

   H. $\frac{3}{5}$ J. $\frac{5}{8}$

3. What is the next number in the sequence?
   $-1, 4, -16, 64, -256, \ldots$

   A. 1,024   B. –512
   C. 648   D. –1,024

4. The number of grams in 5.4 kilograms is

   F. 0.54   G. 0.054
   H. 540   J. 5400

5. With a sales tax of 8%, what is the total cost of a suit priced at $162?

   A. $170   B. $174.96
   C. $178.24   D. $185.60

6. The height of a building is 300 feet and its width is 200 feet. A scale model of the building is 6 inches high. How wide is the scale model?

   F. 4 inches   G. $4\frac{1}{2}$ inches

   H. 5 inches   J. $5\frac{1}{4}$ inches

7. What is the value of $2x^2 - 4x + 5$, when $x = -3$?

   A. –1   B. 11
   C. 26   D. 35

8. If three times a number is decreased by 8, the result is 34. What is the number?

   F. $8\frac{2}{3}$ G. $11\frac{1}{3}$

   H. 14   J. 18

9. Find the value of $36 - 18 \div 9 \times 2$.

   A. 4   B. 32
   C. 35   D. 37

10. How much greater is $36\frac{1}{10}$ than $24\frac{3}{5}$?

   F. $11\frac{1}{2}$ G. $11\frac{3}{4}$

   H. $12\frac{1}{4}$ J. $12\frac{1}{3}$

Short-Response Questions

11. Find the next two numbers in the pattern.
    3, 3, 6, 18, 72, ____, ____

12. At Kenny's Kitchen, 4 dinner rolls sell for $1 and pies sell for $4 each. Mr. Danza bought 20 items that cost $20. What did he buy?

13. Find the answer.
    $-|-19 + 23 - 8| + |31 - 17 + 14|$

Extended-Response Questions

14. Hillary, Jim, Nancy, Leon, and Theo are in these positions at the end of a race:
    • Hillary is 15 meters behind Jim.
    • Theo is 25 meters ahead of Hillary.
    • Nancy is 45 meters behind Theo and 10 meters ahead of Leon.

    a. Write the order of finish from first to last place.
    b. How many meters apart were the first and the last racer?

15. Each line of a computer printout contains 72 characters.

    a. How many characters can print in a 75-line report?

    b. The computer printer can print 135 characters each second. How long will it take to print this report?

16. Solve and graph the solution: $3x + 2 \leq 20$.

17. A plumber charges \$45 for the first half hour of work and \$35 for each half hour after the first. How much does the plumber charge for a job that lasted from 10:10 A.M. until 12:40 P.M.?

18. Ms. Ortiz is planning a sales trip that will require her to drive about 2,685 miles. If her average speed is 55 miles per hour, estimate the number of hours she will spend driving. Explain how you estimated.

19. The numbers 253 and 194 are multiplied. The hundreds digit and units digit of 253 and 194 are then interchanged and the new numbers are multiplied. What is the difference between the two products?

20. The table shows the average number of hours and minutes of weekly television viewing for different groups during prime-time hours.

Group	Age Range	Mon.–Sun. (8–11 P.M.)
Women	18–24	6 h 13 min
	25–54	3 h 2 min
	55 or older	10 h 38 min
Men	18–24	4 h 56 min
	25–54	7 h 29 min
	55 or older	10 h 3 min
Teens	12–17	5 h 15 min
Children	6–11	2 h 45 min

    a. How many more hours per week do men ages 25–54 watch than women in the same age group?

    b. Approximately how many hours of prime-time television would an 8-year-old be likely to watch per year?

CHAPTER 8

Geometry

Chapter Vocabulary

angle	vertex	complementary
supplementary	perpendicular	vertical angles
parallel	transversal	congruent
exterior angle	interior angle	remote interior angles
polygon	regular polygon	diagonal
congruent figures	corresponding parts	similar
perimeter	circle	center
radius	chord	diameter
central angle	circumference	area
altitude	surface area	prism
pyramid	cylinder	volume
hypotenuse	legs	Pythagorean Theorem
Pythagorean triple	trigonometric ratios	opposite leg
adjacent leg	sine	cosine
tangent		

8.1 Lines and Angles

An **angle** ($\angle$) is the figure formed by two rays with a common endpoint called the **vertex**. Angles are measured in degrees. There are 360° in a complete rotation. A protractor is an instrument used to measure angles.

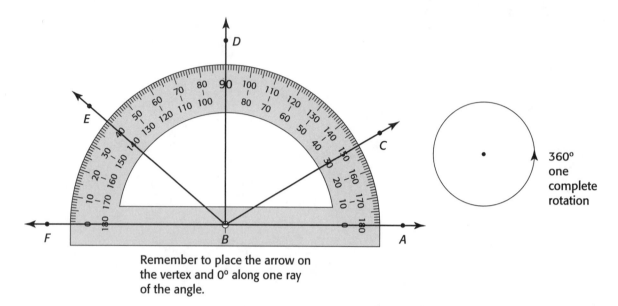

Remember to place the arrow on the vertex and 0° along one ray of the angle.

The protractor has two rows of degree measures so that it can be used to measure angles that open to the left or to the right.

For example, to measure $\angle LMN$ below, you would use the top row of degree measures. To measure $\angle QRS$, you would use the bottom row of degree measures. Both angles measure 60° up from 0°.

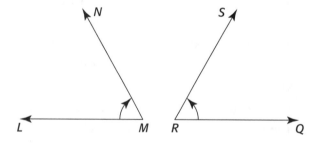

Angles are classified according to their measures.

Angle Type	Definition	Examples
Acute	Measures less than 90°	$\angle ABC$ measures 30°
Right	Measures 90°	$\angle ABD$ measures 90°
Obtuse	Measures greater than 90°, but less than 180°	$\angle ABE$ measures 140°
Straight	Measures 180°	$\angle ABF$ measures 180°

Angle Pairs

Two angles are **complementary** if the sum of their measures is 90°.

Two angles are **supplementary** if the sum of their measures is 180°.

Model Problems

1. Find the complement and supplement of each.

   a. 42°   b. 17°   c. 81°

   Solution

Complement	*Supplement*
a. 90° − 42° = 48°	180° − 42° = 138°
b. 90° − 17° = 73°	180° − 17 = 163°
c. 90° − 81° = 9°	180° − 81° = 99°

2. Two angles are complementary. The measure of one angle is 4 times the measure of the other angle. Find the measures.

   Solution   Let x represent the measure of the smaller angle.
   Then $4x$ represents the measure of the larger angle.
   The angles are complementary, so

   $x + 4x = 90$

   $\quad x = 90$ Combine like terms.

   $\dfrac{5x}{5} = \dfrac{90}{5}$ Divide both sides of the equation by 5 to solve.

   $\quad x = 18$, so $4x = 4(18) = 72$

   Answer   The measures of the angles are 18° and 72°.

Two lines that intersect to form right angles are **perpendicular**.
The symbol ⊥ means *is perpendicular to*.

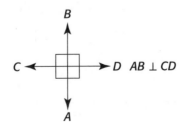

Two lines can also intersect to form two pairs of congruent angles called **vertical angles**.

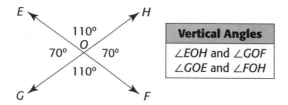

Vertical Angles
∠EOH and ∠GOF
∠GOE and ∠FOH

Two lines in the same plane that never intersect are **parallel**. The symbol ‖ means *is parallel to*.

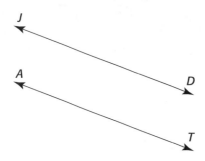

A line that intersects two or more parallel lines is a **transversal**. In the figure below, $\overline{TU}$ is a transversal. Two parallel lines intersected by a transversal form pairs of congruent angles. Angles that are **congruent** have the same degree measure. The symbol ≅ means *is congruent to*.

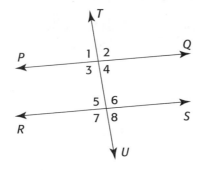

Pairs of congruent angles in the figure are listed in the table below.

Corresponding Angles	Alternate Interior Angles	Alternate Exterior Angles
∠1 ≅ ∠5	∠3 ≅ ∠6	∠1 ≅ ∠8
∠3 ≅ ∠7	∠4 ≅ ∠5	∠2 ≅ ∠7
∠2 ≅ ∠6		
∠4 ≅ ∠8		

Practice

Multiple-Choice Questions

1. Which pair of angles is complementary?

   A. 70°, 70°
   B. 123°, 57°
   C. 45°, 90°
   D. 16°, 74°

2. Which is the measure of an obtuse angle?

   F. 19°
   G. 102°
   H. 180°
   J. 76°

3. An angle measures 33°. What is the measure of its supplement?

   A. 57°
   B. 123°
   C. 147°
   D. 157°

4. What is the measure of ∠ABD?

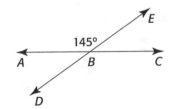

   F. 35°   G. 65°
   H. 85°   J. 145°

5. The number of degrees in $\frac{5}{6}$ of a straight angle is

   A. 75°   B. 105°
   C. 150°   D. 210°

6. The measure of an angle is 50° more than the measure of its supplement. What is the measure of the smaller angle?

   F. 45°   G. 65°
   H. 70°   J. 80°

7. $\overline{LM} \perp \overline{NP}$ What is the measure of ∠QON?

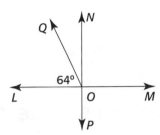

   A. 26°
   B. 54°
   C. 116°
   D. 154°

8. $\overline{VW} \parallel \overline{XY}$. If the measure of ∠3 is 80°, which of the following is NOT true?

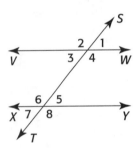

   F. ∠4 measures 100°
   G. ∠7 measures 80°
   H. ∠5 measures 100°
   J. ∠1 measures 80°

9. Which of the following is always true of lines in the same plane?

   A. If two lines intersect, they are perpendicular.
   B. If two lines intersect, they are parallel.
   C. If two lines are not parallel, then they are perpendicular.
   D. If two lines do not intersect, then they are parallel.

10. Which represents the greatest number of degrees?

    F. $\frac{1}{10}$ of a complete rotation

    G. $\frac{1}{2}$ of a right angle

    H. $\frac{2}{3}$ of a straight angle

    J. 15° more than a right angle

Short-Response Questions

11. Use a protractor to measure each angle. Then classify the angles as acute, right, or obtuse.

    a.
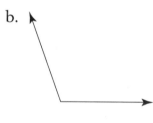

    b.
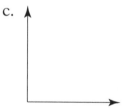

    c.

12. a. Draw an angle that measures 25°.
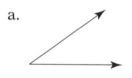
    b. Draw and identify the complement of the angle in part a.
    c. Draw and identify the supplement of the angle in part a.

13. Find the complement and supplement of each.

    a. 50°
    b. 8°
    c. 77°

14. Find the measure of the angle formed by the hands of a clock at each given time. Classify each angle as acute, right, obtuse, or straight.

    a. 1 P.M.   b. 4 P.M.
    c. 6 P.M.

15. Two angles are supplementary. One angle is 8 times larger than the other. Find the measure of the two angles.

16. $\overline{AB} \perp \overline{CD}$ and the measure of $\angle CEF = 43°$. Find the measure of each angle.

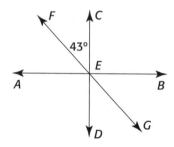

    a. $\angle AEF$ b. $\angle DEG$
    c. $\angle AED$ d. $\angle GEC$

17. $\overline{JK} \parallel \overline{LM}$ and the measure of $\angle 6 = 148°$.

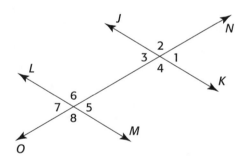

Find the measures of the other seven angles.

Extended-Response Questions

18. a. Draw and label a diagram that fits the following description.
    $\overline{AB}$ intersects $\overline{CD}$ at E and the measure of $\angle AED = 24°$.
    b. Find the measure of each angle: $\angle CEB$, $\angle BED$, $\angle CEA$.

19. Without using a protractor, find the measure of angles a, b, c, and d.

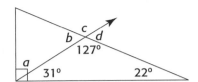

20. Two lines intersect and each pair of vertical angles is supplementary. What can you conclude about the lines? Explain.

8.2 Triangles and Angles

- Triangles can be classified by the types of angles they contain.

Acute Triangle
80°
55° 45°
Three acute angles

Right Triangle
50°
40°
One right angle and two acute angles

Obtuse Triangle
40°
110° 30°
One obtuse angle and two acute angles

- Triangles can also be classified by the number of congruent sides.

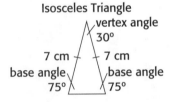

Scalene Triangle
25° 9 cm
6 cm
120°
35°
4 cm
No congruent sides
No congruent angles

Isosceles Triangle
vertex angle
30°
7 cm   7 cm
base angle   base angle
75°   75°
At least two congruent sides
At least two congruent angles

Equilateral Triangle
8.4 cm 60° 8.4 cm
60° 60°
8.4 cm
Three congruent sides
Three congruent angles

- The sum of the measures of the angles of any triangle is 180°.

In △ABC below, ∠BCE is an **exterior angle**. An exterior angle of a triangle is an angle that forms a straight angle with one of the angles of the triangle. ∠ACB is the **interior angle** of the triangle adjacent to ∠BCE. The two angles at the remaining vertices, ∠B and ∠A are called **remote interior angles**.

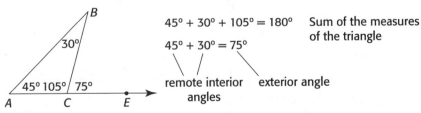

$45° + 30° + 105° = 180°$ Sum of the measures of the triangle

$45° + 30° = 75°$

remote interior   exterior angle
angles

The measure of an exterior angle of a triangle is equal to the sum of the measures of the two remote interior angles. That is:
∠BCE = ∠B + ∠A.

Model Problems

1. Classify △DEF according to its sides and its angles.

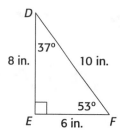

Solution   The triangle has one right angle and two acute angles, so it is a right triangle.

There are no congruent sides, so the triangle is also scalene.

Answer   △DEF is a right scalene triangle.

2. Find the missing angle measures.

a.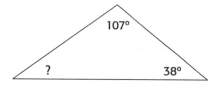

b.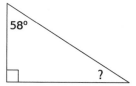

Solution

a. 107° + 38° = 145°
   180° − 145° = 35°
   The missing angle measure is 35°.

b. 90° + 58° = 148°
   180° − 148° = 32°
   The missing angle measure is 32°. Note that the acute angles of a right triangle are complementary.   58° + 32° = 90°

3. Find the number of degrees in the value of *x*.

a.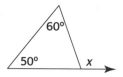

b.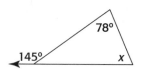

Solution

a. The measures are given for the two remote interior angles corresponding to the exterior angle with measure *x*.
   60° + 50° = *x*
         110° = *x*
   The value of *x* is 110°.

b. The measures are given for the exterior angle and one remote interior angle.

$$x + 78° = 145°$$
$$x + 78° - 78° = 145° - 78°$$ Subtract 78° from both sides to solve for x.

$$x = 67°$$
The value of x is 67°.

Practice

Multiple-Choice Questions

1. Which group could be the angles of an acute triangle?

   A. 30°, 60°, 90°
   B. 40°, 45°, 65°
   C. 15°, 15°, 150°
   D. 75°, 20°, 85°

2. The measures of the sides of a triangle are 7.2 cm, 4.6 cm, and 7.2 cm. The triangle is

   F. scalene
   G. isosceles
   H. equilateral
   J. obtuse

3. The measures of two angles of a triangle are 29° and 47°. The measure of the third angle is

   A. 14°   B. 76°
   C. 104°   D. 166°

4. The measure of two angles of a triangle are 36° and 54°. The triangle is

   F. right   G. acute
   H. obtuse   J. isosceles

5. The maximum number of obtuse angles that a triangle can have is

   A. 0   B. 1
   C. 2   D. 3

6. In △XYZ, the measure of angle Y is twice the measure of angle X and the measure of angle Z is 6 times the measure of angle X. What is the measure of angle Z?

   F. 20°   G. 60°
   H. 90°   J. 120°

7. What is the measure of ∠S in △STU?

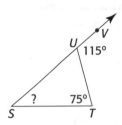

   A. 40°   B. 65°
   C. 80°   D. 180°

8. Which of the following triangles is impossible?

   F. acute scalene triangle
   G. obtuse isosceles triangle
   H. right equilateral triangle
   J. right isosceles triangle

9. If the measure of an angle of an equilateral triangle is represented by $2x + 24$, what is the value of x?

   A. 12   B. 18
   C. 33   D. 78

10. What is the measure of an exterior angle of an equilateral triangle?

F. 60°   G. 90°
H. 120°   J. 150°

Short-Response Questions

11. Classify each triangle according to its angles.

a.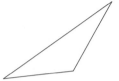

b.

c.

12. Measure the sides and angles of △XYZ. Classify the triangle according to its sides and its angles.

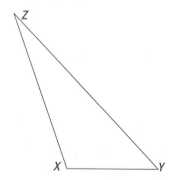

13. Find the measure of the third angle of a triangle if the measures of the other two angles are

a. 47° and 100°
b. 63° and 29°
c. 8° and 114°

14. In an isosceles triangle, the measure of a base angle is 42°. Find the measure of the vertex angle.

Extended-Response Questions

15. a. Draw and label a diagram to fit the following conditions, and then find the value of x.
In △PQR, the measure of ∠P = x, the measure of ∠Q = x + 10, and the measure of an exterior angle at R is 80°.
b. Classify △PQR according to its angles.

16. $\overline{WXYZ}$ is a straight line, the measure of ∠WXV is 132°, and the measure of ∠ZYV = 144°.

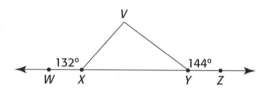

a. Find the measures of ∠VXY, ∠XYV, and ∠XVY.
b. What kind of triangle is △VXY?

17. a. In △ABC, the measure of ∠B is twice the measure of ∠A and ∠C is three times the measure of ∠A. Find the measure of each angle of the triangle.
b. What kind of triangle is △ABC?

18. Cecilia said that a triangle can have two obtuse angles. Explain why you agree or disagree with Cecilia.

19. A triangle has vertices C, A, and T. $\overline{CA}$ is the same length as $\overline{AT}$. $\overline{CA}$ is 3 cm and the measure of ∠ACT is 50°.

a. Draw △CAT and label all known measures.
b. Find the missing angle measures. Show your work.
c. Classify the triangle by angles and by sides.

20. If a right triangle is also isosceles, what is the measure of each of its congruent angles? Explain how you arrived at your answer.

8.3 Polygons and Special Quadrilaterals

A **polygon** is a plane closed figure made up of line segments. Any two sides of a polygon intersect at a vertex of the polygon. The name of each type of polygon tells how many sides and how many angles it has. In a **regular polygon**, all sides and all angles are congruent.

Triangle
3 sides
3 angles

Quadrilateral
4 sides
4 angles

Pentagon
5 sides
5 angles

Hexagon
6 sides
6 angles

Octagon
8 sides
8 angles

Decagon
10 sides
10 angles

A **diagonal** is a segment joining any two nonadjacent vertices of a polygon. The sum of the angles of a polygon can be found by dividing the polygon into triangles using the diagonals and then multiplying the number of triangles by 180°.

Example A hexagon can be divided into 4 triangles, so the sum of the angle measures is 4(180°) = 720°.

In general, the sum of the angle measures S of a polygon with n sides is given by the formula:
$$S = (n - 2)180°$$
Note: $(n - 2)$ is the number of triangles formed by drawing the diagonals. For a regular n-sided polygon, the measure of each angle is, therefore,

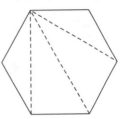

$$\frac{\text{The sum of the angle measures}}{\text{The number of sides}} = \frac{S}{n} = \frac{(n - 2)180°}{n}$$

Model Problems

1. Name each polygon.

   a.

   b.

   c.

   Solution
   a. A 5-sided figure is a pentagon; this is a regular pentagon.
   b. A 6-sided figure is a hexagon; this is an irregular hexagon.
   c. An 8-sided figure is an octagon; this is a regular octagon.

2. Find the measure of each angle of a regular decagon.

Solution A decagon has 10 sides, so use $n = 10$ in the formula.

$$\frac{S}{n} = \frac{(n - 2)180°}{n}$$

$$\frac{S}{n} = \frac{(8)180°}{10} = \frac{1,440°}{10} = 144°$$

Answer Each angle of a regular decagon measures 144°.

3. The sum of the measures of the angles of a regular polygon is 540°. Name the polygon.

Solution Since $S = (n - 2)180°$, then

$$(n - 2)180° = 540°$$

$$\frac{(n - 2)180°}{180°} = \frac{540°}{180°} = 3 \quad \text{Divide both sides by 180°.}$$

$$n - 2 = 3 \qquad \text{Add 2 to both sides to solve for } n.$$
$$n - 2 + 2 = 3 + 2$$
$$n = 5$$

Answer A 5-sided polygon is a pentagon.

Special Quadrilaterals

There are some special quadrilaterals.

Trapezoid
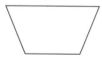
One pair of opposite sides parallel

Parallelogram

Both pairs of opposite sides parallel

Opposite sides and opposite angles congruent

Rectangle

A parallelogram with 4 right angles

Rhombus

A parallelogram with 4 congruent sides

Square

A rectangle with 4 congruent sides

Using the formula for the sum of the angle measures with $n = 4$ shows that the sum of the measures of the angles of a quadrilateral is 360°. For parallelograms, consecutive angles are supplementary.

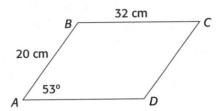

Consecutive angles
120° + 60° = 180°

4. Give the missing measures for all sides and angles of the parallelogram.

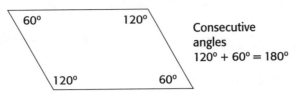

Solution Opposite sides of a parallelogram are congruent.
$\overline{CD}$ = 20 cm
$\overline{AD}$ = 32 cm
Consecutive angles are supplementary.
$$\angle A + \angle B = 180°$$
$$53° + \angle B = 180°\quad \text{Subtract 53° from both sides to solve for } \angle B.$$
$$53° + \angle B - 53° = 180° - 53°$$
$$\angle B = 127°$$
Answer Opposite angles are congruent, so $\angle C$ measures 53° and $\angle D$ measures 127°.

5. Find the measures of $\angle A$ and $\angle D$.

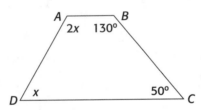

Solution The figure is a trapezoid, which is a quadrilateral. The sum of the angles in a quadrilateral is 360°. Write an equation and solve for x.
$$x + 2x + 130° + 50° = 360°$$
$$x + 2x + 180° = 360°$$
$$x + 2x + 180° - 180° = 360° - 180°\quad \text{Subtract 180° from both sides.}$$
$$x + 2x = 180°$$
$$3x = 180°\quad \text{Combine like terms.}$$
$$\frac{3x}{3} = \frac{180°}{3}\quad \text{Divide both sides by 3 to solve for } x.$$
$$x = 60°$$
Answer So, the measure of $\angle D = 60°$ and the measure of $\angle A = 2(60°) = 120°$.

Multiple-Choice Questions

1. A polygon with 8 sides is called a(n)

    A. decagon   B. octagon
    C. hexagon   D. pentagon

2. Name the polygon shown.

    F. quadrilateral G. pentagon
    H. hexagon   J. decagon

3. The measure of each angle of a regular polygon is 120°. Name the polygon.

    A. quadrilateral B. pentagon
    C. hexagon   D. octagon

4. How many diagonals can be drawn from each vertex of a decagon?

    F. 7   G. 8
    H. 10   J. 12

5. What is the missing angle measure?

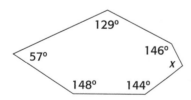

    A. 36°   B. 76°
    C. 96°   D. 106°

6. Which name does NOT describe *WXYZ*?

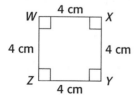

    F. parallelogram G. rhombus
    H. rectangle   J. trapezoid

7. Which statement is true?

    A. A parallelogram is a rectangle.
    B. A trapezoid is a parallelogram.
    C. Some trapezoids have 4 right angles.
    D. A parallelogram with one angle measuring 90° is a rectangle.

8. *PQRS* is a parallelogram. What is the value of *x*?

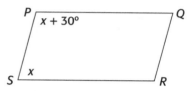

    F. 45°
    G. 75°
    H. 115°
    J. 165°

9. A polygon has vertices *M*, *N*, *O*, and *P*. $\overline{MP} \parallel \overline{NO}$. $\overline{MP} = \overline{NM} = \overline{NO} = \overline{PO} = 5$ cm. The measures of all four angles are equal. The most descriptive name for *MNOP* is

    A. trapezoid
    B. square
    C. parallelogram
    D. rhombus

10. Figure *FGHI* is a rectangle. What is the value of *x*?

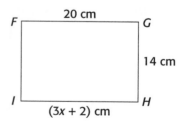

    F. 12
    G. $7\frac{1}{3}$
    H. 6
    J. 4

Short-Response Questions

11. Name the polygons. If the figure is NOT a polygon, explain why.

a.

b.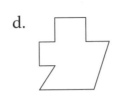

c.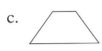

d.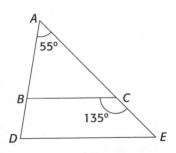

12. Write *sometimes*, *always*, or *never* for each statement.

    a. A rectangle is a parallelogram.
    b. A rhombus is a trapezoid.
    c. A rectangle is a square.
    d. A square is a rhombus.

13. For a regular polygon with 12 sides:

    a. Find the sum of the angle measures.
    b. Find the measure of each angle.

14. The sum of the measures of the angles of a regular polygon is 1,080°.

    a. Name the polygon.
    b. Find the measure of each angle.

Extended-Response Questions

15. For the given figure, $\overline{BC}$ is parallel to $\overline{DE}$. Find the measures of $\angle ADE$, $\angle AED$, and $\angle DBC$. Show your work.

16. Figure *QRST* is a parallelogram.

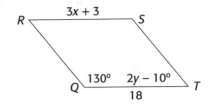

    a. Find the value of x.
    b. Find the value of y.

17. *WXYZ* is a square. If $\overline{WX} = 3x$ and $\overline{XY} = 2x + 8$, find the length of each side of the square.

18. a. Use a ruler and protractor to draw the figure described. Write the name of the figure.
A quadrilateral has $\overline{KL}$ parallel to $\overline{NM}$, all sides measure $1\frac{1}{2}$ in., and $\angle KLM = 60°$.

    b. Find the measure of $\angle LMN$.

19. Complete the table for a heptagon (7 sides) and an octagon. Describe any patterns you see.

Polygon	Number of Diagonals from One Vertex	Total Number of Diagonals
triangle	0	0
quadrilateral	1	2
pentagon	2	5
hexagon	3	9
heptagon	?	?
octagon	?	?

20. In a heptagon, the sum of the measures of 4 of the angles is 478°. What is the sum of the measures of the other 3 angles? Show your work.

Polygons and Special Quadrilaterals **241**

8.4 Congruent Figures

Congruent figures have the same size and the same shape. The symbol ≅ means *is congruent to*. Parts of congruent figures that match are called **corresponding parts**. Corresponding parts of congruent figures are congruent.

- Corresponding sides are congruent (equal in length).
- Corresponding angles are congruent (equal in measure).

Model Problems

1. State if each pair of figures is congruent. If not, explain why.

   a.   b.   c.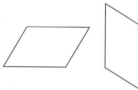

   Solution
   a. Not congruent. The figures are different sizes.
   b. Not congruent. The figures are different shapes.
   c. Congruent. The positions of the figures do not matter. One figure can be moved to fit exactly over the other.

2. △ABC ≅ △DEF. Identify the corresponding sides and corresponding angles of the triangles.

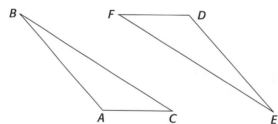

   Solution

Corresponding Sides	Corresponding Angles
$\overline{AB} \cong \overline{DE}$	∠A ≅ ∠D
$\overline{AC} \cong \overline{DF}$	∠B ≅ ∠E
$\overline{BC} \cong \overline{EF}$	∠C ≅ ∠F

3. Quadrilateral $MNOP \cong$ quadrilateral $WXYZ$. Find each measure.

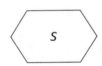

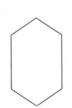

a. $\overline{MP}$ b. $\overline{XY}$ c. $\angle P$ d. $\angle Y$

Solution
a. $\overline{MP} \cong \overline{WZ}$, $\overline{MP}$ = 12 cm
b. $\overline{XY} \cong \overline{NO}$, $\overline{XY}$ = 16 cm
c. $\angle P \cong \angle Z$, measure of $\angle P$ = 110°
d. $\angle Y \cong \angle O$, measure of $\angle O$ = 70°

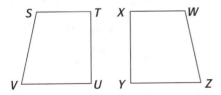

Practice

Multiple-Choice Questions

1. Which figure is congruent to figure S?

A.

B.

C.

D.

2. Quadrilateral $STUV \cong$ quadrilateral $WXYZ$. Which side corresponds to $\overline{UV}$?

F. $\overline{XW}$ G. $\overline{WZ}$
H. $\overline{YZ}$ J. $\overline{XY}$

3. Figure $ABCDE \cong FGHIJ$. Which angle corresponds to $\angle E$?

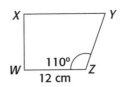

A. $\angle F$
B. $\angle H$
C. $\angle I$
D. $\angle J$

4. $\triangle PQR \cong \triangle SRQ$. Which angle corresponds to $\angle QRP$?

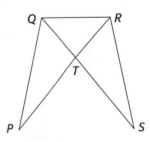

F. $\angle QPR$
G. $\angle RQS$
H. $\angle STR$
J. $\angle PQR$

5. $\triangle JKL \cong \triangle MNO$. What is the measure of $\angle L$?

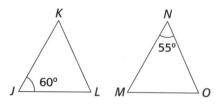

A. 55°   B. 60°
C. 65°   D. 70°

6. An obtuse triangle and a right triangle are

F. sometimes congruent
G. always congruent
H. never congruent
J. not enough information

7. Parallelogram $QRST \cong$ parallelogram $WXYZ$. Find the value of x.

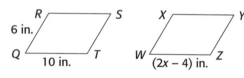

A. 5   B. 6
C. 7   D. 10

8. Which figures are congruent?

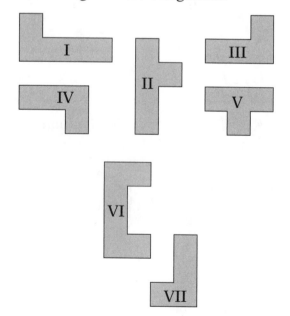

F. I and IV only
G. III and V only
H. II and VI only
J. III, IV, and VII

Short-Response Questions

9. Tell if the figures in each pair are congruent. If not, explain why.

a.   b.

c.

10. Quadrilateral $QRST \cong$ quadrilateral $WXYZ$. Name all the corresponding sides and all the corresponding angles.

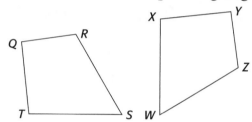

11. Write *sometimes*, *always*, or *never* for each.

a. Two squares are congruent.
b. An acute triangle and an obtuse triangle are congruent.
c. Two regular pentagons are congruent.
d. A trapezoid and rectangle are congruent.

Extended-Response Questions

12. $\triangle ABC \cong \triangle DEF$. Find the value of x, y, and z. Show your work.

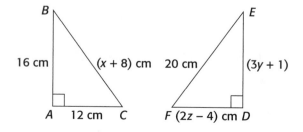

13. If △PQR ≅ △XYZ, find the measure of ∠Q. Show your work.

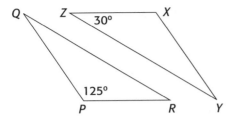

14. Identify all the pairs of congruent triangles in the figure. Use a ruler and protractor to verify your answer.

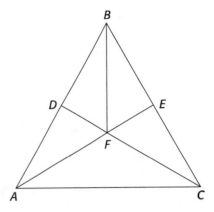

8.5 Similar Figures

Figures that have the same shape are **similar**, whether or not they are the same size. The symbol ~ means *is similar to*.

In similar figures:

- Corresponding angles are congruent (the same measure).
- Corresponding sides are proportional.

To find the length of an unknown side in a pair of similar figures, write and solve a proportion.

Note: All circles are similar because they have the same shape. All squares are similar because they have the same shape.

 **Model Problems**

1. For each pair of figures, tell if they are similar, congruent, or neither.

a.    b.    c.

Solution
a. The figures are the same shape, but not the same size. The figures are similar.
b. The figures are the same shape, so they are similar. The figures are the same size, so they are also congruent.
c. The figures are different shapes. They are neither similar nor congruent.

2. Quadrilateral *ABCD* ~ quadrilateral *EFGH*.
   a. What is the length of $\overline{HG}$?
   b. Find the measure of $\angle C$.

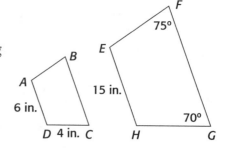

Solution

a. Write a proportion comparing
   the lengths of the sides.

$$\frac{\overline{DC}}{\overline{AD}} = \frac{\overline{HG}}{\overline{EH}}$$

$$\frac{4}{6} = \frac{\overline{HG}}{15}$$

$$6(\overline{HG}) = 60$$

$$\frac{6(\overline{HG})}{6} = \frac{60}{6} \quad \text{Divide both sides by 6 to solve.}$$

$$\overline{HG} = 10 \text{ in.}$$

b. Corresponding angles of similar figures are congruent. $\angle C$ corresponds $\angle G$, so the measure of $\angle C$ is 70°.

3. At the same time of day, a 6-foot-tall person casts a shadow 8 feet long and a nearby tree casts a shadow 36 feet long. What is the height of the tree?

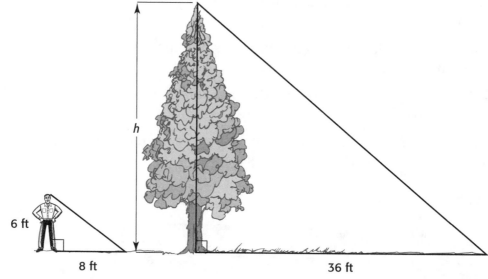

Solution When the sun is shining, nearby vertical objects and their shadows form similar right triangles. Write and solve a proportion to find the height of the tree.

$$\begin{array}{c}\text{height} \rightarrow \\ \text{shadow} \rightarrow\end{array} \frac{6}{8} = \frac{h}{36} \begin{array}{c}\leftarrow \text{height} \\ \leftarrow \text{shadow}\end{array}$$

$$8 \cdot h = 6 \cdot 36$$

$$8h = 216$$

$$\frac{8h}{8} = \frac{216}{8}$$

$$h = 27$$

Answer The tree is 27 feet tall.

Multiple-Choice Questions

1. Which of these figures are similar but NOT congruent?

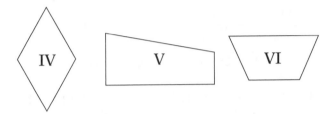

A. III and VI
B. I and V
C. II and IV
D. III and V

2. Which of the following is true?

F. All squares are similar.
G. All rectangles are similar.
H. All parallelograms are similar.
J. All right triangles are similar.

3. Which of the following sets of lengths could be the lengths of the sides of a triangle that is similar to △XYZ?

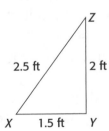

A. 3.5 ft, 4 ft, 4.5 ft
B. 6 ft, 8 ft, 10 ft
C. 15 ft, 25 ft, 35 ft
D. 16.5 ft, 22 ft, 32.5 ft

4. The lengths of the sides of a triangle are 4 in., 14 in., and 16 in. If the longest side of a similar triangle measures 40 in., what is the length of the shortest side of this new triangle?

F. 10 in.
G. 12 in.
H. 28 in.
J. 35 in.

5. Which of the triangles shown is similar to △MNO?

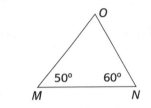

A.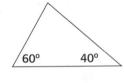

B.

C.

D.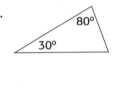

6. Two regular hexagons are

F. always congruent
G. always similar
H. never similar
J. never congruent

7. A building casts a shadow 24 feet long. At the same time, Andy, who is 5 feet tall, casts a 3-foot shadow. What is the height of the building?

A. 14.4 ft
B. 30 ft
C. 40 ft
D. 60.5 ft

8. $\triangle LMN \sim \triangle PQR$. Which of the following proportions can NOT be used to find x?

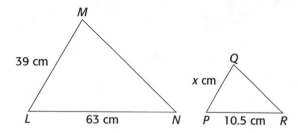

F. $\dfrac{x}{39} = \dfrac{10.5}{63}$

G. $\dfrac{x}{10.5} = \dfrac{39}{63}$

H. $\dfrac{39}{x} = \dfrac{63}{10.5}$

J. $\dfrac{63}{39} = \dfrac{x}{10.5}$

9. Rectangle $GHIJ \sim$ rectangle $KLMN$. Find the length of $\overline{LK}$.

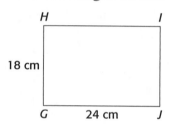

A. 3.6 cm
B. 6.4 cm
C. 7.2 cm
D. 12.8 cm

10. A right triangle and an equilateral triangle are

F. sometimes similar
G. sometimes congruent
H. always similar
J. never similar

Short-Response Questions

11. Write *sometimes*, *always*, or *never*.

a. Congruent figures are similar.
b. Similar figures are congruent.
c. Figures that are not similar are congruent.

12. Are rectangles $ABCD$ and $EFGH$ similar? Explain why or why not.

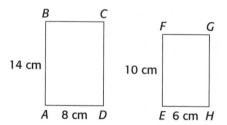

13. The quadrilaterals are similar. Find the values of x and y.

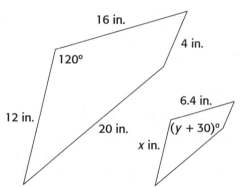

14. The lengths of the sides of a triangle are 25 cm, 60 cm, and 65 cm. If the longest side of a similar triangle measures 39 cm, what is the length of the shortest side of this triangle?

15. A 9-foot utility pole casts a shadow that is 22 feet long. At the same time of day, a nearby building casts a shadow that is 99 feet long. What is the height of the building?

16. A 3-foot sunflower casts a shadow that is 4.8 feet long. At the same time of day, how long would the shadow be of a 40-foot tree?

Extended-Response Questions

17. The rectangles are similar.

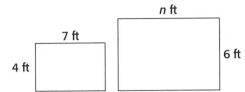

a. Find the value of n.
b. Give the dimensions of another rectangle that would be similar to the two shown.

18. Are all regular pentagons similar? Explain why or why not.

19. Use a ruler and a protractor to draw the triangles.

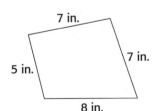

a. Draw right triangle ABC in which $\angle B$ is right, the measure of $\angle A$ is $60°$, and $\overline{AB} = 1\frac{1}{2}$ in.

b. Draw $\triangle DEF \sim \triangle ABC$, so that $\overline{DE}$ corresponds to $\overline{AB}$ and $\overline{DE} = 2$ in.

20. $\triangle MNO \sim \triangle QPO$. Find each measure.

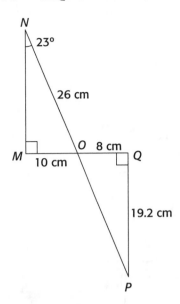

a. $\angle POQ$
b. $\angle QPO$
c. $\overline{MN}$
d. $\overline{OP}$

8.6 Perimeter and Circumference

The **perimeter** of a polygon is the distance around the polygon. To find the perimeter, add the lengths of the sides.

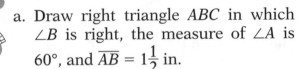

P (perimeter) = 5 in. + 7 in. + 7 in. + 8 in.
$P = 27$ in.

You can use a formula to find the perimeter of some polygons.

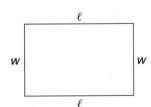

Rectangle
$P = 2\ell + 2w$

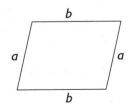

Parallelogram
$P = 2a + 2b$

To find the perimeter of a regular polygon, multiply the length of each side by the number of sides.

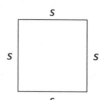

Equilateral
Triangle
$P=3s$

Square
$P=4s$

Regular
Pentagon
$P=5s$

Model Problems

1. Find the perimeter of the isosceles triangle shown.

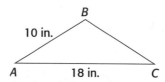

Solution The triangle is isosceles, so $\overline{BC} = 10$ in.
   $P = 10$ in. $+ 10$ in. $+ 18$ in.
Answer 38 in.

2. The perimeter of a rectangular room is 32 ft. The length of the room is 9 ft. Find the width of the room.

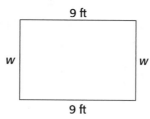

Solution For a rectangle, $P = 2\ell + 2w$.
   $32 = 2 \cdot 9 + 2w$ Since the length is 9, substitute 9 for ℓ.
   $32 = 18 + 2w$
   $32 - 18 = 18 - 18 + 2w$ Subtract 18 from both sides to solve for w.
   $14 = 2w$
   $\dfrac{14}{2} = \dfrac{2w}{2}$ Divide both sides by 2.
   $7 = w$
Answer The width of the room is 7 ft.

Circles

A **circle** is a flat, closed curve that has all its points the same distance from an inside point called the **center**. A circle is not a polygon because it is not made of straight line segments. A **radius** is a line segment that has the center and a point on the circle as endpoints.

A **chord** is a line segment that has two points on the circle as endpoints. A **diameter** is a chord that has both endpoints on the circle and that passes through the center. The length of the diameter d is twice the length of the radius r.

A **central angle** is an angle whose vertex is at the center of the circle.

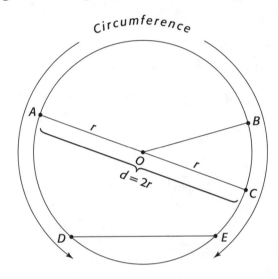

Center: O
Radius: $\overline{OA}$ or $\overline{OB}$ or $\overline{OC}$
Chord: $\overline{AC}$ or $\overline{DE}$
Diameter: $\overline{AC}$
Central Angle: $\angle AOB$ or $\angle BOC$

The distance around a circle is called the **circumference** (C). The circumference is about 3 times longer than the circle's diameter. The exact ratio of the circumference to the diameter is represented by the Greek letter π (pi). π is an irrational number whose value is approximately $\frac{22}{7}$ or 3.14. To find the circumference of a circle, use a formula.

Circumference = $\pi \times$ length of diameter
$C = \pi d$, or
$C = 2\pi r$ (since $d = 2r$)

3. Find the circumference of a circle if the diameter measures 20 cm. Use $\pi \approx 3.14$.

> **Solution**   The diameter is given, so use the formula $C = \pi d$.
> $C = \pi d$
> $C \approx 3.14 \times 20$
> $C \approx 62.8$
>
> **Answer**   The circumference is 62.8 cm.
> *Note*: The symbol $\approx$ means *is approximately.*

4. Find the circumference of the circle shown. Use $\pi \approx \frac{22}{7}$.

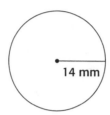

14 mm

> **Solution**   The radius is given so use $C = 2\pi r$.
> $C = 2\pi r$
> $C \approx 2 \times \frac{22}{7_1} \times \cancel{14}_2$
> $C \approx 2 \times 44$
> $C \approx 88$
>
> **Answer**   The circumference is 88 mm.

5. Using $\pi \approx \frac{22}{7}$, find the measure of the diameter of a circle whose circumference is 132 in.

> **Solution**   Rewrite the formula $C = \pi d$ to find an expression for d.
> $C = \pi d$
> $\frac{C}{\pi} = \frac{\pi d}{\pi}$ Divide both sides by π.
>
> $\frac{C}{\pi} = d$
>
> $132 \div \frac{22}{7} = d$ Substitute known values for C and π.
>
> $\cancel{132}_6 \times \frac{7}{22_1} = d$ Multiply by the reciprocal.
> $42 = d$
>
> **Answer**   The diameter measures 42 in.

Multiple-Choice Questions

1. Find the perimeter of the polygon shown on the grid.

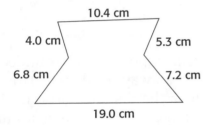

A. 20 units   B. 22 units
C. 23 units   D. 24 units

2. Find the perimeter of the polygon.

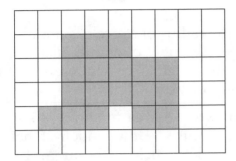

F. 42.3 cm   G. 48.7 cm
H. 52.7 cm   J. 58.0 cm

3. Find the perimeter of a regular hexagon with each side measuring 14 in.

A. 56 in.   B. 84 in.
C. 98 in.   D. 112 in.

4. Find the length of the side of a square whose perimeter is 300 yd.

F. 75 yd   G. 150 yd
H. 600 yd   J. 1,200 yd

5. A park that is shaped like a rectangle is 210 meters long and has a perimeter of 600 meters. How wide is the park?

A. 90 m   B. 105 m
C. 140 m   D. 390 m

6. Determine the perimeter of the polygon shown. (All angles are right angles.)

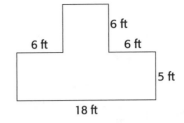

F. 41 ft   G. 46 ft
H. 58 ft   J. 64 ft

7. In circle O, $\overline{OF}$ is

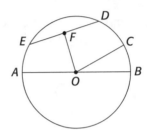

A. a chord
B. a radius
C. a diameter
D. no special line segment

8. Find the circumference of the circle. Use $\pi = 3.14$.

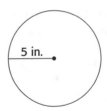

F. 15.7 in.   G. 22.5 in.
H. 31.4 in.   J. 47.1 in.

9. A circular flower bed has a diameter of 25 feet. How much fencing is needed to enclose the flower bed? Use 3.14 for π.

A. 39.25 ft   B. 78.5 ft
C. 117.75 ft   D. 157 ft

10. The length of a rectangle is twice its width. The perimeter of the rectangle is 162 in. How long is the rectangle?

F. 27 in.   G. 36 in.
H. 54 in.   J. 60 in.

Short-Response Questions

11. Find the perimeter of the polygon.

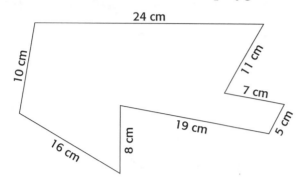

12. Find the perimeter of a regular octagon with each side measuring 7.6 in.

13. Find the perimeter of the polygon. (All angles are right angles.)

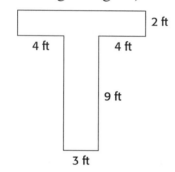

14. The ratio of the length of a rectangular meeting room to its width is 4:3. The perimeter of the room is 238 feet. What are the dimensions of the room?

15. a. Draw a circle with center *O*.
   b. Draw radius $\overline{OA}$.
   c. Draw chord $\overline{AB}$.
   d. Draw diameter $\overline{BC}$.
   e. Identify two central angles.

Extended-Response Questions

16. How far will the wheel shown travel when it makes one rotation? Use 3.14 for π.

17. If the radius of a circle is doubled, how does the new circumference compare to the original circumference? Justify your answer.

18. An exercise area for dogs is shown in the figure. How much fencing is needed to enclose the area? Use 3.14 for π. Show your work.

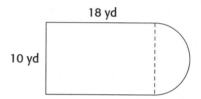

19. A circular pool is surrounded by a paved walkway as shown. How much longer is the distance around the outside of the walkway than around the inside? Use 3.14 for π. Show your work.

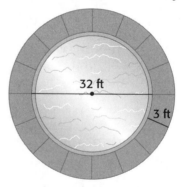

20. a. What is the circumference of the largest circle that can be cut from a square measuring 30 cm on a side? Use 3.14 for π.
   b. What percent of the perimeter of the square is the circumference of the circle?

8.7 Area

The **area** of a closed plane figure is the number of square units that are contained within its interior. Some units of area are: square inch (in.²), square foot (ft²), square centimeter (cm²), and square meter (m²).

These formulas can be used to find area.

Rectangle

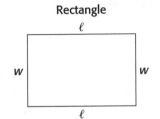

The area of a rectangle is its length times its width.

$$A = \ell w$$

Square

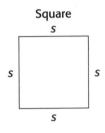

A square is a special rectangle where $\ell = w = s$.

$$A = s^2$$

Parallelogram

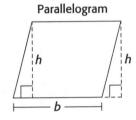

The area of a parallelogram is its base times its height.

$$A = bh$$

Triangle

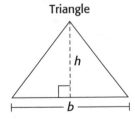

The area of a triangle is one half its base times its height. Another name for the height of a triangle is **altitude**.

$$A = \frac{1}{2}bh$$

Trapezoid

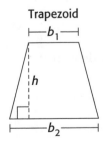

The area of a trapezoid is one half its height times the sum of its bases.

$$A = \frac{1}{2}h(b_1 + b_2)$$

The area of a circle is the square of its radius times pi (π).

$$A = \pi r^2$$

1. How much will it cost to cover a rectangular room that is 7 yards long and 5 yards wide with carpet that costs $22 per square yard?

   Solution Use the formula for the area of a rectangle.

   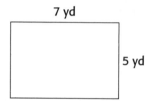

   7 yd

   5 yd

   $A = \ell w$
   $A = 7 \times 5$
   35 yd² of carpet are needed.
   Find the cost. $35 \times \$22 = \$770$

   Answer The carpet will cost $770.

2. Find the area of each figure.

   a.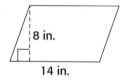

   8 in.

   14 in.

   b.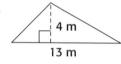

   4 m

   13 m

   c.

   6 cm

   7 cm

   9 cm

   Solution

   a. Use the formula for the area of a parallelogram.
   $A = bh$
   $A = 14 \times 8$
   $A = 112$ in.²

   b. Use the formula for the area of a triangle.
   $A = \frac{1}{2}bh$
   $A = \frac{1}{2} \times 13 \times 4$
   $A = 26$ m²

   c. When applying the formula for a trapezoid, be careful to follow the order of operations.
   $A = \frac{1}{2}h(b_1 + b_2)$
   $A = \frac{1}{2} \times 7(6 + 9)$
   $A = \frac{1}{2} \times 7(15)$
   $A = 52.5$ cm²

3. A tabletop is a circular piece of glass that has a diameter of 60 in. Find the area of the glass. Use 3.14 for π.

Solution   To use the formula for the area of a circle, first find the radius.

60 in.

$r = \frac{1}{2}d$, so $r = \frac{1}{2} \times 60 = 30$

$A = \pi r^2$ Substitute the length of the radius in for r and 3.14 for π.
$A \approx 3.14 \times 30^2$
$A \approx 3.14 \times 900 \approx 2{,}826$ in.^{2}

Answer   The area of the glass is 2,826 in.^{2}.

4. Find the area of the figure. Assume that all angles are right angles.

4 in.

14 in.

20 in.

12 in.

6 in.

16 in.

Solution   Divide the figure into polygons that do not overlap. Sometimes there is more than one way to do this. Find the area of each polygon, then add.

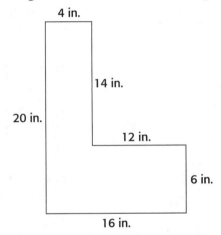

4 in.

14 in.   **II**   14 in.

12 in.

6 in.   **I**   6 in.

16 in.

A (rectangle I) $= \ell w = 16 \times 6 = 96$ in.^{2}
A (rectangle II) $= \ell w = 14 \times 4 = 56$ in.^{2}

$$\begin{array}{r} 96 \text{ in.} \\ \text{Add.} \underline{+\ 56 \text{ in.}} \\ 152 \text{ in.} \end{array}$$

Answer   The area of the figure is 152 in.^{2}.

Practice

Multiple-Choice Questions

1. The area of a square rug is 144 ft². What is the length of a side of the rug?

   A. 6 ft   B. 12 ft
   C. 24 ft   D. 36 ft

2. The area of a rectangle whose dimensions are 30 inches by 4 feet is

   F. 10 ft²   G. 30 ft²
   H. 60 ft²   J. 120 ft²

3. The sides of a square are tripled in length. What effect does this have on the area of the square?

   A. The area is tripled.
   B. The area is multiplied by 6.
   C. The area is multiplied 9.
   D. The area is multiplied by 12.

4. Which figure has the greatest area?

   F.
   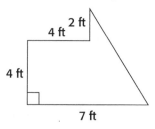
   10 in.
   12 in.

   G.
   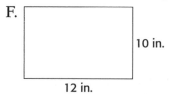
   18 in.
   20 in.

   H.
   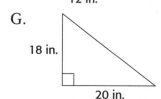
   8 in.
   16 in.

   J.
   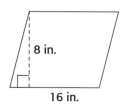
   14 in.
   10 in.
   6 in.

5. Find the area of the figure.

   A. 22 ft²
   B. 25 ft²
   C. 28 ft²
   D. 34 ft²

6. A circular linen tablecloth has a diameter that measures 80 inches. How much linen was used for the tablecloth? Use 3.14 for π.

   F. 251.2 in.²
   G. 502.4 in.²
   H. 5,024 in.²
   J. 20,096 in.²

7. The radius of circle I is one half the radius of circle II. What is the ratio of the area of circle I to the area of circle II?

   A. 1:4
   B. 2:4
   C. 1:8
   D. 1:16

8. The perimeter of a square is 36 in. What is the area of this square?

   F. 81 in.²
   G. 144 in.²
   H. 324 in.²
   J. 1,296 in.²

9. If the radius of a circle is 5, the area of the circle is

   A. exactly 314
   B. between 314 and 315
   C. exactly 78.5
   D. between 78 and 79

Short-Response Questions

10. What is the area of the figure when $x = 6$ and $y = 8$?

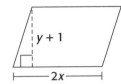

11. The area of a triangle is 32.4 cm² and the height of the triangle is 5.4 cm. What is the length of the base of the triangle?

12. A rectangle is twice as long as it is wide. The area of the rectangle is 50 ft². What is the length of the rectangle?

13. The Wilsons decided to enlarge their rectangular patio. They doubled the length and tripled the width of the original patio. What is the ratio of the area of the new patio to the area of the original patio?

Extended-Response Questions

14. The drawing shows the plan for a garden. If sod costs $6 per square yard, how much will it cost to cover the whole garden with sod? Show your work. Use $\frac{22}{7}$ for π.

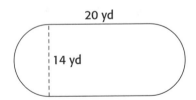

15. Find the area of the hallway shown in the figure. Show your work.

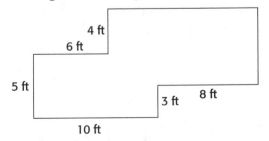

16. A shape was cut from a piece of cardboard as shown. Find the area of the cardboard that remains. Show your work.

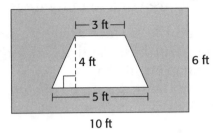

17. What is the area of this kite? Show your work.

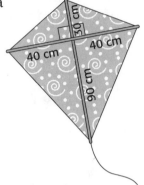

18. What is the area of the sidewalk around Circle Pond? Round your answer to the nearest square foot.

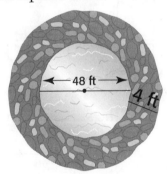

19. Nikolas has 40 feet of fencing to enclose a rectangular space for a vegetable garden. He wants the garden to have the greatest area possible and he wants the dimensions to be whole number lengths.
a. What dimensions should Nikolas use?
b. What is the greatest area possible for his garden?

20. A horse is tied to a post in a field. The length of the rope holding the horse is 24 ft. To the nearest square foot, over how much area can the horse graze? Use 3.14 for π.

8.8 Surface Area

The **surface area** of a solid figure is the sum of the areas of all its faces.

- To find the surface area of a **prism**, find the total area of its two congruent bases and its rectangular sides. A prism is named for the shape of its bases.
- To find the surface area of a **pyramid**, find the total area of its base and its triangular sides. A pyramid is named for the shape of its base.
- To find the surface area of a **cylinder**, find the total area of its two congruent circular bases and its curved side, which flattens into a rectangle.

Rectangular Prism
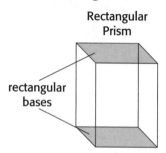
rectangular bases

Triangular Prism
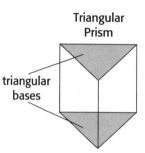
triangular bases

Square Pyramid
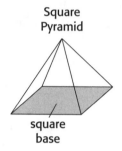
square base

Cylinder
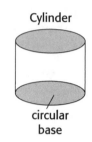
circular base

Model Problems

1. Find the surface area of the rectangular prism.

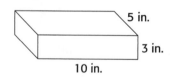

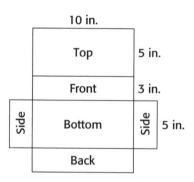

Solution Draw a pattern that when folded would form the rectangular prism. Use the fact that congruent faces have equal areas to simplify your work.

Area of top = Area of bottom = $10 \times 5 = 50$
Area of front = Area of back = $10 \times 3 = 30$
Area of left side = Area of right side = $3 \times 5 = 15$
The sum of all areas = $2(50) + 2(30) + 2(15) = 190$ in.^{2}

Answer The surface area is 190 in.^{2}.

2. Find the surface area of the square pyramid.

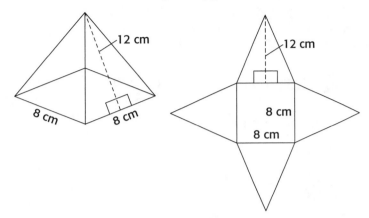

> **Solution**   The square pyramid has four congruent triangular faces.
>   Area of square base = $s^2 = 8^2 = 64$
>
>   Area of one triangular face = $\frac{1}{2}bh = \frac{1}{2} \times 8 \times 12 = 48$
>
>   The sum of all areas = $64 + 4(48) = 256$ cm^2
> **Answer**   The surface area is 256 cm^2.

3. How much aluminum is needed to make the coffee can shown? Use 3.14 for π.

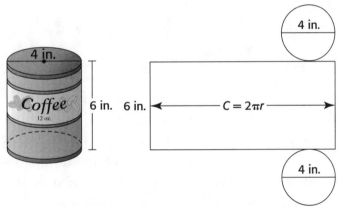

> **Solution**   The height of the cylinder is the width of the rectangle.
> The length of the rectangle is the circumference of one of the bases.
>   Area of one circular base = $\pi r^2 \approx 3.14 \times 2^2 \approx 3.14 \times 4 \approx 12.56$
>   Area of rectangle = $\ell w = (2\pi r)h \approx 2 \times 3.14 \times 2 \times 6 \approx 75.36$
>   Sum of the areas = $2(12.56) + 75.36 = 100.48$ in.^{2}
> **Answer**   The coffee can requires 100.48 in.^{2} of aluminum.

Multiple-Choice Questions

1. What is the surface area of the cube?

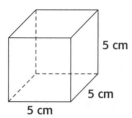

5 cm
5 cm
5 cm

A. 100 cm² B. 125 cm²
C. 150 cm² D. 225 cm²

2. If the length of a side of a cube is doubled, the surface area of the cube is

F. doubled
G. multiplied by 4
H. multiplied by 8
J. multiplied by 12

3. Which of the patterns shown can be folded and taped to form a cube?

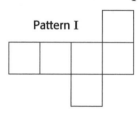

Pattern I

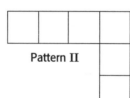

Pattern II

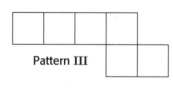

Pattern III

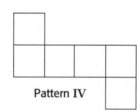

Pattern IV

A. I only
B. I and III only
C. II, III, and IV
D. I and IV only

4. A prop in a play is a giant wedge of Swiss cheese. How much yellow cardboard will be needed to make the prop?

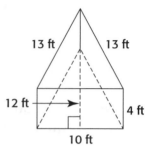

13 ft   13 ft
12 ft   4 ft
10 ft

F. 60 ft²   G. 112 ft²
H. 164 ft²   J. 264 ft²

5. Find the surface area of the cylinder. Use $\frac{22}{7}$ for π.

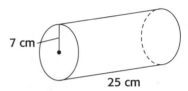

7 cm
25 cm

A. 1254 cm² B. 1408 cm²
C. 2508 cm² D. 7700 cm²

6. What is the ratio of the surface area of a cube with sides measuring 6 in. to the surface area of a cube with sides measuring 2 in.?

F. 9:1   G. 6:1
H. 3:2   J. 3:1

7. The cylinder with which set of dimensions has the greatest surface area?

A. $r = 10$ cm, $h = 8$ cm
B. $r = 5$ cm, $h = 16$ cm
C. $r = 8$ cm, $h = 10$ cm
D. $r = 4$ cm, $h = 20$ cm

8. A bedroom is 18 ft long, 15 ft wide, and 10 ft high. If the walls and ceiling of the bedroom are given one coat of paint, what is the total area to be painted?

    F. 600 ft²
    G. 660 ft²
    H. 930 ft²
    J. 1,200 ft²

Short-Response Questions

For 9–12, find the surface area of each figure. Use 3.14 for π. Show all work.

9.

9 ft

4 ft   4 ft

10.

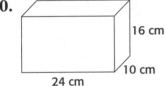

16 cm

10 cm

24 cm

11.

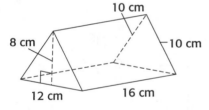

10 cm

8 cm

10 cm

12 cm   16 cm

12.

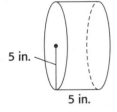

5 in.

5 in.

13. The surface area of a cube is 216 in.². What is the length of each side of the cube?

Extended-Response Questions

14. The inside of a rectangular swimming pool will be resurfaced. The pool is 40 feet long, 18 feet wide, and 7 feet deep. What is the total area to be resurfaced? Show your work.

15. Mrs. Evans is sending her son a tin of cookies. The tin is cylindrical and it just fits in the carton shown. What is the surface area of the tin? Use 3.14 for π. Round to the nearest square inch.

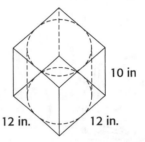

10 in

12 in.   12 in.

16. Find the outside surface area of the wooden storage shed shown.

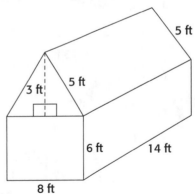

5 ft

3 ft   5 ft

6 ft   14 ft

8 ft

8.9 Volume

Volume is the measurement of the amount of space that a solid figure contains. Volume is measured in cubic units such as cubic inches (in.^{3}), cubic feet (ft.^{3}), cubic centimeters (cm^3), or cubic meters (m^3). A cubic inch is the space contained by a cube that measures 1 inch on each side.

To find the volume of a figure, use the appropriate formula.

- The volume of a prism or cylinder is the area of the base (B) multiplied by the height of the prism.
  V (prism or cylinder) $= Bh$

- The volume of a cube with side s is s^3.
  V (cube) $= s^3$.

- The volume of a pyramid or cone is one third the area of the base (B) multiplied by the height.

  V (pyramid or cone) $= \frac{1}{3}Bh$

Model Problems

1. Find the volume of each figure.

   a.
   4 ft   3 ft   5 ft

   b.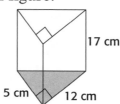
   17 cm   5 cm   12 cm

   c.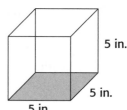
   5 in.   5 in.   5 in.

   Solution

   a. For this rectangular prism,
      $V = Bh$ *Think:* The area of the rectangular base equals length times width.
      $V = \ell wh$
      $V = 4 \times 3 \times 5$
      $V = 60 \text{ ft}^3$

   b. For this triangular prism,
      $V = Bh$ and $B = \frac{1}{2} \times 5 \times 12 = 30$
      $h = 17$
      $V = 30 \times 17 = 510 \text{ cm}^3$

c. For this cube,
$$V = s^3$$
$$V = 5^3$$
$$V = 125 \text{ in.}^3$$

2. A cylinder and a cone each have a radius of 3 in. and a height of 10 in. Find their volumes. Use 3.14 for π.

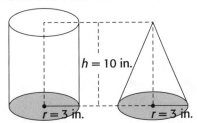

Solution   Since the figures have the same radius and height, the volume of the cone is one third the volume of the cylinder.

V (cylinder) $= Bh$ and V (cone) $= \frac{1}{3}Bh$ Where $B = \pi r^2 \approx 3.14 \times 3^2$
                                                            ≈ 28.26

V (cylinder) $\approx 28.26 \times 10 \approx 282.6 \text{ in.}^3$

V (cone) $\approx \frac{1}{3} \times 282.6 \approx 94.2 \text{ in.}^3$

Answer   The volume of the cylinder is 282.6 in.³ and the volume of the cone is 94.2 in.³.

3. Find the volume of the pyramid.

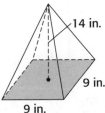

Solution   To remember the formula, keep in mind that the volume of a pyramid is one third the volume of a rectangular prism with the same base and height. Here, the base is a square.

V (pyramid) $= \frac{1}{3}Bh$ $B = s^2 = 9^2 = 81$

$V = \frac{1}{3} \times 81 \times 14 = 378 \text{ in.}^3$

Answer   378 in.³

![Practice](puzzle piece graphic)

Multiple-Choice Questions

1. The volume of the figure is

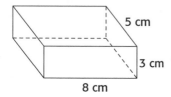

    A. 40 cm³   B. 120 cm³
    C. 136 cm³   D. 158 cm³

2. The volume of a rectangular prism is 1,001 in.³. The height of the prism is 13 in. and its width is 7 in. What is the length of the prism?

    F. 11 in.   G. 14 in.
    H. 17 in.   J. 70 in.

3. The length of each side of a cube is tripled. How many times greater is the volume of the new cube than the original?

    A. 3 times   B. 6 times
    C. 9 times   D. 27 times

4. The volume of a cylinder is 216 in.³. What is the volume of a cone with the same radius and height as the cylinder?

    F. 54 in.³   G. 72 in.³
    H. 108 in.³   J. 648 in.³

5. Find the volume of chili that the can holds. Leave your answer in terms of π.

    A. 9π in.³   B. 11.25π in.³
    C. 12.5π in.³   D. 45π in.³

6. From least to greatest volume, the figures are:

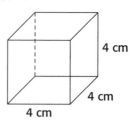

I

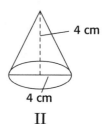

II

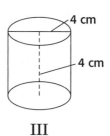

III

    F. I, II, III
    G. III, I, II
    H. II, I, III
    J. II, III, I

7. Find the volume of a pyramid that has a height of 8 inches and a rectangular base 6 inches long and 4 inches wide.

    A. 64 in.³   B. 96 in.³
    C. 192 in.³   D. 576 in.³

8. The length of the edge of a cube is represented as $x + 3$. Find the volume of the cube when $x = 2$.

    F. 11 cubic units
    G. 35 cubic units
    H. 75 cubic units
    J. 125 cubic units

Short-Response Questions

For 9–12, find the volume of each figure. Use 3.14 for π.

9.

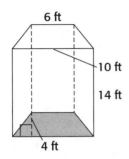

10.

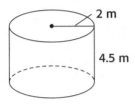

11.

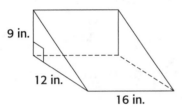

12.

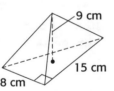

13. The volume of a cylinder is 1632 cm³. The height of the cylinder is 24 cm. Find the area of its base.

14. Find the volume of the cube when $x = 4$.

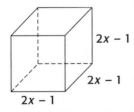

Extended-Response Questions

15. A cereal manufacturer needs a box that will have 60 in.³ of space inside.

   a. Give the dimensions of two possible boxes the manufacturer can use.
   b. Which of the two boxes you suggested will use less cardboard? Show how you determined your answer.
   c. Based on your findings, what general statement can you make about boxes with the same volume?

16. a. If the height of a cylinder is doubled, what effect does this have on the volume?
   b. If the radius of a cylinder is doubled, what effect does this have on the volume?
   c. If both the height and radius of a cylinder are doubled, what effect does this have on the volume?

17. A cylinder with height of 10 cm and diameter of 10 cm is placed inside a 10-cm cube. What percent of the volume of the cube is NOT taken up by the cylinder? Show your work. Use 3.14 for π.

18. A straight driveway leading to a hotel is 150 feet long and 12 feet wide. It is paved with concrete 6 inches thick. At a cost of $6.25 per cubic foot, how much did the concrete cost? Show your work.

8.10 The Pythagorean Theorem

The **hypotenuse** of a right triangle is the side opposite the right angle. The hypotenuse is longer than either of the two other sides called **legs**.

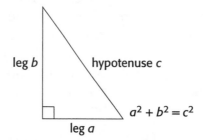

The **Pythagorean Theorem** describes the relationship between the lengths of the legs of a right triangle and its hypotenuse. Understanding this relationship can be very helpful in solving problems about right triangles because it allows you to find the length of one side of the triangle if you know the lengths of the other two.

In a right triangle, the square of the length of the hypotenuse is equal to the sum of the squares of the lengths of the two legs.

Let a and b represent the lengths of the legs and c the length of the hypotenuse. Then the theorem can be expressed as:

$a^2 + b^2 = c$

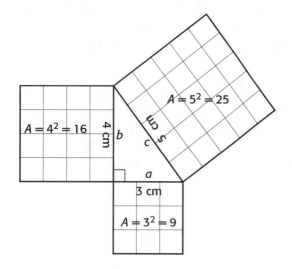

The figure illustrates the Pythagorean Theorem geometrically for $a = 3$, $b = 4$, and $c = 5$.

$(3 \text{ cm})^2 + (4 \text{ cm})^2 = (5 \text{ cm})^2$

$9 \text{ cm}^2 + 16 \text{ cm}^2 = 25 \text{ cm}^2$

Three whole number measures of a right triangle, such as 3, 4, and 5, are called a **Pythagorean triple**.

The Pythagorean Theorem also allows you to determine if a triangle is a right triangle or not.

- If the sides of a triangle satisfy the Pythagorean Theorem, then the triangle is a right triangle.
- If the sides of a triangle do not satisfy the Pythagorean Theorem, then the triangle is not a right triangle.

Model Problems

1. Determine whether each group of measures represents the lengths of the sides of a right triangle.
   a. 9 cm, 12 cm, 15 cm
   b. 8 ft, 11 ft, 14 ft

   Solution Substitute the side lengths into the theorem to check if the Pythagorean relationship holds.
   a. The longest side must be the hypotenuse c.
   $$a^2 + b^2 = c^2$$
   $$9^2 + 12^2 = 15^2$$
   $$81 + 144 = 225$$
   $$225 = 225$$
   The triangle is a right triangle.

   b. $$a^2 + b^2 = c^2$$
   $$8^2 + 11^2 = 14^2$$
   $$64 + 121 = 196$$
   $$185 \neq 196$$
   The triangle is not a right triangle.
   Note: Once you know which side is the hypotenuse, it does not matter which leg is called a and which is called b.

2. For $\triangle ABC$
   a. Find the length of $\overline{AC}$.
   b. Find the area of $\triangle ABC$.

   Solution
   a. Let $a = 9$, b = length of $\overline{AC}$ and $c = 41$.
   $$a^2 + b^2 = c^2$$
   $$81 + b^2 = 41^2$$
   $$81 + b^2 = 1{,}681$$

   $b^2 = 1{,}600$ Subtract 81 from both sides.

   $b = \sqrt{1{,}600}$ Take the square root of both sides.

   $b = 40$

   The length of $\overline{AC}$ is 40 cm.

   b. Area of $\triangle ABC = \frac{1}{2}bh$

   $A = \frac{1}{2} \times 40 \times 9$

   $A = 180$

   The area of $\triangle ABC$ is 180 cm^2.

3. A ladder 13 feet long leans against a wall. If the foot of the ladder is 5 feet from the wall, how far up the wall does the ladder reach?

Solution   Let $c = 13$, $a = 5$, and b = height of wall where ladder reaches.

$$a^2 + b^2 = c^2$$
$$5^2 + b^2 = 13^2$$
$$25 - 25 + b^2 = 169 - 25 \qquad \text{Subtract 25 from both sides.}$$
$$b^2 = 144$$
$$b = \sqrt{144} \qquad \text{Take the square root of both sides of the equation.}$$
$$b = 12$$

Answer   The ladder reaches up 12 ft on the wall.

Practice

Multiple-Choice Questions

1. In $\triangle DEF$, what is the length of $\overline{DE}$?

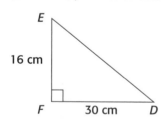

A. 14 cm
B. 25.4 cm
C. 34 cm
D. 46 cm

2. Which of the following could be the lengths of the sides of a right triangle?

F. 2 cm, 4 cm, 7 cm
G. 6 cm, 8 cm, 10 cm
H. 4 cm, 9 cm, 12 cm
J. 5 cm, 10 cm, 15 cm

3. What is the distance from Minton to Newville?

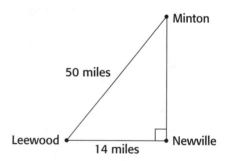

A. 64 miles   B. 48 miles
C. 36 miles   D. 12 miles

4. Find the area of $\triangle XYZ$.

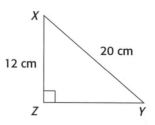

F. 96 cm^2 G. 112 cm^2
H. 120 cm^2 J. 140 cm^2

5. Find the length of diagonal $\overline{PR}$ of rectangle PQRS.

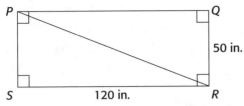

A. 70 in.   B. 109 in.
C. 130 in.   D. 170 in.

6. Find the perimeter of △JKL.

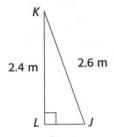

F. 10.0 m   G. 7.4 m
H. 6.8 m   J. 6.0 m

7. Which of the following groups could NOT be the lengths of the sides of a right triangle?

A. $\frac{3}{5}$ in., $\frac{4}{5}$ in., 1 in.

B. $\frac{1}{2}$ yd, $\frac{1}{2}$ yd, $\frac{3}{4}$ yd

C. $1\frac{1}{2}$ in., 2 in., $2\frac{1}{2}$

D. $\frac{3}{10}$ ft, $\frac{2}{5}$ ft, $\frac{1}{2}$ ft

8. The two legs of a right triangle measure 2 cm and $\sqrt{5}$ cm. Find the length of the hypotenuse.

F. 1 cm   G. $\sqrt{3}$ cm
H. 3 cm   J. $\sqrt{7}$ cm

Short-Response Questions

For 9–11, find the missing measure in each right triangle.

9.

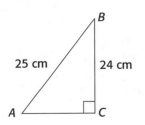

10.

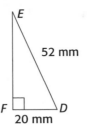

11.

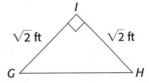

12. Find the area of △TUV. Show your work.

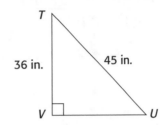

Extended-Response Questions

13.

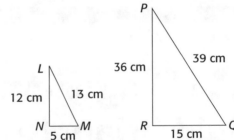

a. What is the relationship between △LMN and △PQR? Justify your answer.

b. Determine if one, both, or neither of the triangles are right triangles. Show your work.

c. Suggest a set of lengths for another triangle that would have the same relationship to △LMN and △PQR.

14. A boat traveled 12 miles north and then 22.5 miles east. What is the shortest distance the boat must travel to return to its starting point?

15. Find the perimeter of rectangle *ABCD*. Show your work.

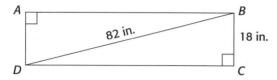

16. A baseball diamond is actually a square with sides 90 feet long. Use a calculator to find the distance from home plate to second base to the nearest tenth of a foot.

17. A cable supports a 28-foot utility pole. The cable is fastened to the ground at a point 21 feet from the base of the pole. Find the length of the cable.

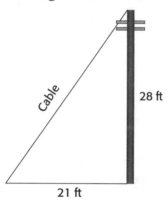

18. The freight entrance to a store is 3 feet above ground level. An access ramp to the entrance is 10 feet long. What is the distance from the bottom of the ramp to the base of the building? Give the exact answer and the answer rounded to the nearest tenth of a foot.

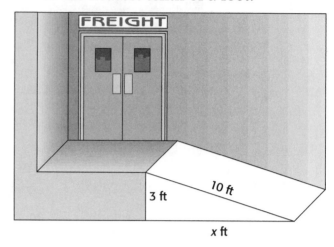

8.11 Trigonometry of the Right Triangle

Right triangles that contain the same acute angles are similar.

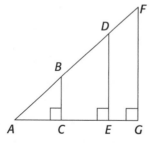

Since corresponding sides of similar triangles are in proportion, some special ratios, called **trigonometric ratios**, are formed to compare the lengths of the sides of these triangles. For a given acute angle, such as 43° or 71° or 80°, these ratios have the same value no matter how large or small the lengths of the sides of the triangle are.

Understanding how to use the trigonometry ratio can be a very powerful tool when solving problems involving right triangles. Trigonometric ratios can be used to find missing side lengths and angle measurements in right triangles.

In these ratios, the legs of the triangle are identified as the **opposite leg** and the **adjacent leg** to describe their positions relative to the acute angles of the triangle. The leg that is directly across from an acute angle is the opposite leg. The leg next to the angle is the adjacent leg.

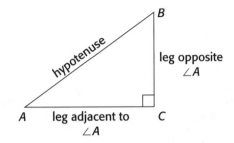

In a right triangle:

The **sine** (sin) of an acute angle is the ratio of the length of the leg opposite the angle to the length of the hypotenuse.

$$\text{sine } A = \frac{\text{length of side opposite } \angle A}{\text{length of hypotenuse}} \qquad \sin \angle A = \frac{\text{opp}}{\text{hyp}} = \frac{BC}{AB}$$

The **cosine** (cos) of an acute angle is the ratio of the length of the leg adjacent to the angle to the length of the hypotenuse.

$$\text{cosine } A = \frac{\text{length of side adjacent to } \angle A}{\text{length of hypotenuse}} \qquad \cos \angle A = \frac{\text{adj}}{\text{hyp}} = \frac{AC}{AB}$$

The **tangent** (tan) of an acute angle is the ratio of the length of the leg opposite the angle to the length of the leg adjacent to the angle.

$$\text{tangent } A = \frac{\text{length of side opposite } \angle A}{\text{length of side adjacent to } \angle A} \qquad \tan \angle A = \frac{\text{opp}}{\text{adj}} = \frac{BC}{AC}$$

SohCahToa

Use a mnemonic (memory) device to remember the trigonometric ratios. The word SohCahToa can help you remember that

$$\text{Sine} = \frac{\text{opposite}}{\text{hypotenuse}}$$

$$\text{Cosine} = \frac{\text{adjacent}}{\text{hypotenuse}}$$

$$\text{Tangent} = \frac{\text{opposite}}{\text{adjacent}}$$

Note: For an explanation of how to use the scientific calculator to find the value of trigonometric ratios, see pages 416–417.

Model Problems

1. $\triangle QRS$ is a right triangle. Write each trigonometry ratio as a fraction.

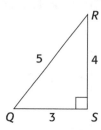

   a. sin $\angle Q$ b. cos $\angle Q$
   c. tan $\angle Q$ d. sin $\angle R$
   e. cos $\angle R$ f. tan $\angle R$

Solution

a. $\sin Q = \dfrac{\text{opp}}{\text{hyp}} = \dfrac{RS}{QR} = \dfrac{4}{5}$ b. $\cos Q = \dfrac{\text{adj}}{\text{hyp}} = \dfrac{QS}{QR} = \dfrac{3}{5}$

c. $\tan Q = \dfrac{\text{opp}}{\text{adj}} = \dfrac{RS}{QS} = \dfrac{4}{3}$ d. $\sin R = \dfrac{\text{opp}}{\text{hyp}} = \dfrac{QS}{QR} = \dfrac{3}{5}$

e. $\cos R = \dfrac{\text{adj}}{\text{hyp}} = \dfrac{RS}{QR} = \dfrac{4}{5}$ f. $\tan R = \dfrac{\text{opp}}{\text{adj}} = \dfrac{QS}{RS} = \dfrac{3}{4}$

2. In right triangle XYZ, sin $\angle X = \dfrac{5}{13}$. Find cos $\angle X$ and tan $\angle X$.

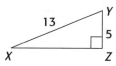

Solution   Use the given information to label the lengths.

$\sin \angle X = \dfrac{\text{opp}}{\text{hyp}} = \dfrac{YZ}{XY} = \dfrac{5}{13}$

Use the Pythagorean Theorem to find XZ.

$$(YZ)^2 + (XZ)^2 = (XY)^2$$
$$5^2 + (XZ)^2 = 13^2$$
$$25 + (XZ)^2 = 169$$
$$(XZ)^2 = 144$$
$$XZ = 12$$

So, $\cos \angle X = \dfrac{\text{adj}}{\text{hyp}} = \dfrac{XZ}{XY} = \dfrac{12}{13}$ and $\tan \angle X = \dfrac{\text{opp}}{\text{adj}} = \dfrac{YZ}{XZ} = \dfrac{5}{12}$

Answer $\cos \angle X = \dfrac{12}{13}$, $\tan \angle X = \dfrac{5}{12}$

3. Find the height of the ski jump to the nearest tenth of a meter.

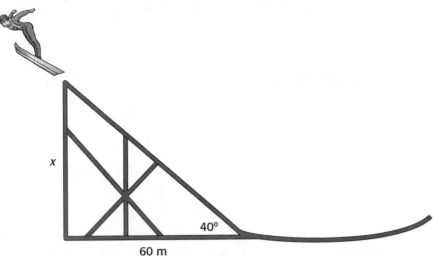

 Solution   Decide which trigonometric ratio to use. The parts of the tri-angle involved are an acute angle of a right triangle and the opposite and adjacent sides. Use the tangent ratio.

$$\tan 40° = \frac{x}{60}$$

To find tan 40°, use the Trigonometric Table on page 393, or a scientific calculator.

$$\tan 40° = 0.8391 = \frac{x}{60}$$

$$60(0.8391) = x \qquad \text{Use cross products.}$$

$$50.346 = x$$

Answer   The ski jump is about 50.3 m high.

4. A ladder 20 ft long leans against a building and reaches a point 18.8 ft above the ground. Find to the nearest degree the angle that the ladder makes with the ground.

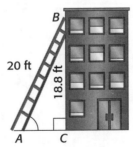

Solution   In relation to ∠A, the given lengths are for the opposite side and the hypotenuse. Use the sine ratio.

$$\sin \angle A = \frac{\text{opp}}{\text{hyp}} = \frac{18.8}{20}$$

$$\sin \angle A = 0.9400$$

Look up the resulting decimal measure on the Trigonometric Table (page 393) or use the inverse ($\sin^{-1}$) function key on a calculator. Round to the nearest degree.

∠A = 70° (rounded to the nearest degree)

Answer   The ladder makes an angle of 70° with the ground.

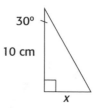

Practice

Multiple-Choice Questions

1. Cos A =

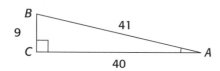

   A. $\dfrac{40}{41}$ B. $\dfrac{9}{41}$

   C. $\dfrac{9}{40}$ D. $\dfrac{41}{40}$

2. Which statement is true?

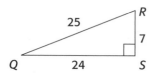

   F. $\sin \angle R > 1$
   G. $\cos \angle Q > 1$
   H. $\tan \angle R > 1$
   J. $\tan \angle Q > 1$

3. Tan $\angle Y$ =

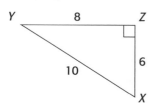

   A. 0.60   B. 0.75
   C. 0.80   D. $1.3\overline{3}$

4. In $\triangle LMN$, sin $\angle L = \dfrac{8}{17}$. What is the value of sin M?

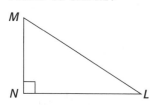

   F. $\dfrac{17}{8}$ G. $\dfrac{15}{8}$

   H. $\dfrac{17}{15}$ J. $\dfrac{15}{17}$

5. Find x to the nearest tenth.

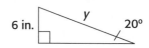

   A. 5.0 cm   B. 5.8 cm
   C. 8.7 cm   D. 10.6 cm

6. Find y to the nearest tenth.

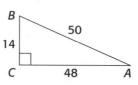

   F. 2.1 in.   G. 6.4 in.
   H. 16.5 in.   J. 17.5 in.

7. Cos $\angle M$ = 0.8192. What is the measure of $\angle M$?

   A. 35°   B. 43°
   C. 55°   D. 67°

8. In $\triangle FGH$, $\angle H$ is a right angle and $\overline{GH} = \overline{FH}$. To the nearest degree, what is the measure of $\angle G$?

   F. 15°   G. 30°
   H. 45°   J. 60°

Short-Response Questions

9. Refer to right triangle ABC. Give the value of each ratio as a fraction in lowest terms.

   a. sin $\angle A$
   b. cos $\angle A$
   c. tan $\angle A$
   d. sin $\angle B$
   e. cos $\angle B$
   f. tan $\angle B$

10. Fill in a degree measure to make each statement true. Use the trigonometric table on page 393.

a. sin 60° = cos ____
b. sin 25° = cos ____
c. cos 50° = sin ____
d. cos 15° = sin ____

11. Refer to right triangle *XYZ*. Find the value of $(\sin \angle X)^2 + (\cos \angle X)^2$. Show your work.

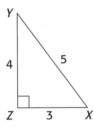

12. a. Find *x* to the nearest tenth of a foot.
b. Find *y* to the nearest tenth of a foot. Show your work.

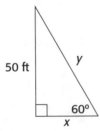

13. Find the height of the tree to the nearest meter. Show your work.

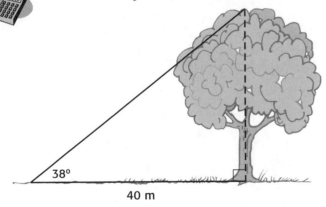

14. For right triangle *DEF*,

a. Find the measures of ∠*D* and ∠*E* to the nearest degree.
b. Show how you can check your answers to part a.

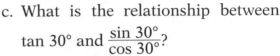

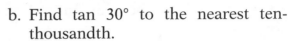

15. a. Find $\dfrac{\sin 30°}{\cos 30°}$ to the nearest ten-thousandth.

b. Find tan 30° to the nearest ten-thousandth.
c. What is the relationship between tan 30° and $\dfrac{\sin 30°}{\cos 30°}$?

16. In rectangle *ABCD*, diagonal $\overline{BD}$ = 20 in. and makes an angle of 35° with base $\overline{CD}$.

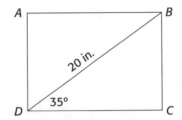

a. Find the length of $\overline{BC}$ to the nearest tenth.
b. Find the length of $\overline{DC}$ to the nearest tenth.
c. Find the area of rectangle *ABCD* to the nearest tenth.

Extended-Response Questions

17. A ladder leans against a building. The top of the ladder reaches a point on the building that is 21 feet above the ground. The foot of the ladder is 6 feet from the building.

a. Find to the nearest degree the measure of the angle that the ladder makes with level ground.
b. Find to the nearest foot the length of the ladder.

18. An airplane climbs at an angle of 12° with the ground. Find to the nearest ten feet the distance the airplane has flown when it has reached an altitude of 500 feet.

19. In isosceles triangle MNO, $\overline{MO} = \overline{NO} = 20$ in. and P is the midpoint of $\overline{MN}$.

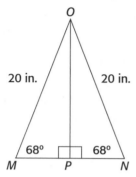

a. Find the length of $\overline{OP}$ to the nearest tenth.
b. Find the length of $\overline{MN}$ to the nearest tenth.
c. Find the area of $\triangle MNO$ to the nearest square inch.

20. While flying a kite, Omar lets out 120 meters of string, which makes an angle of 42° with the ground. There is a strong wind and the string is stretched straight. To the nearest tenth of a meter, how high above the ground is the kite flying?

Chapter 8 Review

Multiple-Choice Questions

1. $\overline{QR} \parallel \overline{ST}$ and the measure of $\angle POS = 130°$. What is the measure of the complement of $\angle QPO$?

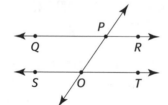

A. 40°   B. 50°
C. 60°   D. 70°

2. Find the measure of $\angle Y$ in $WXYZ$.

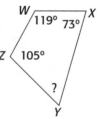

F. 17°
G. 63°
H. 75°
J. 119°

3. Which of the figures below have an area of 36 square units?

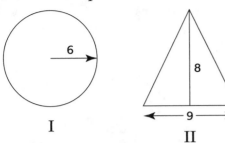

I

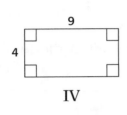

II

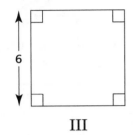

III

IV

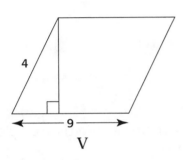

V

A. I, II, III only
B. III, IV, V only
C. II, III, IV only
D. I, II, III, IV, and V

4. Find the perimeter of the figure shown below.

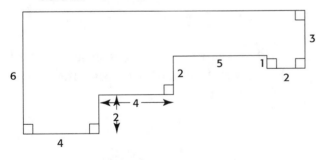

F. 50 units
G. 49 units
H. 44 units
J. 32 units

5. Find the total area of the garden shown below.

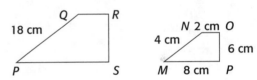

A. 36 ft²
B. 64.26 ft²
C. 100.26 ft²
D. 85.04 ft²

6. What is the measure of each angle of the regular polygon shown?

F. 108°
G. 120°
H. 135°
J. 150°

7. Quadrilateral *PQRS* ~ quadrilateral *MNOP*. What is the perimeter of *PQRS*?

A. 38 cm
B. 66 cm
C. 82 cm
D. 90 cm

8. The storage box shown is 6 times longer than it is wide. What is the volume of the box?

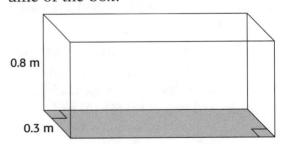

F. 0.432 m³
G. 1.44 m³
H. 2.9 m³
J. 4.32 m³

9. Which pair of figures is similar, but NOT congruent?

A.

B.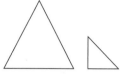

C.

D.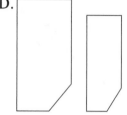

10. The area of a circle with a 12-cm radius is

   F. $6\pi^2 cm^2$

   G. $12\pi cm^2$

   H. $36\pi cm^2$

   J. $144\pi cm^2$

Short-Response Questions

11. A truck traveled steadily at the rate of 30 miles per hour. It traveled south for one half hour, then turned and drove west for 40 minutes and arrived at its destination.

   a. What is the shortest possible distance the truck could travel to get back to its starting point?

   b. How long would it take?

12. A straight trail up a mountain has an incline of 40°. To the nearest foot, how far along the trail must a hiker walk to reach an elevation of 500 ft? Show your work.

13. Find the total area of the two similar triangles below. Show all steps.

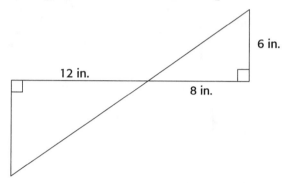

14. Find the perimeter of the figure below. Use 3.14 for π.

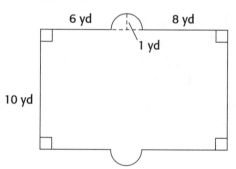

15. Find the volume of the shaded part of the figure. Use 3.14 for π. Round to the nearest tenth.

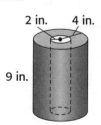

Extended-Response Questions

16. As a craft project, Rosa is covering the closed wooden box shown with a mosaic made from 1-cm² tiles. The tiles come in packages of 100 that cost $2.95 each.

  a. How many tiles does Rosa need to completely cover the box?
  b. How much will Rosa spend for the tiles? Explain how you arrived at your answer.

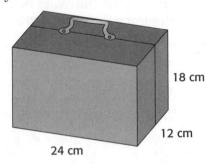

18 cm

12 cm

24 cm

17. In the figure, $\overline{JK} \perp \overline{LM}$ and the measure of LPO is 70°.

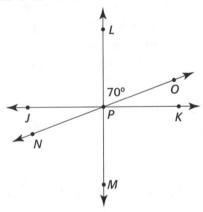

  a. What is the measure of $\angle NPM$?
  b. What is the measure of $\angle JPO$?

  c. What is the measure of $\angle MPO$?
  d. Name an angle that is congruent to $\angle NPL$.
  e. Name an angle that is supplementary to $\angle MPO$.

18. Find the area of the isosceles right triangle shown. Show your work.

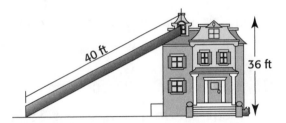

$\sqrt{200}$ in.

19. A 40-foot chute being used to remove roofing material leans against a building that is 36 feet high. Find to the nearest degree the angle formed by the chute and the ground.

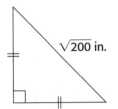

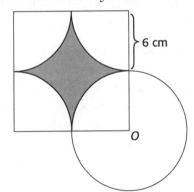

40 ft

36 ft

20. Find the shaded area of the square. Use 3.14 for π. Show your work.

6 cm

O

Multiple-Choice Questions

1. Which is equivalent to a repeating decimal?

   A. $1\frac{1}{5}$ B. $\sqrt{9}$

   C. $\frac{2}{3}$ D. $\frac{3}{8}$

2. Which expression has a value of 180?

   F. $2^2 \cdot 3^2 \cdot 5^1$ G. 0.0180×10^3
   H. $3^2 \cdot 2^0 \cdot 5^2$ J. 18.0×10^{-3}

3. Find the value of $5 \times 6 + 100 \div 2 + 5^2$.

   A. 90   B. 105
   C. 162   D. 290

4. What is the value of x in the equation $14 = 2x - 6$?

   F. 4   G. 6
   H. 10   J. 20

5. At Fleur du Jour, lilacs cost $6.25 a bunch and daffodils cost $5.25 a bunch. If the store sold $113 worth of flowers, how many bunches of each flowers did it sell?

   A. 8 bunches of daffodils and 12 bunches of lilacs
   B. 12 bunches of daffodils and 8 bunches of lilacs
   C. 6 bunches of daffodils and 13 bunches of lilacs
   D. 13 bunches of daffodils and 6 bunches of lilacs

6. A chef opened a 2-liter container of olive oil and, by the end of the day, 35% of the oil had been used. How many milliliters of oil were left?

   F. 650 mL   G. 700 mL
   H. 1050 mL   J. 1300 mL

7. A scale model of a building is $\frac{1}{48}$ the size of the actual building. If the actual building is 30 ft wide, how wide is the scale model?

   A. 625 in.   B. 7.5 in.
   C. 10 in.   D. 1.6 ft

8. A rectangular hallway was carpeted using 374 one-foot squares of carpet. The width of the hallway is 8.5 feet. What is the length of the hallway?

   F. 22 ft   G. 44 ft
   H. 88 ft   J. 158 ft

9. The circumference of a circular pipe is 44 cm. What is the approximate radius of the pipe? Use $\frac{22}{7}$ for π.

   A. 154 cm   B. 22 cm
   C. 14 cm   D. 7 cm

10. Last month, Street Wear had sales of $22,000 and earned a profit of $7,040. At the same rate, how much profit can the store expect to earn this month if it projects sales of $28,000?

    F. $20,960   G. $13,040
    H. $8,960   J. $8,160

Short-Response Questions

11. A train traveled for 3 hours 28 minutes, arriving at its destination at 12:05 P.M. What time did the train leave the station?

12. Find the value of the expression below when $a = -16$, $b = 12$, and $c = -1$.

    $$\frac{a - b^2}{|-2c^2|}$$

13. Karen started the weekend with a full tank of gasoline in her car. On Saturday, she used $\frac{2}{5}$ of the tank. On Sunday, she used $\frac{3}{4}$ of what was left in the tank. What part of the full tank was left after the weekend?

14. The day after the election, the school newspaper headline read: "Donohoe Wins! For every 3 votes for Trusty, there were 5 for Donohoe." If Trusty received 435 votes, how many votes did Donohoe receive?

Extended-Response Questions

15. At temperatures of 100°C or higher, water changes to steam (a gas). At temperatures of 0°C or lower, water changes to ice (a solid). In between these extremes, water is a liquid. Draw number lines showing the temperatures at which water is

a. a solid
b. a gas
c. a liquid

16. A water trough is shaped like a half-cylinder.

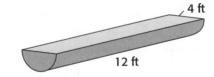

a. Find the outside surface area of the trough.
b. Find the volume of the trough.

Use 3.14 for π. Round your answers to the nearest whole number.

17. Marcia won $16,000 on *Who Wants to Be Rich*. She plans to go on a shopping spree, take a vacation, and put the rest of the money in her college fund. She will spend 3 times the cost of the shopping spree on her vacation. She also plans to put twice as much as she spends on her vacation into her college fund. How much money does she plan to put in her college fund? Show your work.

18. Jim is putting up fence posts around a rectangular garden. He used 50 posts and placed them 2 feet apart. The length of the garden is 30 feet.

a. What is the perimeter of the garden?
b. What is the area of the garden? Show your work.

19. The two triangles below are similar.

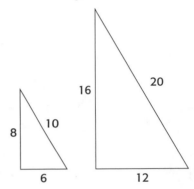

a. Draw a third triangle that is similar to both triangles, but not congruent to either one. Label the lengths of the sides.
b. Write an explanation of how you know that the triangle you drew is similar to the other two triangles.

20. The price of a share of stock rose by 10% on Monday, then fell by 10% on Tuesday. The stock was priced at $115 a share before the Monday increase.

a. What was the share price after the Tuesday decrease?
b. What was the percent change from its opening price on Monday?

Coordinate Graphing and Geometric Constructions

Chapter Vocabulary

coordinate plane	*x*-axis	*y*-axis
origin	ordered pair	coordinates
graph	quadrants	transformation
image	translation	reflection
line of reflection	line symmetry	point symmetry
rotation	rotational symmetry	construction
midpoint	perpendicular bisector	altitude

9.1 Using Coordinates to Graph Points

Two perpendicular number lines can be used to form a system for locating points called a **coordinate plane**. The horizontal line is called the **x-axis**. The vertical line is called the **y-axis**. The point where the axes cross is called the **origin**, and this point is represented by the **ordered pair** (0, 0).

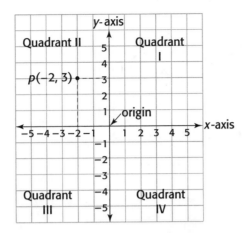

Other points are represented by ordered pairs according to their distance with respect to each axis. The ordered pair $(-2, 3)$ corresponds to -2 on the x-axis and 3 on the y-axis. Each ordered pair is a set of **coordinates** for the point it names. A point is the **graph** of its ordered pair.

To graph an ordered pair

- Start at the origin $(0, 0)$.

- Read the first number (the x-coordinate) and move left or right the number of units indicated.

- Read the second number (the y-coordinate) and move up or down the number of units indicated.

- Place a dot at the location and label the point with a capital letter.

The x-axis and the y-axis divide the graph into four regions called **quadrants**. The signs of the coordinates in each quadrant are:

Quadrant I: $(+, +)$ Quadrant II: $(-, +)$
Quadrant III: $(-, -)$ Quadrant IV: $(+, -)$

Model Problems

1. Use an ordered pair to name the location of each point.

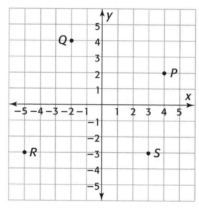

  a. *P*   b. *Q*   c. *R*   d. *S*

Solution
  a. 4 units right and 2 units up   *P*(4, 2)
  b. 2 units left and 4 units up   *Q*(−2, 4)
  c. 5 units left and 3 units down   *R*(−5, −3)
  d. 3 units right and 3 units down   *S*(3, −3)

2. Name the point that has the given coordinates. Give the quadrant for each point.

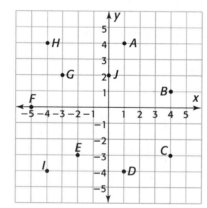

Solution

Coordinates	Point	Quadrant
a. $(-5, 0)$	F	none; x-axis
b. $(4, -3)$	C	IV
c. $(-4, 4)$	H	II
d. $(0, 2)$	J	none; y-axis
e. $(-2, -3)$	E	III
f. $(4, 1)$	B	I

3. Graph each pair of points and connect them with a line segment. Identify the relationship of the segments.

a. $P(3, 2)$ and $Q(-3, 2)$
   $R(-4, -2)$ and $S(4, -2)$
b. $L(2, 2)$ and $M(2, -4)$
   $N(1, 1)$ and $O(5, 1)$

Solution

a. Points P and Q, which have the same y-coordinate, are on a line that is parallel to the x-axis. Points R and S, which have the same y-coordinate, are on a line parallel to the x-axis. Lines parallel to the x-axis are parallel to each other. So, $\overline{PQ} \parallel \overline{RS}$.

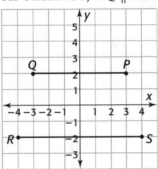

b. Points L and M, which have the same x-coordinate, are on a line parallel to the y-axis. Lines that are parallel to the y-axis are perpendicular to the x-axis and to lines parallel to the x-axis. So, $\overline{LM} \perp \overline{NO}$.

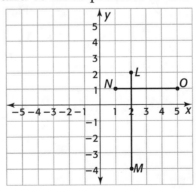

Multiple-Choice Questions

1. Which ordered pair locates a point on the *y*-axis?

   A. (2, 2)
   B. (0, 3)
   C. (−4, −4)
   D. (6, 0)

2. Which ordered pair locates a point on the *x*-axis?

   F. (0, 5)
   G. (−6, 3)
   H. (10, 10)
   J. (8, 0)

3. Points *K*(5, 7) and *L*(−6, 7) lie on a line that

   A. is parallel to the *x*-axis
   B. passes through the origin
   C. is parallel to the *y*-axis
   D. passes through Quadrants III and IV

4. The ordered pair for the point that is 3 units left and 4 units up from point *P* is

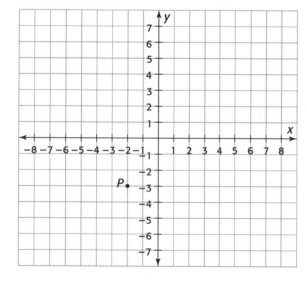

   F. (1, 1)   G. (−5, −7)
   H. (−5, 1)   J. (−3, 4)

5. Which set of points is on a line perpendicular to the *x*-axis?

   A. *P*(3, 6), *Q*(−2, −4), *R*(0, 0)
   B. *S*(6, −9), *T*(6, 11), *U*(6, 4)
   C. *V*(4, 8), *W*(−7, 8), *X*(0, 8)
   D. *K*(10, 5), *L*(−4, −2), *M*(8, 4)

6. Point *Q* is 5 units right and 2 units down from which point?

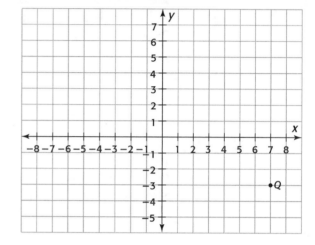

   F. *W*(1, −2)
   G. *X*(12, −5)
   H. *Y*(5, 2)
   J. *Z*(2, −1)

7. Which point does NOT lie on either the *x*-axis or the *y*-axis?

   A. *L*(0, 13)
   B. *M*(−17, 0)
   C. *N*(−15, 15)
   D. *O*(0, 0)

8. Which ordered pair comes next in the pattern?
   (2, −1), (3, 1), (4, 3), (5, 5), (?, ?)

   F. (6, 6)
   G. (6, 7)
   H. (7, 6)
   J. (7, 8)

Short-Response Questions

9. Draw and label a pair of coordinate axes. Graph the point that corresponds to each ordered pair. Label each point with its coordinates.

a. (3, −5)   b. (8, 0)
c. (−6, 2)   d. (0, 4)
e. (−3, −7)   f. (2.5, 2.5)

10. Name the quadrant for each ordered pair.

a. (−9, 6)   b. (−7, −2)
c. (3, 11)   d. (0, 12)
e. (5, −14)

Extended-Response Questions

11. Graph these points and connect them with a line segment: $P(5, 4)$ and $Q(5, −3)$.

a. Give the coordinates of two points R and S that are on a line perpendicular to $\overline{PQ}$. Graph $\overline{RS}$.
b. Give the coordinates of two points V and W that are on a line parallel to $\overline{RS}$. Graph $\overline{VW}$.

12. Write the letter and coordinates of the points graphed that meet each condition given.

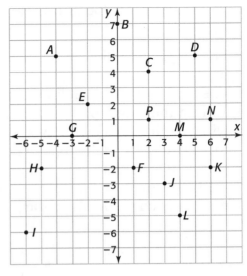

a. The y-coordinate is greater than the x-coordinate.
b. The x-coordinate is the opposite of the y-coordinate.
c. The x-coordinate and the y-coordinate are equal.

13. a. $P(x, y)$ is in Quadrant II. In which quadrant is $Q(−x, −y)$?
b. $S(x, y)$ is in Quadrant I. In which quadrant is $T(−x, −y)$?
c. $M(x, y)$ is in Quadrant III. In which quadrant is $N(−x, y)$?

14. a. Which point does NOT fit the pattern? Explain why the point does not fit and change the coordinates so that it does.
$A(1, −1), B(2, −4), C(3, −7),$
$D(4, −11), E(5, −13), F(6, −16)$
b. Give the ordered pairs for two more points that fit the pattern.

9.2 Distance Between Two Points on a Coordinate Plane

The distance between two points is the length of the line segment that has these points as endpoints. The distance between two points is always a positive number.

Line segments on the coordinate plane are horizontal, vertical, or diagonal.

- The length of a horizontal line segment = |difference of x-coordinates|.
- The length of a vertical line segment = |difference of y-coordinates|.
- To find the length of a diagonal line segment, form a right triangle with the diagonal as its hypotenuse. Then, find the coordinates of the vertex of the right angle. The legs are horizontal and vertical line segments, so their lengths can be found using the rules above. Use the Pythagorean Theorem to find the length of the diagonal line segment that corresponds to the hypotenuse of the right triangle.

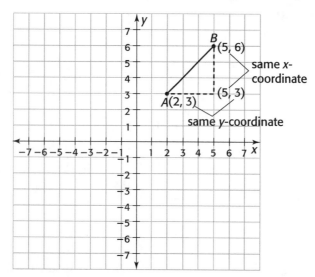

 **Model Problems**

1. Find the distance between each pair of points.

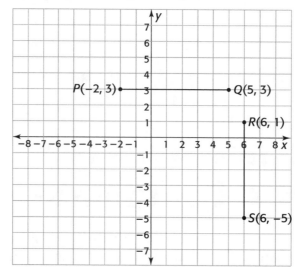

a. $P(-2, 3)$ and $Q(5, 3)$
b. $R(6, 1)$ and $S(6, -5)$

Solution

a. The line segment is horizontal, so:

length = |difference of x-coordinates|

$= |5 - (-2)| = |5 + 2| = |7| = 7$ units

b. The line segment is vertical, so:

length = |difference of y-coordinates|

$= |-5 - 1| = |-6| = 6$ units

You can verify both lengths by counting units on the graph.

2. Find the distance between $A(1, 4)$ and $B(-3, 1)$.

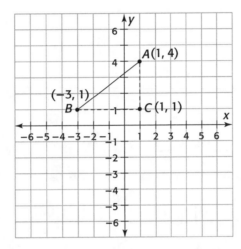

Solution The line segment is diagonal, so you must form right triangle ABC with $C(1, 1)$ and use the Pythagorean Theorem.

$\overline{AC}$ = length of vertical leg = |difference of y-coordinates|

$\overline{AC} = |4 - 1| = |3| = 3$

$\overline{BC}$ = length of horizontal leg = |difference of x-coordinates|

$\overline{BC} = |1 - (-3)| = |1 + 3| = |4| = 4$

Use the Pythagorean Theorem:

$c^2 = 4^2 + 3^2 = 16 + 9$

$c^2 = 25$

$c = 5$

Answer The length $\overline{AB}$ of is 5 units.

3. Graph the following points: $D(5, 2)$, $E(2, 6)$, and $F(-1, 2)$. Then draw $\triangle DEF$ and find its area.

Solution

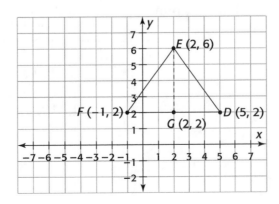

To find the area, calculate the length of the base $\overline{DF}$. Draw $\overline{EG}$ perpendicular to $\overline{DF}$ and find the height of the triangle.

$$\overline{DF} = |5 - (-1)| = |5 + 1| = |6| = 6$$
$$\overline{EG} = |6 - 2| = |4| = 4$$
$$A = \tfrac{1}{2}bh = \tfrac{1}{2}(\overline{DF})(\overline{EF}) \quad \text{Substitute in } \overline{DF} \text{ and } \overline{EG} \text{ for the base and height.}$$
$$A = \tfrac{1}{2} \times 6 \times 4 = 12$$

Answer The area of $\triangle DEF$ is 12 square units.

Multiple-Choice Questions

1. The distance between which pair of points is 7 units?

   A. $(0, 7)$ and $(7, 0)$
   B. $(3, 4)$ and $(3, -3)$
   C. $(-2, -5)$ and $(4, -5)$
   D. $(1, -7)$ and $(1, 7)$

2. What is the distance between $(2, 14)$ and $(7, 2)$?

   F. 21 units
   G. 13 units
   H. 12 units
   J. 9 units

3. What is the area of the figure formed when the points $J(-1, -2)$, $K(-1, -6)$, $L(-5, -6)$, and $M(-5, -2)$ are graphed and connected in order?

   A. 16 units
   B. 20 units
   C. 24 units
   D. 32 units

4. Diego graphed these points: $Q(-3, 2)$, $R(3, 2)$, and $S(3, -2)$. Which point must be graphed to complete a rectangle?

   F. $T(2, 3)$
   G. $T(-2, -3)$
   H. $T(-3, -2)$
   J. $T(-2, 3)$

5. What kind of polygon is formed when these points are graphed and connected in order?
   $W(0, 3)$, $X(5, 3)$, $Y(6, -4)$, $Z(-4, -4)$

   A. rectangle
   B. rhombus
   C. parallelogram
   D. trapezoid

6. What is the area of the polygon formed when these points are graphed and connected in order?
   $J(-8, 7)$, $K(9, 7)$, $L(9, -5)$

   F. 204 square units
   G. 136 square units
   H. 102 square units
   J. 85 square units

7. What is the length of the diagonal of the figure formed when these points are graphed and connected in order? $P(-2, 4)$, $Q(4, 4)$, $R(4, -4)$, $S(-2, -4)$

   A. 7 units
   B. 10 units
   C. 12 units
   D. 15 units

8. The distance from (9, 5) to the origin is

   F. $\sqrt{14}$ units
   G. 14 units
   H. $\sqrt{56}$ units
   J. $\sqrt{106}$ units

Short-Response Questions

9. Find the distance between each pair of points.

   a. (3, −2) and (−11, −2)
   b. (−6, 15) and (−6, −6)
   c. (3, 4) and (10, 28)

10. The distance between $(x, 7)$ and (5, 11) is 5 units. What is the value of x?

11. The distance between (−4, −1) and $(11, y)$ is 17 units. What is the value of y?

Extended-Response Questions

12. a. Graph points $A(-2, 3)$, $B(3, 3)$, $C(5, -2)$, and $D(0, -2)$.
    b. Identify the type of quadrilateral.
    c. Find the area of $ABCD$.

13. a. Give the coordinates for a set of points that form a square with an area of 100 square units. One of the points must be $W(-5, 3)$.
    b. Find the length of the diagonal of the square to the nearest tenth.

14. a. Graph these points: $P(6, 0)$, $Q(3, 6)$, $R(-3, 6)$, $S(-6, 0)$, $T(-3, -6)$, and $U(3, -6)$.
    b. Identify the polygon formed.
    c. Find the lengths of $\overline{QR}$ and $\overline{QP}$. Round to the nearest tenth if necessary.
    d. Is the polygon regular? Explain.
    e. Explain how you could find the area of $PQRSTU$. Carry out your plan, showing all steps.

9.3 Translations

A **transformation** is a way of moving a geometric figure without changing its size or shape. The figure that results after the move is called the **image** of the original figure. For each point of the original figure, there is a corresponding point of the image.

Imagine sliding a chair across the floor so that each leg moves the same distance in the same direction. This is an example of a translation.

A **translation** (or slide) moves every point of a figure the same distance in the same direction. Triangle $A'B'C'$ is the translation image of triangle ABC. $\overline{AA'} = \overline{BB'} = \overline{CC'}$ and the two triangles are congruent and have the same orientation.

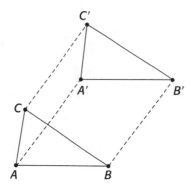

In the figure below, $PQRS$ is translated by moving every point 5 units to the right and 4 units down.

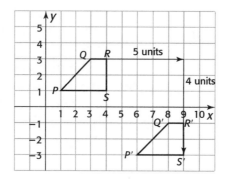

To find the corresponding vertices of the image:
add 5 to each x-coordinate.
add -4 to each y-coordinate.
$P(1, 1) \rightarrow P'(6, -3)$
$Q(3, 3) \rightarrow Q'(8, -1)$
$R(4, 3) \rightarrow R'(9, -1)$
$S(4, 1) \rightarrow S'(9, -3)$

Finding the coordinates of a translation image

Under a translation of a units in the horizontal direction and b units in the vertical direction, the image of $P(x, y)$ is $P'(x + a, y + b)$.

Model Problems

1. Graph the image of $\triangle XYZ$ with vertices $X(3, 3)$, $Y(4, 0)$, and $Z(1, -1)$ after a translation 6 units left and 3 units up.

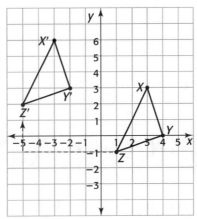

Solution Add -6 to the x-coordinate of each vertex. Add 3 to the y-coordinate of each vertex.

$X(3, 3) \rightarrow X'(3 + -6, 3 + 3) \rightarrow X'(-3, 6)$
$Y(4, 0) \rightarrow Y'(4 + -6, 0 + 3) \rightarrow Y'(-2, 3)$
$Z(1, -1) \rightarrow Z'(1 + -6, -1 + 3) \rightarrow Z'(-5, 2)$

2. The coordinates of $ABCD$ are $A(2, 2)$, $B(4, 2)$, $C(4, -2)$, and $D(2, -1)$. After a translation, the image of A is $A'(6, -5)$. Find the coordinates of B', C', and D' after this same translation and graph $A'B'C'D'$.

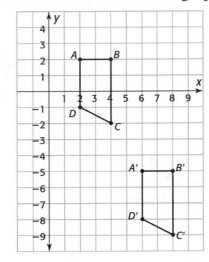

Solution Since $A(2, 2) \rightarrow A'(6, -5)$, you can find the numbers that were added to each coordinate.

$2 + a = 6$, so $a = 4$

$2 + b = -5$, so $b = 7$

Add 4 to each x-coordinate and -7 to each y-coordinate of the other vertices.

$B(4, 2) \rightarrow B'(4 + 4, 2 + (-7)) \rightarrow B'(8, -5)$

$C(4, -2) \rightarrow C'(4 + 4, -2 + (-7)) \rightarrow C'(8, -9)$

$D(2, -1) \rightarrow D'(2 + 4, -1 + (-7)) \rightarrow D'(6, -8)$

Practice

Multiple-Choice Questions

1. $\triangle F'G'H'$ is a translation of $\triangle FGH$

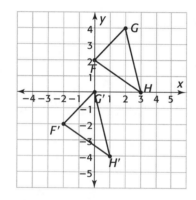

  A. right 2 units, up 4 units
  B. left 4 units, down 3 units
  C. left 2 units, down 4 units
  D. right 3 units, up 5 units

2. The coordinates of *WXYZ* are $W(-6, -1)$, $X(-5, 3)$, $Y(-2, 3)$, and $Z(-3, 1)$. After a translation 8 units right and 3 units down, the coordinates of the image are

  F. $W'(-14, -4)$, $X'(-13, 0)$, $Y'(-10, 0)$, $Z'(-11, -2)$
  G. $W'(2, -4)$, $X'(3, 0)$, $Y'(6, 0)$, $Z'(5, -2)$
  H. $W'(2, 2)$, $X'(3, 6)$, $Y'(6, 6)$, $Z'(5, 4)$
  J. $W'(1, -4)$, $X'(2, 1)$, $Y'(6, -1)$, $Z'(-5, 2)$

3. There are twelve congruent plates. After a translation, the image of plate 1 is plate 8. After the same translation, what is the image of plate 5?

  A. plate 9
  B. plate 10
  C. plate 11
  D. plate 12

4. Which pair of figures shows a translation?

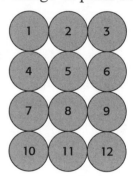

  F.
  G.
  H.
  J.

5. Which of the numbered figures are translations of the shaded figure?

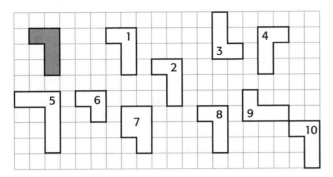

A. 5 only
B. 1, 2, 4, and 8 only
C. 2, 6, 8, and 10 only
D. 1, 2, 8, and 10 only

Short-Response Questions

For 6 and 7, copy each figure onto graph paper. Then graph the image of each figure after a translation 6 units to the right and 3 units down.

6.

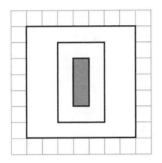

7.

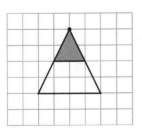

8. Copy the figure shown onto graph paper. Graph the image of the figure on the same set of axes after each translation.

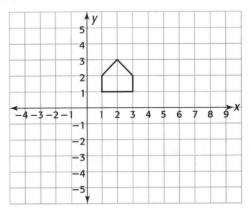

a. 5 units right and 3 units down
b. 4 units left and 2 units up
c. 2 units right and 6 units down

9. △A′B′C′ is the image of △ABC after a translation 7 units left and 3 units up. Graph △ABC before the translation.

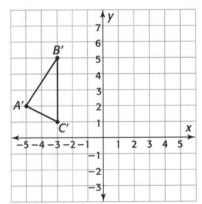

Extended-Response Questions

10. The coordinates of MNOP are M(0, 6), N(2, 6), O(7, 2), and P(2, 2).

a. Give the coordinates of the image after a translation 4 units left and 3 units up.
b. Graph MNOP and M′N′O′P′.

11. The coordinates of △*ABC* are *A*(2, 3), *B*(6, 6), and *C*(7, 2). After a translation, the image of vertex *A* is *A'*(−6, 1).

    a. Give the coordinates of *B'* and *C'* after the same translation.

    b. Graph △*ABC* and △*A'B'C'*.

12. The coordinates of *DEFG* are *D*(2, −1), *E*(5, −2), *F*(4, −5), and *G*(1, −4).

    a. Describe a translation that will move vertex *E* to the origin.

    b. Give the coordinates of *D'*, *E'*, *F'*, and *G'* after the translation described in part a.

    c. Graph *DEFG* and *D'E'F'G'*.

9.4 Reflections and Symmetry

A **reflection** is a transformation in which a figure is flipped or reflected over a **line of reflection** to produce a mirror image. The figure and its image are congruent, but have opposite orientations. Each point and its image are the same distance from the line of reflection.

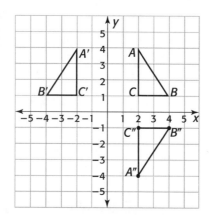

- After reflection in the *y*-axis, the image of *P*(*x*, *y*) is *P'*(−*x*, *y*).
  △*A'B'C'* is the reflection of △*ABC* in the *y*-axis.
  Each *x*-coordinate of the image is multiplied by −1.
  A(2, 4) → *A'*(−2, 4)
  B(4, 1) → *B'*(−4, 1)
  C(2, 1) → *C'*(−2, 1)

- After a reflection in the *x*-axis, the image of *P*(*x*, *y*) is *P"*(*x*, −*y*).
  △*A"B"C"* is the reflection of △*ABC* in the *y*-axis.
  Each *y*-coordinate of the image is multiplied by −1.
  A(2, 4) → *A"*(2, −4)
  B(4, 1) → *B"*(4, −1)
  C(2, 1) → *C"*(2, −1)

A figure has **line symmetry** if it is possible to draw a line that cuts the figure into two parts such that one part is a mirror image of the other. A figure may have one, none, or several lines of symmetry.

$\overline{GH}$ and $\overline{FC}$ are lines of symmetry for hexagon $ABCDEF$.

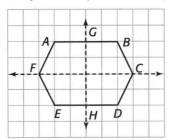

A figure has **point symmetry** if for every point in the figure, there is another point at the same distance from the center on the opposite side. The center is the midpoint of the line segment joining the pair of points.

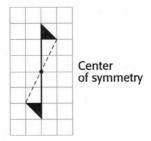

Center of symmetry

Figures may have only line symmetry, only point symmetry, or both, or no symmetry at all.

Model Problems

1. Find the image of $\triangle DEF$ with vertices $D(1, -2)$, $E(4, 2)$, and $F(6, -3)$ after a reflection:
   a. in the y-axis
   b. in the x-axis

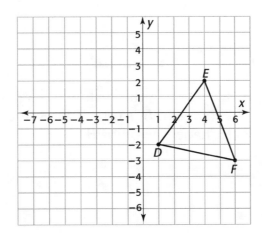

Solution

a. For a reflection in the *y*-axis, each *x*-coordinate is multiplied by –1.

$D(1, -2) \rightarrow D'(-1, -2)$
$E(4, 2) \rightarrow E'(-4, 2)$
$F(6, -3) \rightarrow F'(-6, -3)$

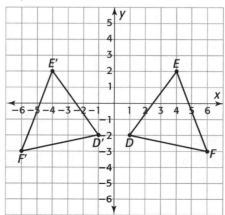

b. For a reflection in the *x*-axis, each *y*-coordinate is multiplied by –1.

$D(1, -2) \rightarrow D''(1, 2)$
$E(4, 2) \rightarrow E''(4, -2)$
$F(6, -3) \rightarrow F''(6, 3)$

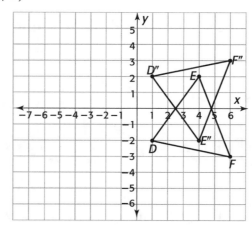

2. Sketch the image of △*ABC* after a reflection in point *A*.

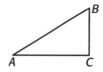

Solution Use a ruler to measure the distance from *B* to *A*. Since *A* is the center of reflection, the image of *B* is on $\overline{AB}$ the same distance as *B* from *A*, but on the opposite side. Label the point *B'*.

Repeat the measuring to locate *C′*, placing the ruler along $\overline{AC}$. Draw △*A′B′C′* the image of △*ABC*.

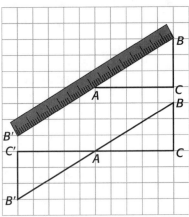

3. Determine what kind of symmetry *KLMN* has.

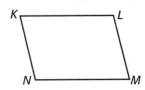

Solution

The parallelogram does not have line symmetry. None of the lines drawn allow the parallelogram to be folded so that points on one side of the line will coincide with points on the other side.

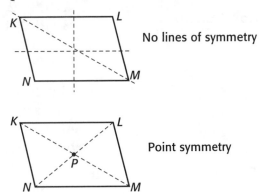

Answer   *KLMN* has point symmetry with *P*, the intersection of its diagonals, as the center of symmetry.

Practice

Multiple-Choice Questions

1. How many lines of symmetry does the figure have?

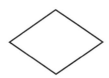

A. 0   B. 1
C. 2   D. 3

2. Which figure does NOT have point symmetry?

F.   G.

H.   J.

3. Describe how $\triangle XYZ$ was transformed to produce $\triangle X'Y'Z'$.

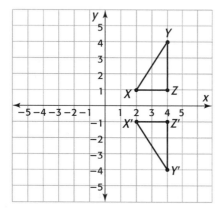

A. reflection in y-axis
B. translation 8 units down
C. reflection in x-axis
D. translation 3 units down

4. Describe how *KLMN* was transformed to produce *K'L'M'N'*.

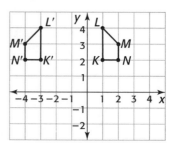

F. reflection in the y-axis
G. translation left 5 units
H. reflection in the y-axis, then translation right 2 units
J. translation right 2 units, then reflection in the y-axis

5. The coordinates of the vertices of $\triangle PQR$ are $P(2, 1)$, $Q(1, 4)$, and $R(-2, 2)$. After a reflection in the x-axis, the coordinates of the vertices are

A. $P'(-2, 1)$, $Q'(-1, 4)$, $R'(2, 2)$
B. $P'(0, 1)$, $Q'(0, 4)$, $R'(0, 2)$
C. $P'(2, -1)$, $Q'(1, -4)$, $R'(-2, -2)$
D. $P'(-2, -1)$, $Q'(-1, -4)$, $R'(2, -2)$

6. The coordinates of the endpoints of a line segment are $S(-4, 7)$ and $T(3, -8)$. After a reflection in the y-axis, followed by a reflection in the x-axis, the coordinates of the image are

F. $S'(4, 7)$, $T'(-3, -8)$
G. $S'(4, -7)$, $T'(3, 8)$
H. $S'(-4, 7)$, $T'(-3, -8)$
J. $S'(4, -7)$, $T'(-3, 8)$

7. Which pair of transformations results in the same image as a translation down 3 units followed by a reflection in the x-axis?

   A. reflection in the x-axis, followed by a translation 3 units up

   B. reflection in the x-axis, followed by a translation 3 units down

   C. translation up 3 units, followed by a reflection in the x-axis

   D. reflection in the y-axis, followed by a translation 3 units left

8. Which figure has both line symmetry and point symmetry?

   F.

   G.

   H.

   J.

Short-Response Questions

9. Copy the figure. Then show how to complete the figure so that $\overline{MN}$ is a line of symmetry.

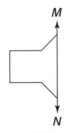

10. Draw a triangle with the given number of lines of symmetry.

   a. 0

   b. 1

   c. 3

11. How many lines of symmetry does each figure have?

   a. square

   b. regular pentagon

   c. regular hexagon

   d. regular polygon of n sides

12. Sketch the image of *JKLM* after a reflection in point *L*.

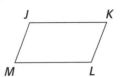

Extended-Response Questions

13. a. Graph the triangle with vertices $D(-5, 2)$, $E(-3, 5)$, and $F(-1, 2)$.

   b. Reflect $\triangle DEF$ in the x-axis. Give the coordinates of D', E', and F'.

   c. Reflect $\triangle D'E'F'$ in the y-axis. Give the coordinates of D'', E'', and F''.

14. Amanda said that all regular polygons have both line symmetry and point symmetry. State whether Amanda is or is not correct. Make drawings to support your conclusion.

15. a. Graph the quadrilateral $A(0, 2)$, $B(1, 1)$, $C(0, -1)$, $D(-1, 1)$.

    b. Reflect $ABCD$ in the y-axis. Describe what you see. Write the coordinates of the vertices of $A'B'C'D'$.

    c. Give a reason for what you observed in part b.

16. a. Draw equilateral triangle ABC.

    b. Draw l, a line of reflection for which the image of A is B.

    c. Draw m, a line of reflection for which the image of A is C.

9.5 Rotations

A **rotation** is a transformation that turns a figure about a point. When rotating a figure, you need:

- a center of rotation about which to rotate the figure.
- a clockwise or counterclockwise direction of rotation.
- a number of degrees of rotation.

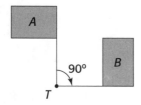

Figure B is the rotation image of figure A. Point T is the center of rotation. Figure A was rotated clockwise 90°. Trace figure A and, without moving the paper, put your pencil point on point T. Turn the paper until figure A matches figure B.

When a figure can be rotated a certain number of degrees about a center point so that the image fits perfectly on top of the original figure, the figure has **rotational symmetry**. Any regular ploygon has rotational symmetry.

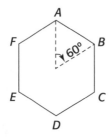

When regular hexagon $ABCDEF$ is rotated $\frac{360°}{6}$ or 60° about its center, the image appears to be in exactly the same position as the original figure. Vertex A has rotated to position B, B to C, C to D, and so on. The hexagon fits over its original position 6 times in the process of a complete rotation (360°).

Model Problems

1. Graph the quadrilateral with vertices $A(5, 4)$, $B(1, 4)$, $C(4, 1)$, and $D(4, 3)$ and its image after a 90° counterclockwise turn about the origin.

   Solution

   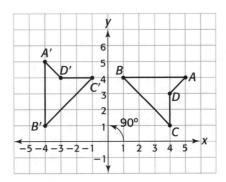

   $A(5, 4) \rightarrow A'(-4, 5)$
   $B(1, 4) \rightarrow B'(-4, 1)$
   $C(4, 1) \rightarrow C'(-1, 4)$
   $D(4, 3) \rightarrow D'(-3, 4)$

2. Give the measure of the smallest angle each figure can be rotated to fit over its original position. Mark the center of rotation.

   a.   b.

   Solution
   a. The parallelogram would fit over itself after a 180° rotation either clockwise or counterclockwise.
   b. The figure would fit over itself after a rotation of $\frac{360°}{3} = 120°$ clockwise or counterclockwise.

   a.   b.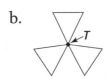

Multiple-Choice Questions

1. Figure *Y* is the image of figure *X*. Identify the transformation.

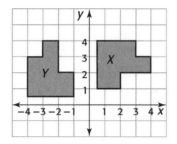

   A. translation left
   B. reflection in the *y*-axis
   C. rotation clockwise 90°
   D. rotation counterclockwise 90°

2. A rotation counterclockwise of 90° is equivalent to

   F. reflection in the *x*-axis
   G. rotation clockwise 270°
   H. reflection in the *y*-axis
   J. translation left and down the same number of units

3. Which is the image of the figure shown after a 90° clockwise rotation about point *P*?

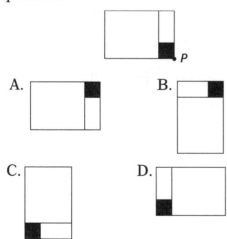

4. Which figure would require a complete turn of 360° to fit over itself?

5. For the figure shown, what is the measure of the smallest angle of rotation about *T* that would allow the image to fit over itself?

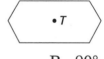

   A. 60°   B. 90°
   C. 120°   D. 180°

6. Which of the following pairs of transformations always brings a figure back to its original position?

   F. reflection in the *x*-axis, then reflection in the *y*-axis
   G. rotation 90° clockwise, translation up
   H. reflection in the *x*-axis, then reflection in the *x*-axis
   J. reflection in the *y*-axis, rotation 90° clockwise

7. $\overline{M'N'}$ is the rotation image of $\overline{MN}$ about *T*. What is the angle of rotation in a clockwise direction?

   A. 90°   B. 180°
   C. 270°   D. 300°

8. The area of $\triangle KLM$ is 32 square units. After a 270° counterclockwise rotation about the origin, the area of image $\triangle K'L'M'$ is

  F. 8 square units
  G. 16 square units
  H. 24 square units
  J. 32 square units

Short-Response Questions

9. Graph the image of the figure shown after a rotation 90° clockwise about (5, 1).

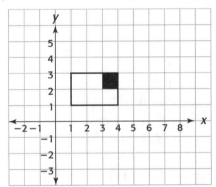

10. Figure *DEFG* has been rotated about *T*. Identify the angle of rotation if the turn was

  a. clockwise   b. counterclockwise

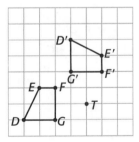

Extended-Response Questions

11. a. Graph the quadrilateral with vertices $A(1, 1)$, $B(4, 1)$, $C(4, 3)$, and $D(2, 3)$.
    b. Graph the image of *ABCD* after a 90° rotation clockwise about the origin.
    c. Graph the image of *ABCD* after a 90° rotation counterclockwise about (1, 1).

12. a. Graph the triangle with vertices $R(1, 2)$, $S(2, 4)$, and $T(4, 1)$.
    b. Graph the image of $\triangle RST$ after a 180° clockwise rotation about the origin. Give the coordinates of R', S', and T'.
    c. Describe another way $\triangle RST$ could have been transformed to produce the same image $\triangle R'S'T'$.

13. What is the image of each of the given points after a rotation 180° clockwise about *T*?

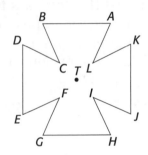

  a. *A*   b. *D*
  c. *F*   d. *K*

14. a. Draw a regular pentagon.
    b. Mark the point about which the figure can be rotated to fit over itself.
    c. Find the measure of the smallest angle of rotation that will allow the figure to fit over itself.

9.6 Constructing and Bisecting Angles

A **construction** is a drawing of a geometric figure that is made using only a compass and an unmarked straightedge. Two of the most fundamental constructions are constructing an angle congruent to a given angle and bisecting an angle (dividing the angle into two congruent parts).

Example 1   Construct an angle congruent to given angle *ABC*.

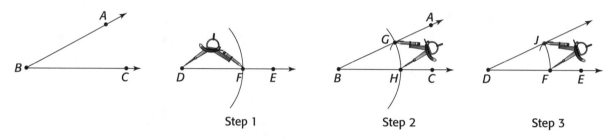

Step 1   Step 2   Step 3

Step 1   Use a straightedge to draw any ray $\overrightarrow{DE}$. Place the compass tip on *D* and draw an arc intersecting $\overrightarrow{DE}$. Label the intersection *F*.

Step 2   Using the same compass width, place the tip on *B*, and draw an arc intersecting both rays. Label the intersections *G* and *H*. Place the tip on *H*. Adjust the compass to draw an arc through *G*.

Step 3   Using the same compass width, place the tip on *F*, and draw a second arc intersecting the first. Label the intersection *J*. Draw *DJ*.

Angle *JDE* is congruent to angle *ABC*. $\angle JDE \cong \angle ABC$.

Example 2   Bisect $\angle XYZ$.

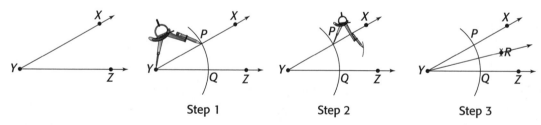

Step 1   Step 2   Step 3

Step 1   With the compass tip on *Y*, draw an arc intersecting both rays. Label the intersections *P* and *Q*.

Step 2   With the compass tip on *P*, draw a second arc inside the angle as shown. (Adjust the compass width if necessary to do this.)

Step 3   Using the same compass width, place the tip on *Q* and draw a third arc as shown. Label the intersection *R*. Draw $\overrightarrow{YR}$.

Ray $\overrightarrow{YR}$ bisects $\angle XYZ$. $\angle XYR \cong \angle RYZ$.

Short-Response Questions

For 1–3, trace each angle, then construct a congruent angle using a compass and straightedge.

1.

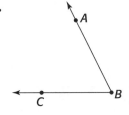

2.

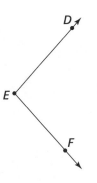

3.

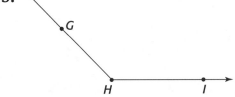

For 4–6, trace each angle, then bisect it using a compass and straightedge.

4.

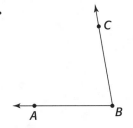

5.

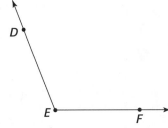

6.

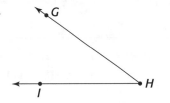

 For 7–9, use a protractor to draw an angle with the given measure. Bisect the angle using a compass and straightedge.

7. 60°

8. 130°

9. 84°

10. Draw a square. Bisect each angle of the square and extend the bisectors. Describe your observations.

Extended-Response Questions

11. a. Draw any obtuse angle. Then use a compass and straightedge to divide the angle into four congruent parts.
   b. Bisect one of the angles that resulted from your work in part a. What is the relationship between the measure of one of the resulting angles and the measure of the original obtuse angle?

12. Draw a parallelogram that is NOT a rectangle. Bisect each angle of the parallelogram and extend the bisectors. Write a description of your observations.

9.7 Constructing Perpendicular Lines

The **midpoint** of a line segment is the point that separates it into two congruent line segments. The **perpendicular bisector** of a line segment is a line, ray, or line segment that is perpendicular to a line segment at its midpoint.

An **altitude** of a triangle is a line segment from a vertex of the triangle perpendicular to the opposite side or to a line containing that side.

Example 1   Construct the perpendicular bisector of $\overline{AB}$.

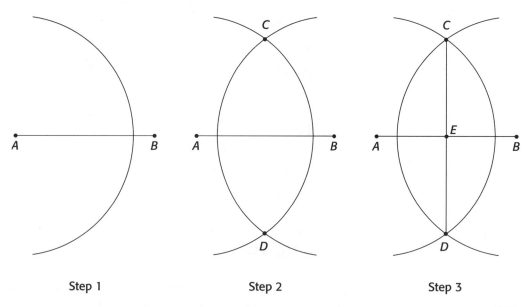

| Step 1 | Step 2 | Step 3 |

Step 1   Open the compass a little more than half the length of $\overline{AB}$. With the compass tip on A, draw an arc intersecting $\overline{AB}$.

Step 2   Use the same compass width. With the tip on B, draw an arc intersecting $\overline{AB}$. Label the points of intersection C and D.

Step 3   Draw $\overleftrightarrow{CD}$ using a straightedge. Label point E.

$\overleftrightarrow{CD}$ bisects $\overline{AB}$. $\overline{AE} \cong \overline{EB}$ and E is the midpoint of $\overline{AB}$.

$\overleftrightarrow{CD}$ is perpendicular to $\overline{AB}$. $\overleftrightarrow{CD} \perp \overline{AB}$.

$\overleftrightarrow{CD}$ is the perpendicular bisector of $\overline{AB}$.

Example 2 Construct a perpendicular to $\overleftrightarrow{RS}$ from point P not on $\overleftrightarrow{RS}$.

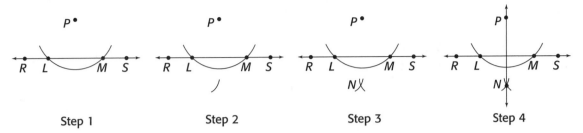

Step 1   Step 2   Step 3   Step 4

Step 1 With the tip of the compass at P, draw an arc that intersects $\overleftrightarrow{RS}$ at L and M.

Step 2 With the compass tip at L, open the compass a little more than half the length of $\overline{LM}$. Draw an arc below $\overleftrightarrow{RS}$.

Step 3 With the compass tip at M and the same width used in Step 2, draw a second arc below $\overleftrightarrow{RS}$ that intersects the arc drawn in Step 2. Label the point of intersection N.

Step 4 Draw $\overleftrightarrow{PN}$. $\overleftrightarrow{PN}$ is perpendicular to $\overleftrightarrow{RS}$. $\overleftrightarrow{PN} \perp \overleftrightarrow{RS}$.

Practice

Short-Response Questions

For 1–3, trace each line segment, then construct its perpendicular bisector.

1.

A————————B

2.

C ... D

3.

E —— F

4. Use a ruler to draw a line segment of length 7 cm. Construct its perpendicular bisector using a compass and straightedge.

For 5–7, trace each figure, then construct a line from point P that is perpendicular to the given line.

5.

P

G   H

6.

•P

I

J

7.

•P

K   L

8. Draw a line, $\overleftrightarrow{XY}$, and a point K not on the line. Construct a line perpendicular to $\overleftrightarrow{XY}$ through K.

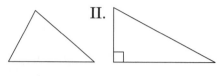

For 9 and 10, use the steps of Example 2 to construct an altitude for each triangle from point B to $\overline{AC}$.

9.

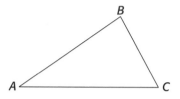

10.

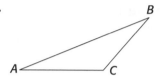

12. Trace each triangle and carry out the following steps for each.

I.

II.

III.

a. Bisect each side of the triangle. The bisectors will meet at a point. Label this point P.
b. Measure the distance from point P to each vertex of the triangle corresponding to P. For each triangle, what do you observe about the measurements?
c. How is the position of the intersection point of the perpendicular bisectors of the sides related to the type of triangle?

Extended-Response Questions

11. Explain how you could use the methods of this section to construct a pair of parallel lines. Complete the construction, showing each step.

Chapter 9 Review

Multiple-Choice Questions

1. The points $P(0, 2)$, $Q(0, -3)$, $R(5, 2)$, and $S(5, -3)$ are graphed on a coordinate plane and the points are connected in order. What is the area of $PQRS$?

  A. 5 square units
  B. 20 square units
  C. 25 square units
  D. 30 square units

2. Which figure is next in the pattern?

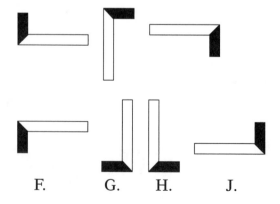

    F.   G.   H.   J.

3. The figure shown is reflected in the *x*-axis. The image is

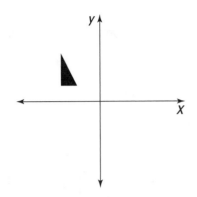

A.

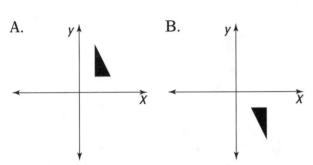

B.

C.

D.

4. What is the distance between points *M* and *N*?

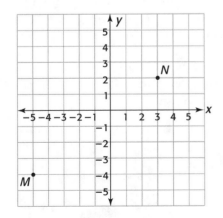

F. 8 units
G. 10 units
H. 14 units
J. 16 units

5. Which figure has exactly 2 lines of symmetry?

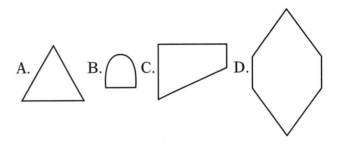

A.   B.   C.   D.

6. Which figure does NOT have point symmetry?

F.   G.   H.   J.

7. For which of the points graphed is the *x*-coordinate less than the *y*-coordinate?

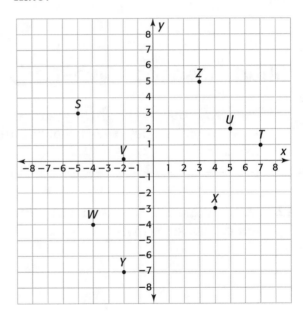

A. *S*, *V*, and *Z* only
B. *S*, *W*, and *X* only
C. *T*, *U*, and *Y* only
D. *V*, *W*, and *Z* only

8. When the *x*-coordinate is positive and the *y*-coordinate is negative, the ordered pair locates a point in which quadrant?

F. I   G. II
H. III   J. IV

9. A point in Quadrant I is reflected in the origin. The image of the point is in which quadrant?

A. I B. II
C. III D. IV

10. $\overline{PQ}$ is translated so that Q' is at the origin. The coordinates of P' are

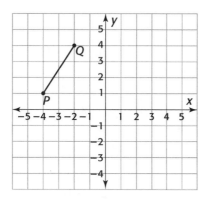

F. $(0, -2)$ G. $(-2, -3)$
H. $(-8, -1)$ J. $(4, -2)$

Short-Response Questions

11. Trace $\overline{AB}$. Use a straightedge and compass to construct the perpendicular bisector.

A •————————————• B

12. Trace angle DEF. Use a straightedge and compass to bisect it.

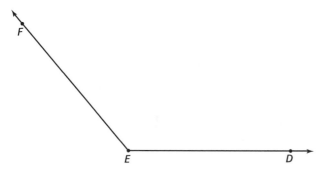

13. Line l contains the points $L(-3, 2)$ and $M(3, 2)$. Line t contains the points $S(-4, -5)$ and $T(4, -5)$.

a. Graph the lines.
b. How are the lines related?

Extended-Response Questions

14. Graph the image of figure S after each transformation.

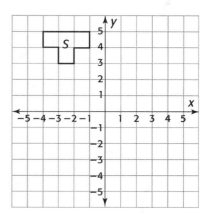

a. translation right 4 units and down 2 units
b. reflection in the x-axis
c. rotation 180° clockwise about the origin

15. A square has a diagonal with endpoints $G(-2, -3)$ and $E(4, 3)$.

a. Graph the square and give the coordinates of its other vertices, D and F.
b. Find the perimeter of the square.

16. An ant was crawling on the lines of a coordinate plane. The ant started at $(-5, 6)$ and went to $(-2, 6)$, turned and went to $(-2, 4)$, then to $(1, 4)$, then to $(1, 0)$, then to $(4, 0)$, then to $(4, -2)$, then to $(0, -2)$, then to $(0, -4)$, and finally to $(3, -4)$.

a. Graph the ant's path.
b. What was the total length of the ant's journey?

17. a. Graph the quadrilateral with vertices $A(3, 1)$, $B(5, 3)$, $C(7, 3)$, and $D(8, 1)$.

b. Graph the image of $ABCD$ after reflection in the line containing $\overline{AB}$. Give the coordinates of each vertex of $A'B'C'D'$.

18. Tell if the capital letter shown has:

a. line symmetry. If so, draw all lines of symmetry.

b. point symmetry. If so, identify the center of symmetry.

c. rotational symmetry. If so, identify the measure of the smallest angle that will rotate the figure to fit over itself.

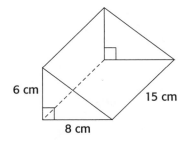

19. Find the coordinates of two points that are the same distance from the origin, but NOT on either axis. Give the distance. Show all work.

20. Examine the dot pattern for the first four triangular numbers.

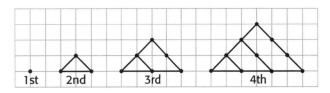

a. Draw the fifth triangular number on a coordinate plane. Place one vertex point at $(0, 0)$. Give the coordinates of the other two vertex points.

b. How many points make up the fifth number?

Chapter 9 Cumulative Review

Multiple-Choice Questions

1. What is the surface area of a cube that has side lengths of 2.2 cm?

A. 6.6 cm²   B. 10.648 cm²
C. 14.52 cm²   D. 29.04 cm²

2. What is the volume of the figure shown?

F. 258 cm³   G. 360 cm³
H. 516 cm³   J. 720 cm³

3. Which is equivalent to 40,000 cm?

A. 4 km   B. 4000 cm
C. 0.4 km   D. 40 m

4. Which measure is the greatest?

F. $4\frac{1}{2}$ gal   G. 74 c

H. $19\frac{1}{2}$ qt   J. 600 fl oz

5. What does this construction show?

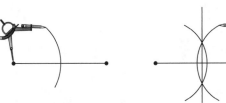

A. congruent segments
B. perpendicular bisector
C. bisected angle
D. parallel lines

6. What is 175% of 180?

    F. 315
    G. 355
    H. 405
    J. 450

7. $\triangle YES$ is similar to $\triangle HOW$. The perimeter of $\triangle HOW$ is

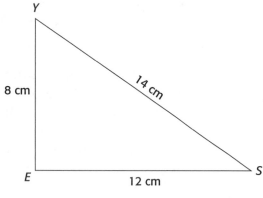

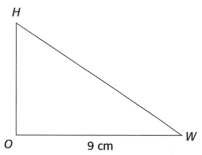

    A. 24 cm
    B. 25.5 cm
    C. 26 cm
    D. 28.5 cm

8. When a number is divided by 10, the result is $\frac{4}{25}$. Find the number.

    F. $\frac{2}{125}$

    G. $\frac{4}{5}$

    H. $1\frac{3}{5}$

    J. $2\frac{3}{10}$

9. Use the price list shown. How much change should you get from $10 if you buy 2 pens, 1 ruler, and 4 folders?

Sam's Supplies (prices include tax)	
Pens	5 for $1.95
Rulers	4 for $2.88
Folders	3 for $3.66
Notebooks	2 for $1.98

    A. $3.62   B. $4.54
    C. $4.84   D. $4.92

10. Find the value of the expression when $x = 8$ and $y = 10$.
$$2x + 3y \div 6 + (x - 5)^2$$

    F. $16\frac{1}{2}$ G. 18

    H. 27   J. 30

Short-Response Questions

11. Find all the whole-number factors less than 11 for the number 6,290,124.

12. The graph shown represents Jay's walk from his home to his friend's house 1 mile away. Write a brief description of Jay's walk that would match the graph.

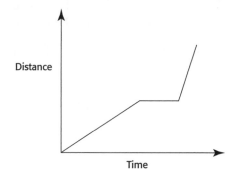

13. How long should an escalator in a new mall be if it is to make an angle measuring 28° with the floor and carry customers a vertical distance of 30 feet between floors? Show your work. Round your answer to the nearest tenth.

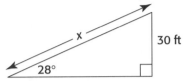

Extended-Response Questions

14. A tractor wheel, including the tire, has a radius of length 27.5 inches.

    a. Find the circumference of the wheel. Use 3.14 for π.

    b. How many times does the wheel go around if the tractor travels 1 mile? Round your answer to the nearest whole number. Show your work.

15. The bar graph shows the number of high school students living in four different towns.

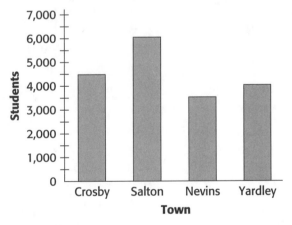

Number of High School Students (2001)

    a. Approximately what percent of the total number of students live in Salton?

b. If the number of high school students living in Yardley is predicted to increase by 15% in 2002, about how many students will be expected to live there?

c. If Crosby expects to have only about 3,800 students in 2002, what is the percent change from 2001 to 2002? Round to the nearest tenth.

16. Five vehicles are in line at a tollbooth. The minivan is paying its toll. The motorcycle is two places behind the truck. The bus is ahead of the sports car, which is fifth in line. Which vehicle will pay its toll next?

17. Points *A* through *J* on the number line represent the following numbers:

$$-4.5,\ 1.8,\ -\pi,\ 2\frac{1}{3},\ 0,\ \sqrt{2},\ \frac{7}{2},\ 2,\ -2\frac{4}{5},\ \sqrt{19}$$

    Point *E* represents 0 and point *G* represents 2.

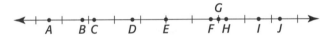

    a. Write the number to which each point corresponds.

    b. Name the rational numbers.

    c. Name the irrational numbers.

    d. Give a rational number that would be graphed between *G* and *H*.

18. a. How many equilateral triangles of all sizes are there in this equilateral triangle of length 3?

    b. How many would there be in an equilateral triangle of length 4?

19. Mr. Wu begins work at 7:30 A.M. and leaves at 3:15 P.M. He works Monday through Friday and takes an unpaid lunch break of $\frac{1}{2}$ hour each day. He is paid for holidays and sick days. His hourly wage is $15.20. How much does Mr. Wu earn on a yearly basis? Show your work.

20. a. Graph the reflection of $\triangle ABC$ in the origin.
   b. Give the coordinates of the vertices of the image.

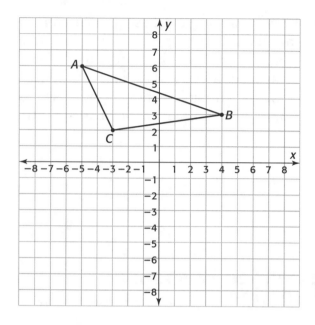

Data Analysis and Probability

Chapter Vocabulary

statistics
line graph
intervals
range
median
box-and-whisker plot
outcomes
independent events
Counting Principle
experimental probability
relative frequency

data
frequency
histogram
measures of central
  tendency
mode
probability
experiment
dependent events
permutation
random sampling

bar graph
frequency table
cumulative frequency
mean
stem-and-leaf plot
sample space
favorable outcomes
compound event
combination
population
theoretical probability
cumulative relative
  frequency

10.1 Displaying Data with Graphs

Statistics is the branch of mathematics that deals with the collection and organization of numerical facts, or **data**, in order to analyze them and draw conclusions. Data are often organized using tables and represented visually with graphs. Different kinds of graphs are used for different purposes.

A **bar graph** is used to show comparisons. The data are presented using bars of different lengths, which correspond to the numbers they represent. The numbers are determined by the scale marked on the horizontal or vertical axis.

The bar graph below shows airfares between New York City and some popular destinations.

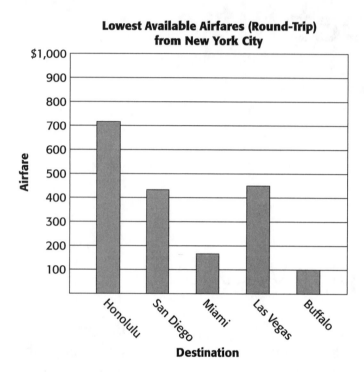

A **line graph** is used to show trends or changes over time. The line graph below shows the number of tornadoes reported, decade by decade. The jagged line indicates that part of the scale is omitted.

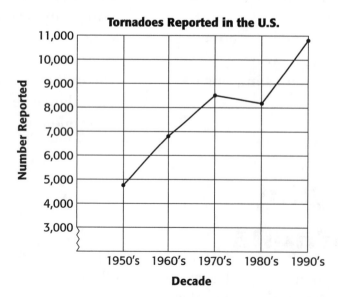

A line graph is read from left to right.

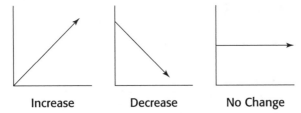

Increase   Decrease   No Change

Although graphs cannot always present data as accurately as tables can, they make it easy to grasp relationships at a glance.

Model Problems

1. Refer back to the graph showing airfares.
   a. About how much more does it cost to fly from New York to Honolulu than to Miami?
   b. Which two destinations have about the same airfares?

   Solution
   a. The cost for Honolulu is about $700 and the cost for Miami is about $170.
      $700 − $170 = $530
   b. Look for bars that are about equal in height.
      San Diego and Las Vegas have about the same airfare.

2. Refer back to the graph showing tornadoes reported.
   a. About how many tornadoes were reported in the 1950's?
   b. During which period did the number of tornadoes reported change the least?
   c. Describe the general trend shown by the graph.

   Solution
   a. The point for the 1950's is a little more than three-fourths of the way between 4,000 and 5,000 on the vertical scale. A good estimate would be about 4,800 tornadoes reported.
   b. Since a horizontal line indicates that there is no change, look for a line segment that is closest to horizontal. The number changed the least during the 1970's and 1980's.
   c. With the exception of the small decrease during the 1970's and 1980's, the graph shows that the number of tornadoes being reported is increasing each decade.

Practice

For 1–6, display each set of data with a bar graph or a line graph. Give a reason for your choice. It may be necessary to use rounded numbers for some graphs.

1.

Daily Circulation of Newspapers	
Newspaper	Circulation (in thousands)
The New York Times	1,160
The New York Daily News	716
The Wall Street Journal	1,820
USA Today	1,853
The New York Post	487

2.

Number of Cable TV Systems	
Year	Number
1978	3,875
1983	5,600
1988	8,500
1993	11,108
1998	10,845

3.

World Weather in May (°F)	
City	Average High/Low
New York	68/53
Hong Kong	82/74
Cairo	91/63
London	62/47
Moscow	66/46
Rome	74/56

4.

The Value of a Dollar Compared to 1978	
Year	Value
1978	$1.00
1980	$0.76
1982	$0.59
1984	$0.55
1986	$0.50
1988	$0.48
1990	$0.43
1992	$0.39
1994	$0.37
1996	$0.35
1998	$0.33

5.

Total Sales for Recording Artists: 1990–1999 (millions of CD's and cassettes)	
Artist	Sales
Garth Brooks	60.0
The Beatles	30.2
Mariah Carey	38.3
N'Sync	34.6
Pearl Jam	22.3
Metallica	33.6

6.

Unemployment Rates in U.S. and Canada (per 100 workers)		
Year	U.S.	Canada
1992	7.5	11.3
1993	6.9	11.2
1994	6.1	10.4
1995	5.6	9.5
1996	5.4	9.7
1997	4.9	9.2

Use the data from the graphs you made for 1–6 to answer questions 7–18.

7. Rank the newspapers by circulation from greatest to least.

8. Estimate the difference between the greatest and least newspaper circulation.

9. During which 5-year period did the number of cable systems show the greatest increase?

   A. 1978–1983   B. 1983–1988
   C. 1988–1993   D. 1993–1998

10. Do you think that when data for 2003 is available, it will show an increase or decrease in the number of cable systems? Give reasons for your answer.

11. How much less was $1.00 worth in 1992 than in 1982?

12. a. Describe the trend in the value of a dollar shown by the graph.
    b. Based on the graph, predict the value of the dollar for 2000 and 2002.

13. Which two recording artists were closest in sales?

14. Which city has the greatest difference between its high and low temperatures?

    A. New York   B. Cairo
    C. Moscow   D. Rome

15. Which two cities have the most similar weather in May? Give reasons for your choice.

16. During which 2-year period did the unemployment rate in Canada show the greatest change?

    A. 1992–1993   B. 1993–1994
    C. 1994–1995   D. 1996–1997

17. In which year was the difference between the rates in the U.S. and Canada the least?

18. Write a paragraph summarizing the conclusions you can draw by looking at the graph of unemployment rates.

10.2 Frequency Tables and Histograms

One way to organize data is to record the number of times, or **frequency**, that each value appears. A tally mark is used each time a particular value appears, and then the marks are counted. The results are shown in a **frequency table**.

Data in a frequency table can be condensed by grouping into **intervals** of equal size. The grouped data can then be displayed in a type of bar graph called a **histogram**.

Model Problems

1. Make a frequency table for the following data on the size of women's shoes sold on one Saturday at ShoeBiz.

   8   6   7   9   7   7   8   10   5   8   9   6
   7   10   11   8   7   6   7   7   8   7   10   8

   Solution   Use the whole numbers from 5 through 11 and count how many times each appears.

Sizes of Shoes Sold		
Size	Tally	Frequency
5	I	1
6	III	3
7	++++ III	8
8	++++ I	6
9	II	2
10	III	3
11	I	1
Total frequency 24		

   Answer   When the data is organized, it is easy to see that size 7 is the most popular.

2. The scores earned by students on a 100-point biology test were as follows:

   83   81   75   72   79   89   86   79   58   95   78   70   74   82   73
   79   84   66   75   90   81   97   71   77   80   68   85   96   93   88

a. Make a frequency table for the scores. Group the data in 5-point intervals.
b. Make a histogram using the grouped data.

Solution

a. Every score must fall into one and only one interval. The lowest score is 58, so begin the first interval at 55.

Interval	Tally	Frequency
55–59	I	1
60–64		0
65–69	II	2
70–74	⊤⊥⊥	5
75–79	⊤⊥⊥ II	7
80–84	⊤⊥⊥ I	6
85–89	IIII	4
90–94	II	2
95–99	III	3
Total frequency 30		

b. Each bar of a histogram represents one interval of a frequency table. There is no space between the bars. The horizontal scale shows the intervals and the vertical scale shows the frequency.

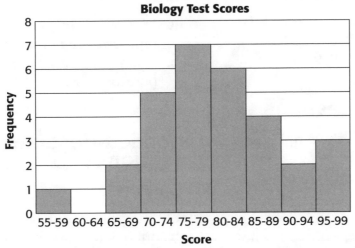

Since no student scored in the interval 60–64, there is no bar on the graph. The number of students who scored at or below a certain number is the **cumulative frequency** for that interval. For example, the number of students who scored 79 or less is the sum of the frequencies for the five intervals.

$1 + 0 + 2 + 5 + 7 = 15$

The cumulative frequency for the interval 75−79 is 15.

Multiple-Choice Questions

For 1–2, use the frequency table showing the dinner times of a survey group.

Dinner Times		
Times (P.M.)	Tally	Frequency
5:00–5:59	IIII	4
6:00–6:59	HHI I	6
7:00–7:59	HHI HHI I	11
8:00–8:59	HHI III	8
9:00–9:59	IIII	4
10:00–10:59	II	2

1. How many people were in the survey group?

   A. 11   B. 25
   C. 35   D. 40

2. Which is the most frequent dinner hour?

   F. 6:00–6:59   G. 7:00–7:59
   H. 8:00–8:59   J. 9:00–9:59

Short-Response Questions

The data gives the age of all the employees working at Romtex, Inc.

  22 31 37 29 40 35 52 26
  24 38 39 44 27 36 27 29
  33 35 28 48 39 36 25 37
  41 50 26 38 29 40 21 33

3. Make a frequency table for the data using intervals of length 5. Begin with the interval 20–24 years.

4. Draw a histogram using the table from question 3.

5. Use the histogram from problem 4 to answer these questions.

  a. How many Romtex employees are under 35 years of age?
  b. To the nearest tenth, what percent of Romtex employees are 40 or older?
  c. What is the cumulative frequency for the interval 35–39?

6. The data below gives the daily high temperature in °F for different cities across the U.S. on one day in April.

  48 53 55 48 50 46 64 70 61
  36 69 51 45 31 48 45 41 38
  50 52 51 54 70 54 39 84 57
  56 63 72 45 48 56 47 38 80
  74 66 61 63

Make a grouped frequency table for the temperatures. Choose an interval length so that you will have between 6 and 10 intervals.

7. Use the frequency table from question 6 to make a histogram.

8. Use the grouped data to answer these questions.

  a. In how many cities was the high temperature between 40° and 69°?
  b. What percent of the cities had temperatures below 40° or above 79°?

Extended-Response Questions

9. Explain why each of the following sets of intervals could NOT be used for a frequency table or histogram.

a.	Interval
	33–45
	46–58
	59–70
	71–73

b.	Interval
	100–115
	115–130
	130–145
	145–160

10. Below are the scores for a particular class on a standardized achievement test. The test is scored on a curve with 200 as the lowest possible score and 800 as the highest possible score.

537	528	379	296	730	541	425
660	612	282	598	655	540	460
436	710	492	584	370	466	390
485	575	395	505	384	590	626
752	608					

Make a grouped frequency table for the scores. Use intervals of length 100 beginning with 200–299.

11. Make a histogram for the test scores in problem 10.

12. Use the frequency table and histogram to answer these questions.

a. In which interval did the greatest number of scores fall?

b. To the nearest tenth, what percent of the students scored between 400 and 599?

c. Standardized tests are structured so that most of the scores fall somewhere in the middle and the extreme low or high scores have about the same frequency. Did this class score as expected? Explain?

10.3 Finding the Range, Mean, Median, and Mode

There are some important statistics that are used to analyze the data collected in a study.

- The **range** is the difference between the highest and lowest values in a set of data. The range tells if the data are closely grouped together or widely spread out.

The next three statistics describe the center of a set of data and are known as **measures of central tendency**.

- The **mean** of a set of data is the arithmetic average. To find the mean, add all the data values and divide by the number of values.
- The **median** is the middle value or the average (mean) of the two middle numbers when the data are arranged in order.
- The **mode** is the data value that appears most often in the set. A data set may have no mode, one mode, or more than one mode.

Model Problems

1. Mike Riley is the star player on the school basketball team. The number of points scored by Mike in his first 10 games this season are:

   30, 23, 34, 31, 34, 28, 33, 25, 38, 24

   Find the range, mean, median, and mode of Mike's scores.

   Solution Arranging the scores in order is useful for several of the calculations.

   23, 24, 25, 28, 30, 31, 33, 34, 34, 38

   range = highest score − lowest score
   range = 38 − 23
   range = 15

   mean = sum of the scores ÷ number of scores
   mean = (23 + 24 +25 + 28 + 30 + 31 + 33 + 34 + 34 + 38) ÷ 10
   mean = 300 ÷ 10
   mean = 30

   There is an even number of scores, so the median is the average of the two middle numbers.

   23, 24, 25, 28, 30, 31, 33, 34, 34, 38
   　　　　　　　　↑　↑ two middle numbers

   $$\text{median} = \frac{30 + 31}{2} = 30.5$$

   The mode is the score that appears most often. Since 34 appears twice and no other score appears two or more times, the mode is 34.

   Answer Mike's statistics: range = 15, mean = 30, median = 30.5, mode = 34.

2. The heights of the players on the basketball team are shown in the frequency table.

Height (inches)	Frequency (number of players)
71	1
72	2
73	3
74	3
75	4
76	0
77	2

a. Find the range, mean, median, and mode of the players' heights.

b. Suppose a new player who is 76 inches tall joins the team. Which of the four statistics above would change and how?

Solution

a. Since the intervals used in the table are of length 1, it is possible to reason with the data as if the individual heights were given.

range = 77 − 71 = 6 in.

To find the mean for the group data, multiply each interval value by its frequency. Add the products, then divide by the total frequency.

$$\text{mean} = \frac{1(71) + 2(72) + 3(73) + 3(74) + 4(75) + 2(77)}{1 + 2 + 3 + 3 + 4 + 2}$$

$$\text{mean} = \frac{1,110}{15} = 74 \text{ in.}$$

To find the median, identify the interval that contains the middle value. Since there are 15 players, the middle value is the 8th. The intervals for 71, 72, and 73 contain a total of 6 players. Therefore, the 8th value occurs in the next interval, 74.

median = 74 in.

To find the mode, identify the interval with the greatest frequency. Since the greatest frequency, 4, appears for a height of 75 in.,

mode = 75 in.

b. A height of 76 in. is not lower than 71 in. or higher than 77 in., so the range of heights would not change.

new range = 77 − 71 in. = 6 in. = original range

A new value different from the original mean will change the mean of the set.

$$\text{new mean} = \frac{1,186}{16} = 74.125 \text{ in.} > 74 \text{ in.}$$

Since the new player's height is greater than the original mean, the new mean is greater than the original.

new mean > original mean

The new median will be the average of the 8th and 9th data values. Since the 8th and 9th values occur in the interval for 74 in., the new median is the same as the original median.

new median = 74 in. = original median

Finally, the change in frequency for the interval for 76 in., from 0 to 1, does not change the interval with the greatest frequency.

new mode = 75 in. = original mode

Multiple-Choice Questions

1. Judy has grades of 91, 73, and 86 on three mathematics tests. What grade must she obtain on the next test to have an average of exactly 85 for all four tests?

A. 85
B. 88
C. 90
D. 97

2. The smallest of five consecutive integers is 42. What are the mean and median of this set of integers?

F. mean = 44, median = 44
G. mean = 43.5, median = 43
H. mean = 44, median = 43.5
J. mean = 44.5, median = 44

3. The table shows the donations collected for a charity. What was the mean donation?

Help All Fund Donations	
Amount	Number
$10	3
$15	6
$25	8
$50	2
$100	1

A. $20
B. $25
C. $26
D. $90

4. For which set of data is there no mode?

F. 12, 11, 23, 11, 29, 15, 16, 9
G. 28, 31, 17, 31, 28, 19, 37, 26
H. 34, 45, 66, 53, 72, 81, 63, 53
J. 41, 44, 47, 42, 58, 61, 45, 51

5. Horatio recorded the number of minutes he spent talking on the telephone each day.

Day	Sun.	Mon.	Tues.	Wed.	Thu.	Fri.	Sat.
Minutes	136	47	38	55	110	24	129

What are his mean and median number of minutes spent talking?

A. mean = 73.5, median = 55
B. mean = 77, median = 82.5
C. mean = 77, median = 55
D. mean = 82.5, median = 47

6. The scores for a science quiz are shown in the table. What are the median and mode(s) of the scores?

Science Quiz Scores	
Score	Number of Students
30	1
40	2
50	4
60	5
70	8
80	11
90	1
100	2

F. median = 68.8, modes = 40 and 100
G. median = 70, mode = 80
H. median = 70, no mode
J. median = 75, mode = 80

7. The mean weight of the first 6 apricots a farmer picked was 1.6 ounces. The next apricot weighed 2.3 ounces. What was the mean weight of all the apricots?

A. 0.7 ounces   B. 1.6 ounces
C. 1.65 ounces   D. 1.7 ounces

8. Andrea had the following bowling scores in five games:

229, 195, 137, 208, 146

Which statement is NOT true?

F. The range of the scores is 92.
G. The median score is higher than the mean score.
H. There is no mode for the scores.
J. A score of 175 in the next game will raise her mean score.

9. Terry had the following test scores: 84, 71, 93, 84, 79, 87. If the teacher offers to erase the lowest score for each student, which of the following will change?

A. the range and mean only
B. the range and median only
C. the range, mean, and mode only
D. the range, mean, median, and mode

10. For which set of data is the mean less than the median?

F. 45, 46, 46, 47
G. 42, 49, 43, 44
H. 47, 49, 40, 48
J. 47, 47, 47, 47

Short-Response Questions

11. Find a set of 5 different numbers such that the mean and median are equal.

12. If the range of a set of numbers is 0, what is true about the numbers in the set?

Extended-Response Questions

13. A travel agent has 7-day vacation packages to Hawaii for the following prices per person: $699, $1,015, $1,255, $1,759, $1,667, and $1,429.

a. Find the range of the prices.
b. Find the mean vacation price.
c. Find the median vacation price.

14. The numbers of season home runs hit by players on a baseball team are shown in the table.

a. Find the mean number of runs.
b. Find the median number of runs.

Player	Number of Runs
Briggs	11
Velez	17
O'Neil	0
Aziz	9
Gold	5
Williams	14
Sanders	2
Lin	8
Trask	15

15. The weekly salaries of ten employees in a small firm are $420, $480, $570, $570, $585, $595, $610, $660, $1,490, and $1,600. For these salaries find

a. the mean
b. the median
c. the mode
d. Is the mean a representative measure of the average salary? Why or why not?

16. A set of data consists of the values 27, 33, 29, 27, x, and 35. Find the possible value of x so that

    a. there is one mode
    b. there are two modes

17. A library received a shipment of new novels containing the following numbers of pages:

277 534 672 404 489 208 274
228 322 592 327 272 327 175
463 321 405 240 208 212

Find the range, mean, median, and mode for the books.

18. A store manager made a frequency table of the sizes of men's shirts sold in one week. Which measure—the mean, median, or mode—is most useful for the manager when ordering new shirts? Why?

Shirts Sold	
Size	**Frequency**
14	2
$14\frac{1}{2}$	5
15	10
$15\frac{1}{2}$	19
16	7
$16\frac{1}{2}$	4
17	1

19. The heights of a group of students are shown in the table.

    a. Find the range, mean, median, and mode of these heights.

    b. A student who had been absent was measured the next day. When the statistics were recomputed, the mean height had increased. What is the shortest the absent student could have been? Explain.

Student Heights	
Height (nearest cm)	**Number of Students**
150	2
151	0
152	1
153	2
154	3
155	1
156	0
157	3
158	2
159	2
160	4

20. The test scores of two students are shown. Use one or more of the statistics discussed in this section to evaluate the students' records. Which student do you think is performing better and why?

	Xavier	Zarah
Test 1	97	78
Test 2	68	80
Test 3	81	80
Test 4	90	84
Test 5	74	88

10.4 Stem-and-Leaf Plots and Box-and-Whisker Plots

Two other useful methods of displaying data are stem-and-leaf plots and box-and-whisker plots.

In a **stem-and-leaf plot**, each data value is separated into two parts. The stems are usually the digits in the greatest common place of each data value. The leaves are the remaining digits of each data value. For example, if the greatest common place value is tens, then the stem for 84 is 8 and the leaf is 4. The advantage of stem-and-leaf plots is that the individual data values can still be identified on the plot.

A **box-and-whisker plot** uses the least and greatest data values along with three special midpoints to represent how the data is spread out.

Model Problems

1. A group of 25 students recorded their pulse rates during a biology experiment. Organize the data into a stem-and-leaf plot.

67 72 86 90 82 78 75 88 75 77 68 84 92
91 85 73 64 76 81 78 74 81 79 67 80

Solution   Since every value is a two-digit number, use tens as the stem. The stems that will be needed are 6, 7, 8, and 9 to include all the data values from least to greatest. List the stems, one under another, with a vertical line to the right.

Stems → 6 | 7 ← Leaves
         7 |
         8 |
         9 |

Write a leaf (the ones digit) for each data value. For example, enter 67 by writing 7 to the right of the vertical line after stem 6.

Rearrange the leaves if necessary so they are in increasing order. Write a key for the stem-and-leaf plot and a title.

Student Pulse Rates

6	4 7 7 8
7	2 3 4 5 5 6 7 8 8 9
8	0 1 1 2 4 5 6 8
9	0 1 2

Key: 6 | 4 represents 64

2. Construct a box-and-whisker plot for the pulse rate data.

Solution Begin by writing the data values in order from least to greatest. Find the median of the whole set.

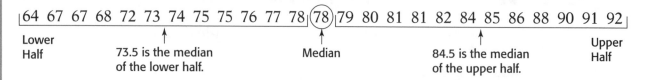

64 67 67 68 72 73 74 75 75 76 77 78 (78) 79 80 81 81 82 84 85 86 88 90 91 92

Lower Half

73.5 is the median of the lower half.

Median

84.5 is the median of the upper half.

Upper Half

The median divides the data into a lower half and an upper half. Find the median of each of these groups of data.

Draw a number line that includes the least and greatest values of the data set. Place a point for the least value, the lower median, the median, the upper median, and the greatest value. Draw a box between the point for the lower and upper medians. Draw a vertical line through the point for the median. Add whiskers by drawing a line segment connecting the points for the least value and the lower median and a line segment connecting the points for the upper median and the greatest value.

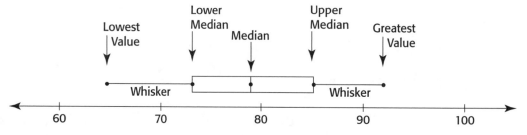

Each whisker and each part of the box represents about one-fourth or 25% of the data values. The whole box represents the middle 50% of the set of data. The longer the whisker or part of the box, the more spread out the values are. If a whisker or part of the box is short, it means the data values represented are close together. In the plot above, the left whisker is a little longer than the right whisker, indicating the data is more spread out at the lower end.

Multiple-Choice Questions

1. On a stem-and-leaf plot, 15 | 8 represents 158. How would 236 be represented?

   A. 2 | 36   B. 23 | 6
   C. 20 | 36   D. 20 | 6

Use the stem-and-leaf plot below for questions 2 and 3.

```
3 | 4 6 7
4 | 0 1 3 3 5 8
5 | 2 2 4 7 7 7
6 | 0 5 6 7 8 9
```
3 | 4 = 34

2. What is the median of the data shown in the stem-and-leaf plot?

F. 35   G. 52
H. 57   J. 58.5

3. How many of the data values are less than 45?

A. 7   B. 8
C. 9   D. 13

Use the box-and-whisker plot to answer questions 4–6.

Hours of TV Watched Weekly

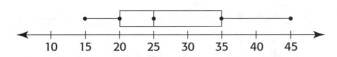

4. What is the range of the data values represented by the box-and-whisker plot?

F. 45   G. 30
H. 25   J. 15

5. What is the median of the data values represented?

A. 15   B. 20
C. 25   D. 35

6. What does the box-and-whisker plot tell you about people's viewing habits?

F. About half the viewers watched 45 hours per week.
G. More viewers watched 35 hours per week than 20 hours per week.
H. Most of the viewers watched 25 hours per week.
J. About half the viewers watched between 20 and 35 hours per week.

7. The weights of pumpkins sold at a farm stand are given below. Which stem-and-leaf plot correctly shows the same data?

Pumpkin Weights (lb)

19	11	10	8	12	15
	7	22	17	23	25	17

A.
```
1 | 0 1 2 5 7 7 7 8 9
2 | 2 3 5            1 | 5 = 15
```

B.
```
0 | 7 8
1 | 0 1 2 5 7 9
2 | 2 3 5            1 | 5 = 15
```

C.
```
10 | 0 1 2 5 7 7 7 8 9
20 | 2 3 5           10 | 5 = 15
```

D.
```
0 | 7 8
1 | 0 1 2 5 7 7 9
2 | 2 3 5            1 | 5 = 15
```

8. Which of the following cannot be determined from a box-and-whisker plot?

F. the mean only
G. the mode only
H. the mean and the range only
J. the mean and the mode only

Use the box-and-whisker plot for questions 9 and 10.

Number of E-mail Messages Received in One Day

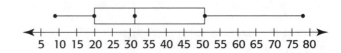

9. The median number of messages received is

A. 20
B. 31
C. 51
D. 77

10. About what percent of the people surveyed received more than 50 messages in one day?

F. 10%
G. 25%
H. 50%
J. 75%

Short-Response Questions

Use the following set of data for problems 11–13.

85 31 67 58 41 28 39 56 47
92 65 33 71 58 40 37 89 96
43 25 48 51 74 66 45

11. Construct a stem-and-leaf plot.

12. Construct a box-and-whisker plot. Label the five values you use.

13. Answer the questions. Tell which plot you used to determine the answer.

    a. What is the range of the data?
    b. What is the median of the data?
    c. What is the mode?
    d. How many data values are less than 60?
    e. Is the difference between the least value and the median less than or greater than the difference between the median and the greatest value?

Extended-Response Questions

14. The normal monthly temperatures in °F for Albany, New York, and Juneau, Alaska, are given in the table below.

    a. Using the same number line, construct a box-and-whisker plot for each city.
    b. Write a paragraph comparing the weather in the two cities. You may wish to include temperature extremes, range of temperatures, median temperature, and variability.

15. Use the stem-and-leaf plot showing the number of visitors for each day of a craft exhibit.

Visitors to Craft Exhibit

18	3 6 8
19	0 2 4 7
20	5 6 6 8 9
21	2 7 9
22	0 5 8
23	9
24	6

18 | 3 represents 183 visitors

    a. What is the range of the number of visitors?
    b. On how many days did at least 210 people visit the exhibit?
    c. On what percent of the days were there fewer than 200 visitors?

16. The heights, in centimeters, of a class of 30 students were recorded.

Heights of Students (cm)

153	162	160	168	171	180	171
167	159	151	164	162	170	173
157	165	168	170	174	175	180
165	174	176	158	166	175	156
159	177					

    a. Construct a box-and-whisker plot of the data.
    b. Into which interval do the middle half of the students fall?

Month	Jan.	Feb.	Mar.	Apr.	May	June	July	Aug.	Sept.	Oct.	Nov.	Dec.
Albany	21	24	34	46	58	67	72	70	61	50	40	27
Juneau	24	28	33	40	47	53	56	55	49	42	32	27

10.5 The Meaning of Probability

A **probability** (*P*) is a number that describes how likely it is that a particular event (*E*) will occur. A **sample space** is the set of all possible results or **outcomes** for some activity or **experiment**. Each of the outcomes is equally likely. An event is any part of the sample space. An event may consist of 0 outcomes, one outcome, or more than one outcome. The outcomes of an event are called the **favorable outcomes**.

The probability that an event will occur is the ratio of the number of favorable outcomes to the number of possible outcomes in the sample space.

The probability of event *E* = *P*(*E*) =
number of favorable outcomes
number of possible outcomes

- The probability of an event that is certain to occur is 1.

- The probability of an event that is impossible is 0.

- The sum of the probability of all the outcomes in a sample space is always 1. So, *P*(*E*) = 1 − *P*(not *E*).

- Probabilities can be expressed as fractions, decimals, or percents.

 Model Problems

1. A number cube is rolled once.

   a. List the outcomes in the sample space.

   For b–e, find each probability.
   b. rolling a 2
   c. rolling an odd number
   d. rolling a 9
   e. rolling a number less than 7
   f. In 120 rolls of the cube, how many times can a number less than 5 be expected?

Solution

a. There are 6 possible outcomes: 1, 2, 3, 4, 5, and 6.

b. The only favorable outcome is 2. $P(2) = \frac{1}{6}$

c. There are 3 favorable outcomes: 1, 3, and 5.

$P(\text{odd number}) = \frac{3}{6} = \frac{1}{2}$

The probability could also be given as .5 or 50%.

d. There are 0 favorable outcomes since no face on the cube shows 9.

$P(\text{rolling } 9) = \frac{0}{6} = 0$ The event is impossible.

e. Since all the faces on the cube show numbers less than 7, there are 6 favorable outcomes.

$P(\text{number} < 7) = \frac{6}{6} = 1$ The event is certain.

f. The numbers less than 5 are 1, 2, 3, and 4, so there are 4 favorable outcomes.

$P(\text{number} < 5) = \frac{4}{6} = \frac{2}{3}$

To predict the number of times the event can be expected to happen, multiply the probability by the number of rolls.

Expected number of rolls less than 5 in 120 rolls $= \frac{2}{3} \times 120 = 80$.

In 120 rolls, a number less than 5 can be expected about 80 times.

2. The cards shown below are placed in a box and one is chosen at random.

a. What is the probability that the card is striped and has A?
b. What is the probability that the card is striped or has A?
c. What is the probability that the card does not have a C?

Solution

a. Count how many selections would make the statement true. There are 2 striped cards with A.

$P(\text{striped and A}) = \frac{2}{8} = \frac{1}{4}$

b. Count how many selections would make the statement true. Do not count any card twice.
There are 4 striped cards and 2 other cards (that are not striped) with A.

$P(\text{striped or A}) = \frac{4+2}{8} = \frac{6}{8} = \frac{3}{4}$

Note: The probability is different when the cards are striped *and* A than when the cards are striped *or* A.

c. *Method I*   There are 7 cards that do not have C.

$P(\text{not C}) = \dfrac{7}{8}$

Method II   There is 1 card with a C.

$P(\text{not C}) = 1 - P(\text{C}) = 1 - \dfrac{1}{8} = \dfrac{7}{8}$

3. A box contains only 4 red marbles, 3 blue marbles, and 5 yellow marbles. If one marble is picked at random from the box, what is the probability of choosing
   a. a red marble
   b. a blue or yellow marble
   c. a marble that is not red

Solution

a. The total number of marbles is 4 + 3 + 5 = 12.
   There are 4 red marbles.

$P(\text{red}) = \dfrac{4}{12} = \dfrac{1}{3}$

b. There are 3 blue and 5 yellow marbles. Since a marble cannot be both blue and yellow at the same time, no outcome is counted twice.

$P(\text{blue or yellow}) = \dfrac{3+5}{12} = \dfrac{8}{12} = \dfrac{2}{3}$

c. $P(\text{not red}) = 1 - P(\text{red}) = 1 - \dfrac{1}{3} = \dfrac{2}{3}$

 **Practice**

Multiple-Choice Questions

1. A box contains only 9 nickels, 7 dimes, and 4 quarters. If a coin is picked at random, what is the probability it is a dime?

   A. $\dfrac{1}{5}$ 　　　　 B. $\dfrac{7}{20}$

   C. $\dfrac{7}{13}$ 　　　　 D. $\dfrac{13}{20}$

2. A box contains only red, blue, and purple marbles. If one marble is picked at random, it is known that the probability of picking red is $\dfrac{1}{4}$ and the probability of picking blue is $\dfrac{1}{3}$. What is the probability of picking purple?

   F. $\dfrac{1}{12}$ 　　　　 G. $\dfrac{1}{6}$

   H. $\dfrac{5}{12}$ 　　　　 J. $\dfrac{7}{12}$

3. If the spinner shown is spun 400 times, how many times can you expect it to land on an odd number?

   A. 125 times
   B. 160 times
   C. 200 times
   D. 250 times

4. A standard number cube is rolled once. Which event has a probability of 0?

   F. rolling a 6
   G. rolling a 2 or a 5
   H. rolling a 3 or an 8
   J. rolling a number greater than 7

5. The chips shown are placed in a box and one chip is drawn randomly. What is the probability that the number on the chip is divisible by 9?

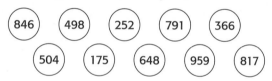

   A. .4
   B. .5
   C. .6
   D. .7

6. The tiles shown are scrambled and placed face down. If one tile is picked at random, what is the probability that the letter on the tile is A or E?

   F. $\frac{2}{11}$

   G. $\frac{3}{11}$

   H. $\frac{2}{9}$

   J. $\frac{3}{8}$

7. In a class of 25 students, 19 take science, 23 take math, and all take either science or math. If one of the students from the class is chosen at random, what is the probability that the student takes science and math?

   A. 24%
   B. 45%
   C. 68%
   D. 92%

8. From a standard deck of 52 cards, one card is drawn. What is the probability that the card is a red face card?

   F. $\frac{1}{13}$ G. $\frac{3}{26}$

   H. $\frac{3}{13}$ J. $\frac{5}{26}$

9. The ATM keypad contains the digits 0, 1, 2, 3, 4, 5, 6, 7, 8, and 9. Jessica is entering a 4-digit PIN code. What is the probability that the last digit of her code is odd or greater than 5?

   A. $\frac{1}{5}$ B. $\frac{3}{10}$

   C. $\frac{1}{2}$ D. $\frac{7}{10}$

10. In basketball practice, the probability that Nell will make a basket is .68. How many baskets can Nell be expected to make in 150 attempts?

   F. 68
   G. 96
   H. 102
   J. 116

Short-Response Questions

11. A box contains 40 marbles that are red, blue, or green. The probability of drawing a red marble is $\frac{2}{5}$ and the probability of drawing a blue marble is $\frac{1}{4}$. How many green marbles are in the box?

12. Find the probability that the spinner will stop on a name that

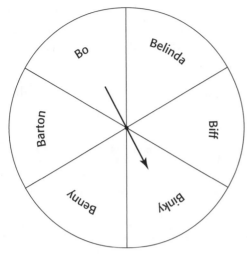

a. begins with B
b. contains N
c. has more than 5 letters
d. has exactly 8 letters

13. The ratio of apples to pears in a fruit basket is 7 to 5. If a piece of fruit is selected at random, what is the probability it is a pear?

14. The probability of getting heads with a magician's weighted coin is $\frac{3}{5}$. If you flip this coin 500 times, about how many times can you expect to get tails?

15. The cards shown are shuffled and turned face down. One of the cards is selected randomly. Find the probability that the figure on the card

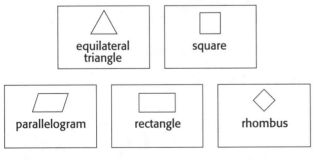

a. is a parallelogram
b. is a rhombus
c. has all sides congruent
d. has angles whose measures total 180°

16. If one card is drawn from a standard deck of 52 cards, what is the probability that it is

a. 7 or a diamond
b. 7 and a diamond

Extended-Response Questions

17. A dime falls out of Jim's pocket and lands at random on the rug shown below. What is the probability that the dime lands in the shaded region? Show your work and express the probability as a decimal.

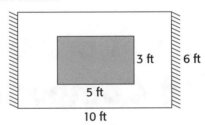

18. A coin bank contains 4 pennies, 2 nickels, 3 dimes, and 1 quarter. If the bank is turned upside down and a coin falls out at random, write a probability question about the coin that will have the given answer.

a. $\frac{3}{10}$ b. $\frac{1}{2}$

c. $\frac{9}{10}$ d. 0

e. 1

19. Laura's CDs are either rock or hip-hop music. She has 3 more rock CDs than hip-hop CDs. If a CD is taken at random from her collection, the probability that it will be hip-hop is $\frac{2}{5}$. How many rock CDs and how many hip-hop CDs does she have? Explain how you found your answer.

20. In a class of 35 students, 17 have ice skates, 31 have a bicycle, **and every student has at least one of the items.** A student from the class is selected at random. Find the probability that the student

a. has ice skates, but not a bicycle
b. has ice skates and a bicycle

10.6 Counting Outcomes and Compound Events

When the outcome of one event has no effect on the outcome of a second event, then the two events are **independent events**. Events that do have an effect on each other are **dependent events**. A **compound event** consists of two or more events.

To determine the number of outcomes in a sample space for a compound event:

- Make a tree diagram or
- Use the **Counting Principle**: If one activity can occur in any of *m* ways and, following this, a second activity can occur in any of *n* ways, then both activities can occur in the order given in *m* × *n* ways.

To find the probability of two independent events E and F, multiply the probabilities of the independent events.

$$P(E \text{ and } F) = P(E) \times P(F)$$

The Counting Principle and probability rule can be extended to include three or more events.

Model Problems

1. Tell whether the events described are independent or dependent.

Solution

Event	Independent or Dependent?
a. Getting heads on a coin toss and rolling a 5 on a number cube	Independent
b. Drawing a 4 from a deck of cards, then drawing another 4 without replacing the first card	Dependent; the number of favorable outcomes for the second draw is affected by the outcome of the first draw.
c. Drawing a queen from a deck of cards, replacing the card, then drawing another queen	Independent

2. On Saturday night, Liza can go to a movie (*M*) or a concert (*C*). She can go with any one of three friends: Jane (*J*), Sarah (*S*), or Toni (*T*). Find the number of different possibilities using
   a. a tree diagram
   b. the Counting Principle

Solution
   a. From a single point, draw a branch for each of the possible outcomes for the first choice. Then for *each* outcome of the first choice, draw as many branches as there are outcomes for the second choice.

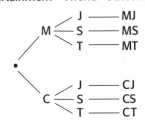

There are 6 possible outcomes.

   b. Number of choices of entertainment × Number of choices of friend = Total outcomes possible
   2 × 3 = 6
   There are 6 possible outcomes.

Using the Counting Principle is less work, but a tree diagram not only gives the number of outcomes, it shows exactly what the outcomes are.

3. A number cube is rolled and a coin is tossed. Find the probability of each compound event, $P(E \text{ and } F)$.
  a. E: Rolling a 2
     F: Tossing heads
  b. E: Rolling a number ≤ 4
     F: Tossing tails

Solution
  a. The events are independent, so the individual probabilities can be multiplied.

$P(\text{rolling a 2}) = \frac{1}{6}$ $\qquad\qquad\qquad$ $P(\text{tossing heads}) = \frac{1}{2}$

$P(\text{rolling a 2 and tossing heads}) = \frac{1}{6} \times \frac{1}{2} = \frac{1}{12}$

  b. The favorable outcomes for the number cube are 1, 2, 3, and 4, so

$P(\text{number} \leq 4) = \frac{4}{6} = \frac{2}{3}$ and $P(\text{tossing tails}) = \frac{1}{2}$.

$P(\text{number} \leq 4 \text{ and tails}) = \frac{2}{3} \times \frac{1}{2} = \frac{2}{6} = \frac{1}{3}$

4. Three fair coins are tossed.
  a. Find the probability of getting all heads.
  b. Find the probability of getting exactly 2 tails.

Solution
  a. The three tosses are independent events. For each coin, $P(\text{H}) = \frac{1}{2}$.

$P(\text{H, H, H}) = \frac{1}{2} \times \frac{1}{2} \times \frac{1}{2} = \frac{1}{8}$

  b. For this problem, make a tree diagram and then count the favorable outcomes.

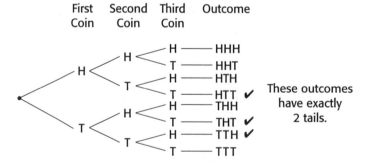

There are 8 possible outcomes. There are 3 outcomes with exactly 2 tails.
$P(\text{exactly 2 tails}) = \frac{3}{8}$

Practice

Multiple-Choice Questions

1. A department store has 7 entrances. There are 2 staircases and 1 escalator going from the first floor to the second floor. How many possible ways are there for a customer to go from outside the store to the second floor?

   A. 10
   B. 14
   C. 15
   D. 21

2. A teacher gives a test consisting of 10 true-or-false questions. Which of the following expressions shows all the possible ways the questions can be answered?

   F. 10×2
   G. 2^{10}
   H. 10^2
   J. 10×20

3. The school cafeteria offers the menu shown. How many different meals consisting of one sandwich, one beverage, and one dessert can be selected from this menu?

Sandwiches	Beverages	Desserts
Egg Salad	Milk	Fruit Salad
Cheese	Chocolate Milk	Cookie
Turkey	Apple Juice	Pudding
Tuna	Orange Juice	Yogurt
Sloppy Joe		

   A. 13
   B. 40
   C. 80
   D. 100

4. A number cube is rolled and a coin is tossed. The probability of rolling a number less than 3 and tossing tails is

   F. $\frac{1}{12}$
   G. $\frac{1}{6}$
   H. $\frac{1}{4}$
   J. $\frac{5}{6}$

5. The spinner shown is spun twice. The probability of spinning two squares is

   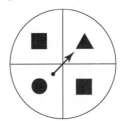

   A. $\frac{1}{16}$
   B. $\frac{1}{8}$
   C. $\frac{1}{4}$
   D. 1

6. A number cube is rolled three times. The probability of getting a 1 on the first roll, a 2 on the second roll, and a 3 on the third roll is

   F. $\frac{1}{216}$ G. $\frac{1}{72}$
   H. $\frac{1}{18}$ J. $\frac{1}{2}$

7. A card is drawn from a standard deck, replaced, then a second card is drawn. The probability of both cards being hearts is

   A. $\frac{1}{169}$ B. $\frac{1}{26}$
   C. $\frac{1}{16}$ D. $\frac{2}{13}$

8. A manufacturer's product code consists of a letter, followed by a digit, followed by a symbol. The choices for each are shown in the chart. What is the probability the code starts with X?

Letters	Digits	Symbols
K	0	*
S	1	#
T	2	&
X	3	
Z		

F. $\dfrac{1}{60}$

G. $\dfrac{1}{30}$

H. $\dfrac{1}{12}$

J. $\dfrac{1}{5}$

9. One box contains 12 cards numbered 1 through 12. A second box contains 10 cards with the letters A through J printed on them. One card is picked at random from the number box and from the letter box. The probability of getting a prime number and a vowel is

A. $\dfrac{3}{20}$

B. $\dfrac{1}{8}$

C. $\dfrac{1}{6}$

D. $\dfrac{4}{11}$

10. Three coins are tossed. The probability of getting at least 1 tail is

F. $\dfrac{1}{3}$

G. $\dfrac{1}{8}$

H. $\dfrac{3}{8}$

J. $\dfrac{7}{8}$

Short-Response Questions

11. Ralph is planning a visit to one of the cities shown in each state.

   a. Make a tree diagram showing all the possible trips Ralph can plan. How many are there?
   b. How many of the possible trips include San Diego?

California	New Mexico	Arizona
San Francisco	Santa Fe	Phoenix
San Diego	Taos	Tucson
Los Angeles		

12. A spinner with five equal sections labeled A, B, C, D, and E is spun and a number cube is tossed.

   a. How many different outcomes are possible?
   b. Let (A, 1) represent getting A on the spinner and 1 on the number cube. Graph the points representing all the possible outcomes. Use the horizontal axis for letters and the vertical axis for numbers.
   c. What is the probability of getting A or B on the spin and 5 or 6 on the toss?

13. Two number cubes are rolled.

   a. How many outcomes are in the sample space?
   b. Display the sample space using ordered pairs.
   c. What is the probability that the two cubes show the same number?

14. Two number cubes are rolled and the sum of the numbers showing is found.

   a. What is the probability of getting a sum of 5?
   b. Which is more likely, a sum of 7 or a sum of 8? Explain.
   c. What is the probability of getting a sum of 13?
   d. Which other sum has the same probability as 12?

15. Each of the spinners shown is spun once. What is the probability of the arrows pointing to:

a. the letter B and the number 7?
b. the letter D and a number less than 10?
c. a consonant and an odd number?
d. a letter before M and a prime number?

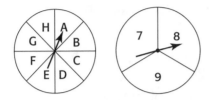

Extended-Response Questions

16. Bath.com offers the choices shown in its towel line.

Bath.com Towels		
Colors	**Sizes**	**Weights**
white	washcloth	regular
yellow	fingertip	extra-thick
pink	hand	
blue	bath	
green		
beige		
multi-stripe		

a. If one of each towel is displayed on the Web site, how many towels are displayed?
b. If a stock person ships a random item to a customer, what is the probability that the item is yellow? What is the probability that it is a yellow bath towel? What is the probability that it is an extra-thick yellow bath towel?

17. An ice cream shop offers 12 different flavors. Noreen orders a different flavor on Sunday than she did on Saturday.

a. Are her flavor choices on Saturday and Sunday independent events?
b. How many choices does she have on Saturday? On Sunday? How many choices in all for the two days?

18. A card is drawn from a standard deck, replaced, the cards are shuffled, and another card is drawn. Find the probability that:

a. both cards are aces
b. both cards are spades
c. both cards are red
d. the first card is the 7 of diamonds and the second card is the 8 of clubs

19. The probability of Carl beating Zack at chess is $\frac{3}{10}$. The probability of Carl beating Leon at chess is $\frac{3}{4}$. (Assume all the games are independent events.) If Carl plays one game with Zack and one with Leon, find the probability that Carl

a. wins both games
b. loses both games
c. Using the results above, how can you find the probability that Carl wins at least one game?

20. A family has four children of different ages.

a. Make a tree diagram showing all the possible outcomes of boys and girls.
b. What is the probability of the children being all boys or all girls?
c. What is the probability that the youngest is a girl?
d. What is the probability that there are exactly two boys?
e. What is the probability that there is at least one girl?
f. What is the probability that the two middle children are girls?

10.7 Permutations and Combinations

A **permutation** is an arrangement of items in some specific order. There are 6 permutations of the 3 letters A, B, C:

ABC   ACB   BAC   BCA   CAB   CBA

The number of permutations can be found by making a tree diagram or by using the Counting Principle:

$$3 \quad \times \quad 2 \quad \times \quad 1 \quad = \quad 6$$

↓	↓	↓
choices of letters for first position	choices of letters for second position	choice of letters for third position

A selection of items, without arranging them in any particular order, is called a **combination**. The combinations of 2 letters chosen from A, B, and C, are:

AB   AC   BC

Combinations *do not* include different arrangements of the same items. AB and BA are the *same* combination.

For a given set of 2 or more items, there are always more permutations than combinations.

Model Problems

1. State whether each situation involves a permutation or combination.
   a. Choose a president, vice-president, and treasurer from a club with 15 members.
   b. Choose a refreshment committee of 3 members from a club with 15 members.

   Solution
   a. The order of the choice is important because Donny (president), Connie (vice-president), Ronny (treasurer) is different from Connie (president), Ronny (vice-president), Donny (treasurer). When order matters, the situation involves a permutation.
   b. Order does not matter. The committee Jan, Dan, Nan is the same as the committee Dan, Nan, Jan. When order does not matter, the situation involves a combination.

2. The director of a film festival wants to select 3 different films from a set of 10 to screen one on Friday, one on Saturday, and one on Sunday. How many different 3-day groups of films are possible?

Solution
The order of the films matters. Only 3 of the possible 10 films will be shown.

$$
\underset{\text{Friday choices}}{10} \times \underset{\text{Saturday choices}}{9} \times \underset{\text{Sunday choices}}{8} = 720
$$

Answer There are 720 ways of presenting the films.

3. A frozen yogurt shop offers these flavors: vanilla (V), chocolate (C), banana (B), strawberry (S), and lemon (L).
   a. How many different choices are there for a 2-flavor cup?
   b. Each week, the store's computer picks 2 flavors at random to be Flavors of the Week. What is the probability the flavors will be banana and strawberry?

Solution
a. Since a vanilla/lemon choice is the same as a lemon/vanilla choice, find the number of combinations of 5 flavors taken 2 at a time. One way is to make a list and check that there are no duplications.
VC   VB   VS   VL
CB   CS   CL
BS   BL
SL
There are 10 possible 2-flavor combinations.

b. There are 10 possible 2-flavor combinations. Of these, only 1 includes banana and strawberry. The probability that the Flavors of the Week will be banana and strawberry is $\frac{1}{10}$.

There is another method for finding the number of combinations of 5 items taken 2 at a time:

$$
\frac{\text{number of permutations of 2 out of 5}}{\text{number of permutations of 2 out of 2}} = \frac{5 \times 4}{2 \times 1} = 10
$$

In other words, the number of permutations of 2 items out of 5 is divided by the number of ways 2 items can be arranged. Similarly, the number of combinations of 6 items taken 3 at a time is:

$$
\frac{\text{permutations of 3 out of 6}}{\text{permutations of 3 out of 3}} = \frac{6 \times 5 \times 4}{3 \times 2 \times 1} = 20 \text{ combinations of 3 out of 6}
$$

Multiple-Choice Questions

1. In how many ways can 6 students be seated in a row of 6 seats?

   A. 30   B. 36
   C. 120   D. 720

2. How many ways are there to arrange 7 books, 4 at a time on a shelf?

   F. 35   G. 120
   H. 840   J. 5,040

3. How many arrangements are possible if 5 letters from the word BIRTHDAYS are used?

   A. 3,024   B. 15,120
   C. 60,480   D. 362,880

4. For a demonstration tape, the band Sleepyhead will choose 3 songs from a group of 8 and record them in order. How many different arrangements are possible?

   F. 56   G. 336
   H. 1,680   J. 40,320

5. Ingrid's cat recently gave birth to 7 kittens. Ingrid has decided to choose 4 of the kittens to give to friends. How many different groups of kittens can she choose?

   A. 35   B. 70
   C. 120   D. 840

6. How many different committees of 3 students can be formed from a group of 9 students?

   F. 27   G. 84
   H. 168   J. 504

7. How many different committees of 11 students can be formed from a group of 11 students?

   A. 1   B. 11
   C. 121   D. 990

8. Romeo Roman has 8 outfits to present at his fashion show. In how many different orders can he present the outfits?

   F. 64   G. 336
   H. 6,720   J. 40,320

9. What is the probability of selecting the number 7,539 at random from all possible arrangements of the digits 3, 5, 7, and 9 (no repetitions)?

   A. $\frac{1}{945}$ B. $\frac{1}{64}$

   C. $\frac{1}{24}$ D. $\frac{1}{4}$

10. Ava, Ben, Celeste, David, and Evan are essay contest winners. Because of time limitations, only 3 of the students can read their essays at an assembly. If the names are chosen out of a hat, what is the probability that Ben, Celeste, and Evan will get to read their essays?

    F. $\frac{1}{20}$ G. $\frac{1}{40}$

    H. $\frac{1}{10}$ J. $\frac{1}{5}$

Short-Response Questions

11. A diner at Five Fountains Restaurant can choose 2 vegetables from among these: string beans, zucchini, broccoli, corn, beets, and peas.

    a. Does the order in which the diner chooses the vegetables matter?
    b. How many choices of 2 different vegetables can the diner make?

12. Four different letters of the alphabet are to be chosen and arranged as identification codes for members of a health club.

    a. How many different codes are possible?
    b. How many different codes are possible if the first letter must be M?

13. Mr. Juarez has to visit the cities of Albany, Rochester, Binghamton, and Utica for business.

a. In how many different orders can he visit these cities?

b. If he plans his route by drawing slips of paper with the city names, what is the probability that his route will be Binghamton-Albany-Rochester-Utica?

c. If he plans his route randomly, as in part b, what is the probability that his route will start in Albany?

14. Albert is a star athlete. He has trophies for baseball (B), basketball (K), swimming (S), and tennis (T). Albert is arranging his trophies on a shelf.

a. How many different arrangements can be made? List them all.

b. In how many arrangements will the swimming trophy be next to the tennis trophy?

15. Nadia must choose 2 of the 50 states and write a report on them.

a. How many different selections can she make?

b. If she chooses randomly, what is the probability that New York or New Jersey (or both) will be included?

16. Joey needed his math book and his biology book for school one day. Because he was late that morning, he grabbed the top two books from a random stack on his desk that included math, history, biology, art, music, health, English, and Spanish. What is the probability that he got the books he needed?

Extended-Response Questions

17. The digits 1, 3, 5, 6, and 9 are written on slips of paper. Two slips are picked at random. What is the probability that the sum of the digits is

a. even   b. odd
c. greater than 10   d. prime

18. a. How many different 3-digit numbers can be made using the digits 2, 3, 5, 7, and 8 if no digit may be used more than once?

b. If all the possible numbers from part a are written on slips of paper and one number is chosen randomly, what is the probability the number will be less than 500?

c. What is the probability the number picked will have 7 as its middle digit?

19. Five cards are chosen randomly from a standard deck of 52 cards.

a. How many different groups of cards are possible?

b. What is the probability that all 5 cards are hearts? (*Hint*: Find the number of different 5-card groups that can be made from hearts.)

c. What is the probability that all 5 cards are of the same suit?

20. If a chimpanzee is given cards with A, C, D, G, O, S, and T printed on them, what is the probability that the chimpanzee will correctly spell DOG or CAT by putting down 3 cards randomly?

10.8 Experimental Probability

The probability of an event based on the results of an experiment or sample is called **experimental probability**. Experimental probabilities are often used when all outcomes are not equally likely (such as for a weighted coin). In **random sampling**, information is collected from a selected group of people or objects and an experimental probability is computed. This probability is used to draw conclusions about the whole **population** from which the sample was taken. For example, 1,000 batteries may be sampled and checked for defects or 1,000 people may be asked for whom they will vote.

Example

Each of five students spun the spinner shown 20 times to determine the probability of getting a circle. Each student determined the **relative frequency** for the experiment:

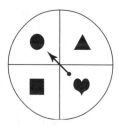

$$\text{relative frequency} = \frac{\text{number of successes}}{\text{number of trials (spins)}} = \text{experimental probability}$$

Student	Number of Circles	Number of Spins	Relative Frequency
Nancy	7	20	$\frac{7}{20} = .35$
Valene	4	20	$\frac{4}{20} = .20$
Lynda	6	20	$\frac{6}{20} = .30$
Jermaine	5	20	$\frac{5}{20} = .25$
Tony	4	20	$\frac{4}{20} = .20$

The results do not mean that only Jermaine was "correct" because .25 is the **theoretical probability** that would have been assigned; the spins just came out that way. The **cumulative relative frequency** can give a better approximation of the probability of spinning a circle. The results for each student are added to the previous results.

Student	Number of Circles	Number of Spins	Cumulative Number of Circles	Cumulative Number of Spins	Cumulative Relative Frequency
Nancy	7	20	7	20	$\frac{7}{20} = .350$
Valene	4	20	11	40	$\frac{11}{40} = .275$
Lynda	6	20	17	60	$\frac{17}{60} = .283$
Jermaine	5	20	22	80	$\frac{22}{80} = .275$
Tony	4	20	26	100	$\frac{26}{100} = .260$

It seems reasonable to expect that as the cumulative number of spins increases, the cumulative relative frequency will get closer to the estimate of the probability, in this case .25 or $\frac{1}{4}$.

Model Problem

1. An appliance manufacturer selected 1,000 toasters at random. Of these, 4 were found to be defective.
   a. What is the probability of a toaster being defective?
   b. In a production run of 7,500 toasters, how many could be expected to have defects?

   Solution

   a. $P(\text{defective}) = \dfrac{\text{number defective in sample}}{\text{total number in sample}} = \dfrac{4}{1,000} = .004$

   b. Expected number defective = $P(\text{defective}) \times$ number of items
   = .004(7,500)
   = 30
   About 30 of the 7,500 toasters could be expected to have defects.

 **Practice**

Multiple-Choice Questions

1. Celia tossed a bottle cap 50 times. It landed up 32 times and down 18 times. The experimental probability that the cap lands down is

   A. .360
   B. .563
   C. .640
   D. 1.778

2. Ricky tossed a penny 100 times and recorded the results shown. Based on his experiment, the probability that the coin he used will land heads is

Heads	56
Tails	44

   F. $\frac{11}{25}$

   G. $\frac{1}{2}$

   H. $\frac{14}{25}$

   J. $\frac{11}{14}$

Use the table below for questions 3 and 4. Four students tossed a number cube and recorded the number of times they each got a 2.

	Number of Tosses	Number of 2's
Taylor	25	3
Nicole	25	2
Tori	25	6
Jordan	25	7

3. Based only on her results, Jordan calculated the experimental probability of tossing a 2 to be

   A. $\frac{2}{25}$

   B. $\frac{7}{25}$

   C. $\frac{9}{50}$

   D. $\frac{18}{25}$

4. Using the cumulative relative frequency, the students calculated the experimental probability of tossing a 2 to be

   F. $\frac{7}{50}$

   G. $\frac{9}{50}$

   H. $\frac{18}{25}$

   J. $\frac{13}{100}$

5. On a production line, 26 leaky batteries were found among 2,000 randomly selected batteries. What is the probability of a battery being leaky?

   A. .0026
   B. .0052
   C. .013
   D. .13

6. A manufacturer randomly selected 5,000 computer chips and found that 85 were defective. In a production run of 100,000 chips, about how many can be expected NOT to be defective?

   F. 1,700
   G. 4,250
   H. 24,575
   J. 98,300

7. Renee can win a prize by picking a black marble from any of the boxes below. From which box should Renee pick to have the best chance of winning a prize?

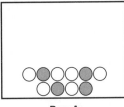
Box A

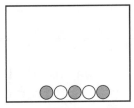
Box B

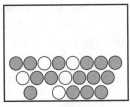
Box C

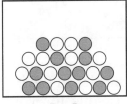
Box D

A. Box A
B. Box B
C. Box C
D. Box D

8. A group of students were asked to name their favorite weekday. The results are shown in the table. If 500 students were asked to name their favorite weekday, how many could be expected to say Friday?

Favorite Weekday	
Monday	4
Tuesday	11
Wednesday	12
Thursday	6
Friday	27

F. 120
H. 225
G. 135
J. 270

Short-Response Questions

Use the data on the table to answer questions 9–11. This table is a summary of the result of an experiment in which a number cube was tossed 50 times.

Number on cube	1	2	3	4	5	6
Number of tosses	10	4	9	14	11	2

9. Find each probability based on the results of the experiment.

a. $P(1)$
b. $P(4)$
c. P(even number)
d. $P(3 \text{ or } 5)$

10. Which experimental probability is closest to the theoretical probability that would be assigned to the outcomes of tossing a number cube? Explain.

11. Explain how a better approximation of the probabilities for each number on the cube could be obtained experimentally.

	Number of Heads	Number of Tosses	Cumulative Number of Heads	Cumulative Number of Tosses	Cumulative Relative Frequency
Gini	9	20			
Pedro	12	20			
Inez	8	20			
Sid	14	20			
Fiona	6	20			

Use the table above for questions 12 and 13. Five students took turns tossing a coin. Each student performed 20 tosses and recorded the number of heads in the table.

12. Copy and complete the table to find the cumulative results of this experiment. Express frequencies to the nearest thousandth.

13. Do you think the coin used in this experiment was a fair coin? Explain. Why or why not?

Extended-Response Questions

The table shows the result of an experiment in which a paper cup was tossed 200 times and its landing position was recorded.

Paper Cup Toss	
Position	Frequency
up	12
down	43
side	145

14. Find each experimental probability:

a. $P(\text{up})$
b. $P(\text{down})$
c. $P(\text{side})$

15. Alex, one of the students working on the experiment, said that since there were 3 possible outcomes, the theoreti-cal probability for each outcome is $\frac{1}{3}$. Do you agree with Alex? Explain.

16. If the same paper cup was tossed 1,000 times, about how many times could the cup be expected NOT to land up?

17. A box contains 6 marbles of equal size and shape. Lucy selected a marble without looking, recorded its color, and then replaced the marble. She did this 200 times and obtained these results:

Color	Red	Blue	Green	Yellow
Number of Picks	63	31	34	72

a. Based on this data, describe how you think the marbles in the box are colored. Give reasons for your answer.
b. Could there be an orange marble in the box? Is it likely there is an orange marble?

18. Four weeks before the election, a radio station surveyed 400 registered voters about which mayoral candidate they preferred. The results are shown in the table.

Candidate	Number
Wang	158
Fields	135
Stanton	33
Undecided	74

a. Based on this data, at the time of the survey what is the probability that a registered voter prefers Wang?
b. By the time of the election, Stanton has dropped out and asked his supporters to vote for Fields. Also, it has been determined that about 80% of the undecided vote will go to Wang. Who seems more likely to win the election, Wang or Fields?

Chapter 10 Review

Multiple-Choice Questions

Use the frequency table showing heights of students for questions 1 and 2.

Student Heights Class 9-201	
Height (in.)	Number of Students
58	2
59	1
60	2
61	0
62	3
63	4
64	2
65	3
66	3

1. The mean height of the students in class 9-201 is

   A. 62 in.
   B. 62.7 in.
   C. 63 in.
   D. 63.6 in.

2. Which of the following is true for the height data?

   F. median = mode
   G. median < mode
   H. median > mode
   J. there is no mode

Use the histogram showing earth science test scores for questions 3 and 4.

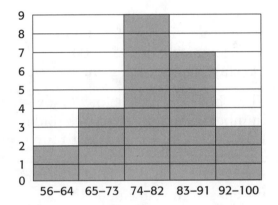

Scores on Earth Science Test

3. What percent of the students scored higher than 82 on the test?

   A. 15%
   B. 35%
   C. 40%
   D. 76%

4. In which interval does the median score fall?

   F. 65–73
   G. 74–82
   H. 83–91
   J. 92–100

5. The graph shows the monthly sales for Sparkle Jewelry. Estimate the change in sales from October to November.

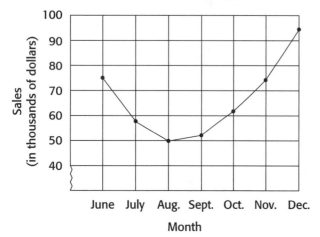

Sparkle Jewelry Monthly Sales

A. increased about $13,000
B. increased about $20,000
C. decreased about $8,000
D. stayed about the same

6. The bar graph below shows the number of new automobiles sold in Jefferson County for the past four years. Jim Savine purchased a car in that time period. What is the probability that he purchased the car in 2002?

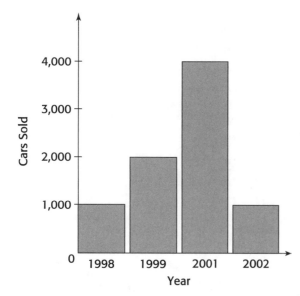

Jefferson Auto Sales

F. $\frac{1}{8}$ G. $\frac{1}{4}$

H. $\frac{1}{3}$ J. $\frac{1}{2}$

7. A bag contains 2 red marbles, 4 blue marbles, 3 green marbles, and 6 yellow marbles. If Rodney picks one marble from the bag without looking, what is the probability that the marble is red or yellow?

A. $\frac{4}{75}$ B. $\frac{2}{15}$

C. $\frac{8}{15}$ D. $\frac{8}{7}$

8. A coin is tossed and a number cube is rolled. What is the probability of getting heads and a number less than 5?

F. $\frac{1}{3}$ G. $\frac{5}{12}$

H. $\frac{1}{2}$ J. $\frac{7}{6}$

9. There are 7 bicyclists in a race. Assuming there are no ties and all the racers finish, how many different finishing orders are possible?

A. 28   B. 49
C. 840   D. 5,040

10. Selena's mother has asked her to pick 3 chores from the following list: dusting, doing laundry, vacuuming, washing windows, grocery shopping. How many different ways can Selena make her selection?

F. 10   G. 15
H. 20   J. 120

Short-Response Questions

11. The ages of people taking a ceramics class are given below.

37	62	19	41	55	32	28	35	24
40	46	45	53	47	32	64	44	23
30	27	48	51	27	36

a. Make a stem-and-leaf plot of the ages.
b. What is the range of ages?
c. What is the median age?
d. If one of the students in the class is picked at random, what is the probability the person is over 45?

Mountain View Diner Menu				
Soup	**Entrée**	**Potato**	**Vegetable**	**Dessert**
Tomato	Chicken	Mashed	Corn	Pie
Minestrone	Duck	Baked	Cabbage	Cheesecake
Split pea	Fish fry	Fried	Broccoli	Pudding
Chicken	Steak		Green beans	Ice cream
	Grilled tofu		Stewed tomatoes	Brownie
	Pasta special		Spinach	Layer cake
	Omelet			

12. At the Mountain View Diner, a complete meal consists of soup, entrée, potato, one vegetable, and dessert.

   a. How many different complete meals can be created from the menu choices shown?

   b. How many different complete meals can be created if a person always orders minestrone soup and cheesecake?

13. The number of cases of salmonella food poisoning for 1985–1996 are given below.

Year	Cases	Year	Cases
1985	5,657	1991	7,755
1986	6,036	1992	6,578
1987	7,052	1993	8,071
1988	7,063	1994	9,866
1989	8,466	1995	10,201
1990	8,734	1996	9,566

   a. Make a graph of the data. Give reasons for your choice of graph. (You may round the data for graphing purposes.)

   b. Write a paragraph describing the trends shown by the graph.

14. Robin wants to determine the probability that an index card folded in half will fall on its side, on its edge, or as a tent. Describe in detail a method she can use to obtain this information.

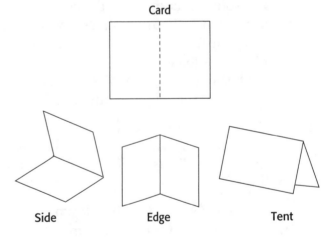

Card

Side   Edge   Tent

15. Ed works part-time for a landscaping service. Over a 20-week period, he earned the following amounts:

$66 $75 $60 $48 $38 $46 $64 $70 $68 $54 $42 $63 $71 $74 $86 $75 $68 $48 $60 $78

   a. Make a box-and-whisker plot for the data. Identify the numbers you use to make the plot.

   b. Write one conclusion that can be drawn from the box-and-whisker plot.

16. a. In how many different ways can the set of cards shown below be arranged?

b. How many different 3-card "hands" can be selected from these cards?

M	O	N	K	E	Y

Extended-Response Questions

17. Alan has two nickels and one dime in his pocket. The pocket has a hole in it and a coin drops out. Alan picks up the dropped coin and puts it back in the pocket. A coin drops out of his pocket again.

  a. Draw a tree diagram and list the possible outcomes to describe the coins that fell out of Alan's pocket.
  b. Find the probability that the dime fell out at least once.
  c. Find the probability that the total value of the dropped coins is 15¢.

18. Each of the four faces of the tetrahedron die shown is an equilateral triangle. The die is rolled twice. Find each probability.

  a. Both rolls show 3 facing down.
  b. Both rolls show an even number facing down.
  c. The sum of the numbers facing down is 4.

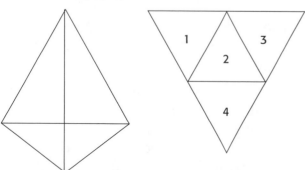

19. The names and ages of the children at a family reunion are

  Boys: Charlie (15), Lionel (13), Steven (12), Mark (13), Theodore (9)

  Girls: Cathy (13), Yasmin (10), Briget (9), Marti (14), Moira (8), Jessica (11), Ashley (7)

  The children's names are written on slips of paper and one name is picked to win a prize. Find the probability that the prize is won by

  a. a girl
  b. a teenager
  c. someone under 12 years old
  d. a 13 year-old boy
  e. someone whose name starts with M

20. The table shows the scores of 25 students on a surprise math quiz.

Score	Frequency	Cumulative Frequency
50	2	
60	3	
70	6	
80	9	
90	4	
100	1	

  a. Find the mean, median, and mode of the scores.
  b. Copy and complete the table.
  c. What is the probability that a student picked at random scored 80 or better?

Multiple-Choice Questions

1. Which is the best buy for oatmeal?

    A. 12 oz for $1.11
    B. 15 oz for $1.43
    C. 1 lb 3oz for $1.75
    D. $1\frac{3}{4}$ lb for $2.59

2. Regis Dental Products produced 7,200 toothbrushes, which are packed 40 to a box. Each box sells for $37.50. The total income for the shipment of toothbrushes is

    F. $1,500   G. $4,050
    H. $5,320   J. $6,750

3. It takes Marla 1 h 10 min to travel to or from work. She spends 8 h 15 min at work. If she leaves home at 7:35 A.M., at what time does she return?

    A. 5:00 P.M.   B. 5:55 P.M.
    C. 6:10 P.M.   D. 6:25 P.M.

4. Which shows the measures of $\angle MOP$ and $\angle NOP$ (in that order)?

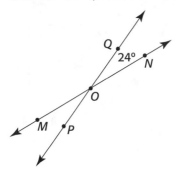

    F. 24°, 66°   G. 24°, 156°
    H. 114°, 156°   J. 66°, 114°

5. Find the volume of the can of paint. Use 3.14 for π.

    A. 5024 cm^3
    B. 1256 cm^3
    C. 628 cm^3
    D. 558.92 cm^3

6. Solve: $\dfrac{1\frac{1}{2}}{n} = \dfrac{6}{2\frac{2}{3}}$.

    F. $\frac{2}{3}$

    G. $\frac{3}{4}$

    H. 4

    J. $10\frac{2}{3}$

7. Which is the graph of $|m| = 3$?

    A.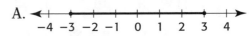

    B.

    C.

    D.

8. △*LMN* is translated 6 units right and 5 units down. The coordinates of the vertices of the image are

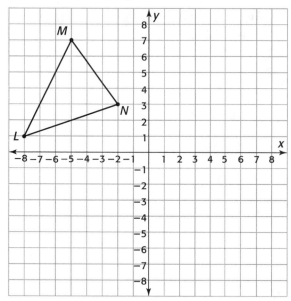

F. $L'(2, 4), M'(-11, 2), N(4, -2)$
G. $L'(-2, -4), M'(2, 1), N'(4, -2)$
H. $L'(-2, -4), M'(1, 2), N'(4, -2)$
J. $L'(-14, 6), M'(1, -2), N'(4, 8)$

9. Find the mean and median of the temperatures shown.

Daily High Temperatures		
11°C	16°C	10°C
12°C	12°C	14°C
14°C	13°C	11°C
15°C	13°C	12°C

A. mean = 12.5°C, median = 12°C
B. mean = 12.75°C, median = 13°C
C. mean = 12.25°C, median = 13.5°C
D. mean = 12.75°C, median = 12.5°C

10. The scale of a map is 2 cm : 5 km. The actual distance from Rock Ridge to Crown corners is 48 km. What is the distance on the map?

F. 24 cm
G. 19.2 cm
H. 15.4 cm
J. 9.6 cm

Short-Response Questions

11. The table shows the results of a school survey that asked the question: Should the school newspaper include restaurant reviews?

Newspaper Survey			
Vote	9th Grade	10th Grade	11th Grade
Yes	176	164	190
No	149	129	116
Undecided	60	68	45

a. To the nearest whole number, what percent of the 9th-grade students voted Yes?
b. To the nearest whole number, what percent of the students who voted No were 11th graders?
c. To the nearest whole number, what percent of the students were undecided?

12. Find the value of $3x^4 + 2x^3 - 4x + 1$, when $x = -4$.

13. The lobby of a hotel is to be carpeted.

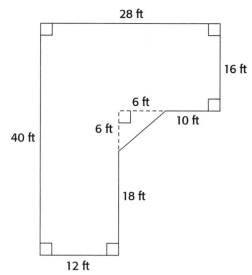

a. Find the number of square feet of carpeting that is needed to cover the whole lobby. Use the diagram to show how you found the area and show all work.

b. The carpet to be used costs $30 per square yard. How much will it cost to carpet the lobby? Explain.

14. Quadrilateral *PQRS* is reflected in the *y*-axis and then translated 5 units down. Graph the resulting image and give the coordinates of the vertices.

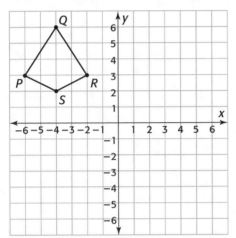

Extended-Response Questions

15. An aquarium is 50 cm long, 30 cm wide, and 40 cm high. The water level is 3 cm from the rim. What percent of the aquarium's volume is filled with water? Show your work.

16. Refer to △*JKL*. Show all work.

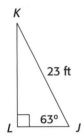

a. Find the measure of ∠*LKJ*.
b. Find $\overline{LJ}$ to the nearest tenth.
c. Find $\overline{KL}$ to the nearest tenth.

17. Assume the pattern shown in Figures 1–3 continues.

Fig. 1

Fig. 2

Fig. 3

a. Describe what Figure 15 will look like.
b. How many dots will be in Figure 15?
c. Which figure will have 420 dots?

18. One card is selected at random. Find each probability.

a. *P*(triangle and even number)
b. *P*(square or odd number)
c. *P*(circle and 2)
d. *P*(prime number)

19. Each of the spinners shown is spun once and the numbers are multiplied.

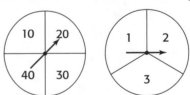

a. Make a tree diagram to show all the possible outcomes.
b. What is the probability that the outcome will be 40?
c. What is the probability that the outcome will be divisible by 3?

20. The stem-and-leaf plot shows the number of sit-ups students were able to do in 1 minute.

```
1 | 6 8 9
2 | 1 2 2 5 7 7 8
3 | 0 3 4 4 5 6 9 9
4 | 2 6 7 8 8
5 | 1 2
```

1|6 = 16 sit-ups

a. What percent of the students were able to do at least 25 sit-ups?
b. What was the median number of sit-ups students completed?
c. Construct a histogram for the data. Use intervals of length 6 starting with 15–20.
d. State one advantage of the stem-and-leaf plot and one advantage of the histogram for displaying data.

Patterns and Functions

Chapter Vocabulary

relation	domain	range
function	direct variation	inverse variation
solution	graph of an equation	linear equation
half-planes	graph of an inequality	slope
linear function	nonlinear function	quadratic function
exponential function		

11.1 Relations and Functions

If a person walks at a steady rate of 4 kilometers per hour, then the distance d that the person travels is related to the number of hours h spent walking. One way to describe the relationship is with the equation $d = 4h$. Another way to show part of the same relationship is with a graph.

A **relation** is a set of ordered pairs (x, y). The **domain** of the relation is the set of all the values of x. The **range** of the relation is the set of all the values of y.

For the relation shown in the graph, each x-value is the time in hours. Each corresponding y-value is the distance traveled in that time. The ordered pairs shown on the graph are $(0, 0)$, $(1, 4)$, $(2, 8)$, $(3, 12)$, $(4, 16)$, $(5, 20)$.

A **function** is a relation in which each value of x is paired with one and only one value of y. That is, each value of x appears only once in the set of ordered pairs. In the above situation, the distance traveled is said to be a function of the time spent walking. In other words, how far you walk depends on how much time you walk for—the distance depends on the time.

Model Problems

1. Tell if each is a relation, a function, or neither.
   a. $(1, 2)$, $(2, 3)$, $(3, 4)$, $(4, 5)$
      Function
   b. $(4, 2)$, $(4, -2)$, $(9, 3)$, $(9, -3)$
      Relation that is not a function. There are two y-values for each x-value.
   c. 10.5, 12.7, 14.9, 17.1
      Neither; there are no ordered pairs.
   d.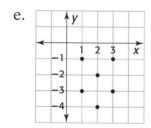

      Function. The ordered pairs are $(1,2)$, $(2,2)$, $(3,2)$. A function may have y-values repeated as long as the x-values are different.

   e.

      Relation that is not a function. The ordered pairs are $(1, -1)$, $(1, -3)$, $(2, -2)$, $(2, -4)$, $(3, -1)$, $(3, -3)$. There are two different y-values for each x-value.

2. The equation $P = 3s$ expresses the relationship between the perimeter P of an equilateral triangle and the length of a side s.
a. Is the relation a function?
b. Graph the relation for lengths of 1, 2, 3, 4, and 5 inches.

Solution
a. The relation is a function. There is only one value for the perimeter for each given length.

b.
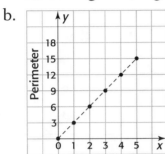

The distance and perimeter functions above are examples of situations in which one variable is a constant multiple of the other. As one variable increases, so does the other. Moreover, the ratio of the y-value compared to the x-value is always the same and is equal to the constant in the equation.

$$d = 4h: \frac{4}{1} = 4, \frac{8}{2} = 4, \frac{12}{3} = 4, \frac{16}{4} = 4, \frac{20}{5} = 4$$

$$P = 3s: \frac{3}{1} = 3, \frac{6}{2} = 3, \frac{9}{3} = 3, \frac{12}{4} = 3, \frac{15}{5} = 3$$

This type of relationship between the variables is called a **direct variation**; y is said to vary directly as x. The points on the graph lie on a straight line

Now, consider the situation where a fixed distance of 20 miles is to be covered. If a bicyclist rides at the rate of 2 miles per hour, it would take 10 hours; at 4 miles per hour, it would take 5 hours; at 6 miles per hour, it would take $3\frac{1}{3}$ hours, and so on. In this case, the relationship between the variables is called an **inverse variation**. As one variable increases (in this case the rate), the other variable decreases (in this case, the time). However, the product of the two variables is constant (in this case, 20).

$$rt = 20$$
$$2 \times 10 = 20$$
$$4 \times 5 = 20$$
$$6 \times 3\frac{1}{3} = 20$$

The graph of an inverse variation is shown below. The points lie on a curve.

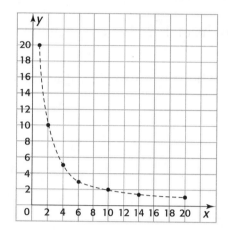

3. If 4 workers can build a skateboard ramp in 10 days, how long will 5 workers take to build the ramp? What type of variation is involved?

Solution As the number of workers increases, the number of days it takes to build the ramp decreases. This is an example of inverse variation.

Think: workers × days = worker-days
The number of worker-days remains constant.
4 workers · 10 days = 40 worker-days
5 workers · d days = 40 worker-days
$$5d = 40$$
$$d = 8$$

Answer If there are 5 workers, the ramp will take only 8 days to complete.

 Practice

Multiple-Choice Questions

1. Which diagram shows the relationship between relations and functions?

A. Relations Functions

B. Functions
Relations

C. Relations Functions

D. Relations
Functions

2. Which set of ordered pairs does NOT describe a function?

   F. $(2, 4), (1, 1), (0, -4), (-2, -6)$
   G. $(1, 6), (2, -5), (1, -3), (3, 8)$
   H. $(3, 7), (5, 7), (7, 7), (9, 7)$
   J. $(-3, 2), (-4, 4), (-6, 10), (2, -10)$

3. Which is NOT the graph of a function?

A.

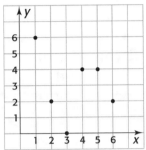

B.

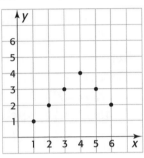

C.

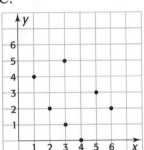

D.

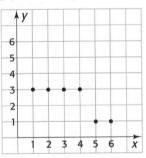

4. What is the domain of the function T given by
T: $(0, -1), (2, 5), (5, 14), (-4, -13),$
   $(-7, -22)$

   F. $-7, -4, -1, 2, 5$
   G. $-22, -13, -1, 5, 14$
   H. $-7, -4, 0, 2, 5$
   J. $-1, 7, 19, -17, -29$

5. What is the range of the function W given by
W: $(-4, 0), (-3, 0), (-1, 0), (2, 1), (5, 1)$

   A. $0, 1$
   B. $-4, -3, -1, 2, 5$
   C. $-4, -3, -1, 3, 6$
   D. $-1, 0$

6. Which statement is true about the ordered pairs (x, y) shown in the table.

Student	x	1	2	3	4	5	6	7
Age	y	13	10	15	14	13	12	14

   F. The set of ordered pairs is neither a relation nor a function.
   G. The set of ordered pairs is a relation but not a function.
   H. The set of ordered pairs is a function but not a relation.
   J. The set of ordered pairs is both a relation and a function.

7. The cost of renting a bus for a picnic is $400. The relationship between the number of students s who go on the bus and the cost c per student is an example of

   A. direct variation
   B. inverse variation
   C. circular variation
   D. increasing variation

8. Kendra earns $8 per hour walking dogs. The relationship between Kendra's total earnings T and the number of hours h she works is an example of

   F. direct variation
   G. uniform variation
   H. inverse variation
   J. parallel variation

9. Which equation expresses the relationship between x and y shown in the table?

x	0	1	3	8
y	0	5	15	40

   A. $x = 5y$
   B. $y = 5x$
   C. $x + y = 6$
   D. $y - x = 4$

10. $A = 10\ell$ is a formula for the area of any rectangle whose width is 10. If ℓ is tripled, what change takes place in A?

    F. A is tripled.
    G. A is multiplied by 6.
    H. A is multiplied by 9.
    J. A is multiplied by $\frac{1}{3}$.

Short-Response Questions

11. If 20 people share the cost of a gift, each will pay $7.50. How much will each pay if 24 people share the cost?

12. Stephanie attends State College. If she can drive from her home to the school in 4 hours at 45 miles per hour, how long will it take her at 50 miles per hour? Show your work.

13. For this set of ordered pairs: (1, 5), $(-2, 8)$, $(x, 15)$, $(9, -20)$

    a. Give a possible value for x that will make the set of ordered pairs a relation but not a function.
    b. Give a possible value of x that will make the set of ordered pairs a function.

14. Given Q: (0, 0), $(-1, 2)$, $(3, -6)$, (5, 12), (10, 10)

    a. Is Q a function?
    b. What is the domain of Q?
    c. What is the range of Q?

15. The domain of a function F is 10, 20, 30, 40, 50, and 60. The range of F is 100. Write the ordered pairs for F.

Extended-Response Questions

16. Linus has $25 at the beginning of a week. He spends $3 each day. Let y rep-

resent the amount of money Linus has left after x days.

    a. Is the relationship between x and y a function? Explain.
    b. Write the ordered pairs for each of the 7 days of the week.
    c. Graph the ordered pairs.

17. Gloria sells cosmetics. She earns a commission of 5% on the items she sells. On Thursday, she worked 5 hours and earned $100. On Friday, she worked 4 hours and earned $70. On Saturday, she worked 8 hours and earned $200. On Sunday, she worked 5 hours and earned $80.

    a. Graph the relation between hours and earnings.
    b. Is the relation a function? Explain.

18. Is the relationship between x and y shown in the table a direct variation? Explain why or why not.

x	1	2	3	4
y	4	6	8	10

19. $C = 6n$ is a formula for the cost of n items that sell for $6 apiece.

    a. How do C and n vary?
    b. How will the cost of 8 items compare with the cost of 4 items?
    c. If the number of items is multiplied by 10, how will the total cost change?

20. Bruno wants to make a rectangular dog pen with an area of 48 square feet.

    a. If the length of the pen is 8 feet, what is the width?
    b. If the length of the pen is 12 feet, what is the width?
    c. In this situation, how do length and width vary?

11.2 Graphing Linear Equations

The equation $y = x + 3$ represents two numbers x and y such that y is 3 more than x. Since the value of y depends on the value of x, y is a function of x. An ordered pair is a **solution** of this equation if a true statement results when the coordinates of the ordered pair are substituted for x and y in the equation. For example, (4, 7) is a solution for $y = x + 3$ because $7 = 4 + 3$.

The **graph of an equation** with two variables is all the points whose coordinates are solutions of the equation. An equation whose graph is a straight line is called a **linear equation**.

Model Problems

1. Tell if the ordered pair is a solution for $y = 2x - 4$.
   a. $(1, -2)$
   b. $(-3, 10)$

   Solution
   a. Substitute 1 for x and –2 for y.
   $$y = 2x - 4$$
   $$-2 = 2(1) - 4$$
   $$-2 = -2 \qquad \text{So, } (1, -2) \text{ is a solution of } y = 2x - 4.$$
   b. $\quad y = 2x - 4$
   $$10 = 2(-3) - 4$$
   $$10 \neq -10 \qquad \text{So, } (-3, 10) \text{ is not a solution for } y = 2x - 4.$$

2. Find three solutions of $y = 3x + 5$.

   Solution Make a table of values by choosing a value for x and finding the corresponding value of y.

x	3x + 5	y
−2	3(−2) + 5	−1
0	3(0) + 5	5
1	3(1) + 5	8

   Answer Three solutions of $y = 3x + 5$ are $(-2, -1)$, $(0, 5)$, and $(1, 8)$.

3. Graph the equation $y = 2x - 1$.

   Solution Find at least three ordered pairs that are solutions of the equation. (Two points determine a line; the third point is a check.) Make a table of values.

x	2x − 1	y
0	2(0) − 1	−1
1	2(1) − 1	1
2	2(2) − 1	3

Graph the points that correspond to each solution. Draw a line through these points and label the graph with the equation.

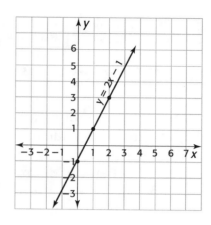

Note: All points on the line, and only those points, are solutions for the equation.

4. The sum of two numbers is 8.
   a. Write an equation to represent the verbal sentence.
   b. Graph the equation.

Solution
   a. Let x = one number and y = other number. Then $x + y = 8$.
   b. Make a table of values for $x + y = 8$. Rewriting the equation so that y is alone on one side makes it easier to work with.

$$x + y = 8$$
$$x - x + y = 8 - x \qquad \text{Subtract } x \text{ from both sides.}$$
$$y = 8 - x$$

x	$8 - x$	y
0	$8 - 0$	8
1	$8 - 1$	7
-1	$8 - (-1)$	9

Graph the points, then draw and label the line.

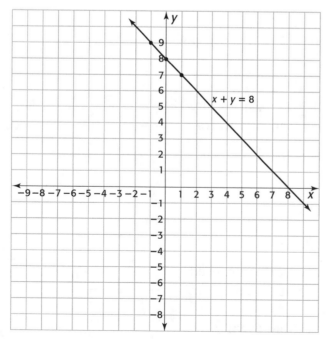

Practice

Multiple-Choice Questions

1. Which ordered pair is a solution of $y = 2x - 3$?

A. $(0, 3)$ B. $(5, 7)$
C. $(-1, 1)$ D. $(6, 8)$

2. Which point lies on the graph of $y = -3x + 5$?

F. $(0, -5)$ G. $(-1, 2)$
H. $(3, -4)$ J. $(2, 1)$

3. Which shows the graph of $x = 3$?

A.
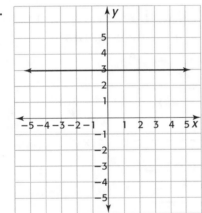

B.
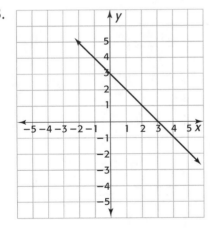

C.
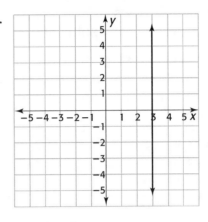

D.
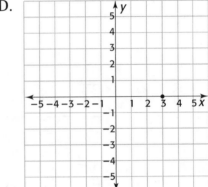

4. What must be the value of k if $(k, 6)$ lies on the graph of $y = 4x - 6$?

F. -3 G. $-\dfrac{3}{2}$

H. 0 J. 3

5. What must be the value of w if (w, w) lies on the graph of $x + y = 0$?

A. 0 B. 2
C. -2 D. -8

6. The graph of $3x - y = 7$ is a

F. point
G. straight line
H. pair of parallel lines
J. rectangle

7. How many solutions for the equation $y = 4 - x$ are there?

    A. none
    B. one solution
    C. two solutions
    D. infinitely many solutions

8. A person's age six years from now (y) is a function of the person's present age (x). Which equation represents this function?

    F. $x + y = 6$
    G. $y = 6x$
    H. $y = x + 6$
    J. $x = y + 6$

9. Which shows the graph of $y = -2$?

A. 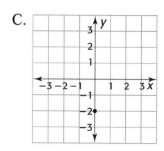 B.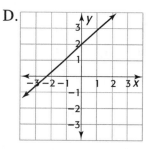

C. 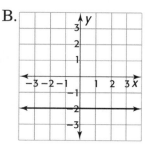 D.

10. Which ordered pair is a solution for both $y = x + 1$ and $y = 5 - x$?

    F. $(0, 1)$
    G. $(-2, 7)$
    H. $(2, 3)$
    J. $(4, 1)$

Short-Response Questions

11. State whether the point whose coordinates are given is a solution for the given equation.

    a. $x + y = 9$; $(4, 5)$
    b. $y = 3x + 1$; $(-2, 7)$
    c. $y = \frac{3}{2} + 2x$; $(6, 10)$
    d. $y = 4x - 5$; $(-8, -27)$

12. Find the number that can replace n so that the resulting ordered pair will be on the graph of the given equation.

    a. $y = 2x - 3$; $(5, n)$
    b. $y = -4x + 7$; $(n, -17)$
    c. $2x + y = 15$; $(10, n)$
    d. $x + y = 18$; (n, n)

For 13–15, find the missing values for each table and graph the equation.

13. $y = -2x$

x	y
−1	?
?	6
4	?

14. $y = x - 1$

x	y
0	?
3	?
?	−3

15. $x + y = 4$

x	y
5	?
?	3
−2	?

16. For each equation, make a table of three solutions and then graph the equation.

a. $y = 2x - 2$ b. $y = -3x + 4$

Extended-Response Questions

17. a. Graph these equations on the same coordinate plane:
$y = 3, x = 4, y = -2, x = -5$
b. Which pairs of lines intersect? What are the coordinates of the point where the lines intersect?
c. Describe the figure with vertices at the points of intersection.

18. a. On a coordinate plane, graph the points $(1, -2)$ and $(-1, -4)$. Draw a line through the points.
b. Make a table of these ordered pairs and three others that are on the line you drew.
c. Look for the pattern. Write the equation for the line.

19. The cost, y, of renting a kayak is $5 plus $2 times the number of hours, x, that the kayak is used. (Assume that x can be a fractional part of an hour.)

a. Write an equation that expresses the relationship between x and y.
b. Find three ordered pairs that are solutions for the equation you wrote in part a. Are there any numbers that do not make sense in this situation?
c. Graph the equation you wrote.
d. Use the graph to find how many hours the kayak was rented if the total cost was $16.

20. a. Graph the following equations on the same coordinate plane: $y = x + 2$ and $y = x - 3$.
b. Do you think there are any ordered pairs that are solutions to both equations? Give reasons for your answer.

11.3 Graphing Inequalities

When a line is graphed in the coordinate plane it separates the plane into two regions called **half-planes**. For example, the half-plane above the line $y = 2$ is the set of all points whose y-coordinate is greater than 2, such as $A(3, 4)$. The half-plane below the line $y = 2$ is the set of all points whose y-coordinate is less than 2, such as $B(3, -1)$. The **graph of an inequality** is the set of all points that are solutions to the inequality. The graph of $y > 2$ consists of the shaded half-plane in the figure. The line $y = 2$ is drawn dashed to show that it is not part of the graph of the inequality.

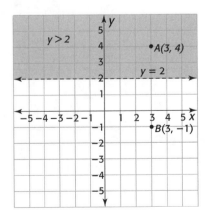

Model Problems

1. Graph the inequality $y \leq x$.

   Solution   Graph the line $y = x$. Use a solid line to show that $y = x$ is part of the inequality.
   Choose two points, one above and one below the line.
   $A(1, 3)$ lies above the line.
   $B(3, 0)$ lies below the line.
   Test both points to find the one that satisfies the inequality.
   For $A(1, 3)$: $3 > 1$.
   For $B(3, 0)$: $0 < 3$. So, $B(3, 0)$ satisfies the inequality.
   Shade the half-plane on the same side of the line as the point that satisfies the inequality. Since $B(3, 0)$ satisfies the inequality, shade below the line.

   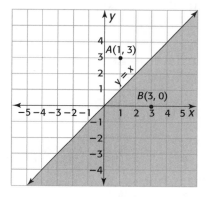

2. Graph the inequality $x > -1$.

   Solution   The graph is the set of all points whose x-coordinate is greater than -1. The line $x = -1$ is not part of the inequality.

Practice

Multiple-Choice Questions

1. Which ordered pair is a solution of $y > x + 2$?

  A. $(0, 0)$
  B. $(-4, -1)$
  C. $(5, 2)$
  D. $(9, 10)$

2. Which ordered pair is NOT a solution of $y \leq 2x - 3$?

  F. $(2, 0)$
  G. $(5, 8)$
  H. $(-3, -12)$
  J. $(7, 4)$

3. Which shows the graph of $y \leq -3$?

  A.

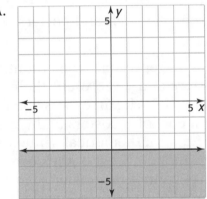

  B.

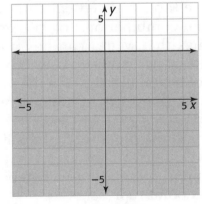

C.

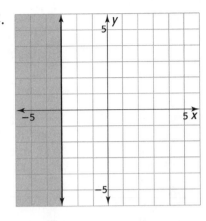

D.

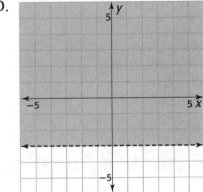

4. Which inequality is graphed below?

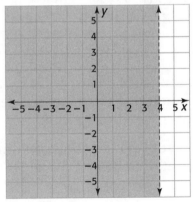

  F. $x \geq -4$
  G. $y < 4$
  H. $x < 4$
  J. $x + y \geq 4$

5. Which inequality is graphed below?

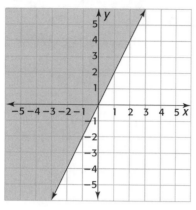

   A. $y \geq x + 2$
   B. $y \leq 2x$
   C. $y \geq 2 - x$
   D. $y \geq 2x$

6. The graph of $y \geq 5$ is a

   F. point
   G. line
   H. half-plane
   J. region between two parallel lines

Short-Response Questions

7. Tell whether the given ordered pair is a solution of the given inequality.

   a. $y < 3x$; (2, 7)
   b. $y \geq 2x + 4$; (−3, 0)
   c. $x \leq 5$; (5, 11)
   d. $y < x - 6$; (0, 0)

8. Paul graphed the line $y = 3x + 4$. Write a mathematical sentence that describes each of the following:

   a. the set of points above the line
   b. the set of points on or below the line

For questions 9–12, graph each inequality on a coordinate plane.

   9. $y \leq x - 1$

  10. $y > 2x - 3$

  11. $y \geq \frac{1}{2}x$

  12. $x < 2$

For questions 13 and 14, write an equation to express each relationship and draw the graph.

13. One number is equal to or greater than 2 more than another number.

14. The sum of two numbers is less than 7.

Extended-Response Questions

15. Marta has $24 to spend on souvenirs. Mugs cost $4 each and posters cost $2 each.

   a. Let x represent the number of posters and y represent the number of mugs. Write an expression for the total cost of x posters and y mugs.
   b. Write a mathematical sentence that says the total cost of the posters and mugs cannot be greater than $24.
   c. Graph the mathematical sentence you wrote in part b.
   d. Give three ordered pairs that represent combinations of posters and mugs that Marta can buy and stay within her budget. Use only numbers that make sense in the problem and explain what each ordered pair represents.
   e. Graph the point (6, 5). Explain how you know from the graph that Marta cannot purchase the combination this point represents.

16. Delray, Inc. manufactures two models of barbecue grills, regular and deluxe. For Delray to make a profit, its daily production must satisfy the inequality $y \geq 2x + 3$, where x represents the number of regular grills and y represents the number of deluxe grills.

a. Graph the inequality.
b. Locate the ordered pair for each day's production. Tell if the company made a profit that day.

	Regular	Deluxe
Monday	8	14
Tuesday	6	16
Wednesday	9	21
Thursday	4	20
Friday	18	10

11.4 Slope

The **slope** of a line is a number that tells how steep a line is and in what direction it slants. The slope of a line may be a positive number, a negative number, or zero. A vertical line has no defined slope.

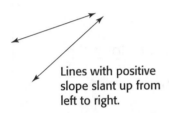

Lines with positive slope slant up from left to right.

Lines with negative slope slant down from left to right.

Lines with a slope of zero are horizontal.

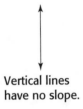

Vertical lines have no slope.

The slope of a line can be expressed as a ratio, using any two points on the line.

$$\text{slope} = \frac{\text{change in } y\text{-value}}{\text{change in } x\text{-value}} = \frac{\text{rise}}{\text{run}}$$

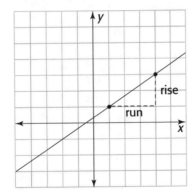

A basic property of a straight line is that its slope is constant. Therefore, any two points on the line may be used to determine the slope of the line.

1. Find the slope of the line that contains the points (2, 1) and (5, 7) and describe the direction of the line.

   Solution

   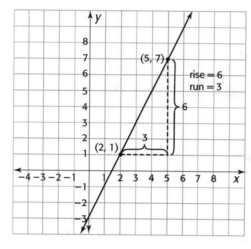

$$\text{slope} = \frac{\text{change in } y\text{-value}}{\text{change in } x\text{-value}} = \frac{7 - 1}{5 - 2} = \frac{6}{3} = 2$$

   Answer A slope of 2 means that for every change of 1 unit along the x-axis, there is a change of 2 units along the y-axis.
   The slope is positive, so the line slants up from left to right.

2. Find the slope of the line $y = -4x + 5$ and describe the direction of the line.

   Solution Make a table of values. Use integral values for x that are easy to work with.

x	−4x + 5	y	(x, y)
0	−4(0) + 5	5	(0, 5)
1	−4(1) + 5	1	(1, 1)
2	−4(2) + 5	−3	(2, −3)

   Use any two points on the line, such as (0, 5) and (1, 1).

   $$\text{slope} = \frac{\text{change in } y\text{-value}}{\text{change in } x\text{-value}}$$

   $$= \frac{1 - 5}{1 - 0} = \frac{-4}{1} = -4$$

   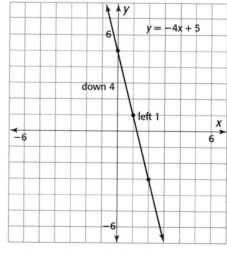

   Answer The slope is negative, so the line slants down from left to right. For every change of 1 unit along the x-axis, there is a change of −4 units along the y-axis.

3. Through point (1, 2) draw the line whose slope is $\frac{3}{2}$.

Solution Graph point (1, 2).

Since slope = $\frac{3}{2}$, when y changes 3, x changes 2. Start at (1, 2) and move 3 units up and 2 units right to locate another point. Then repeat these movements to locate a third point. Draw the line that contains these three points.

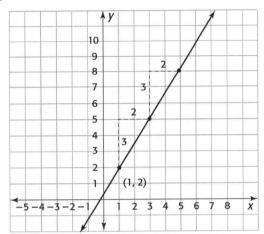

Practice

Multiple-Choice Questions

1. Which line has a negative slope?

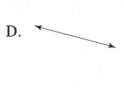

A.

B.

C.

D.

2. The slope of the line that contains the points (7, 3) and (6, −1) is

F. $-\frac{1}{4}$

G. $\frac{1}{4}$

H. −4

J. 4

3. The slope of the line that contains the points (−4, 6) and (5, 6) is

A. 0

B. –2

C. 10

D. undefined

4. The slope of the line that contains the points (3, −8) and (3, 4) is

F. 0

G. $-\frac{4}{3}$

H. 4

J. undefined

5. The line that contains the points (1, 5) and (−2, −5)

A. is parallel to the x-axis

B. is parallel to the y-axis

C. slants up from left to right

D. slants down from left to right

6. Which pair of points lies on a line that slants down from left to right?

F. (0, 0) and (3, 7)

G. (1, 2) and (4, −6)

H. (−2, 3) and (5, 3)

J. (4, −8) and (6, 7)

7. The slope of the line $y = 5x - 7$ is

A. −9   B. $-\dfrac{1}{5}$

C. 5   D. 9

8. The slope of the line shown is

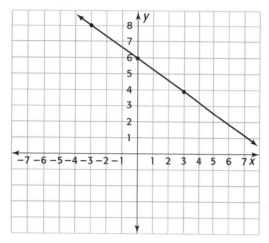

F. $-\dfrac{3}{2}$ G. $-\dfrac{2}{3}$

H. −2   J. −3

9. A hill rises 3 meters vertically for every 30 meters of horizontal distance. The slope of the hill is

A. $\dfrac{1}{10}$ B. $\dfrac{10}{3}$

C. 10   D. 27

10. Which line has a slope of 2?

F.

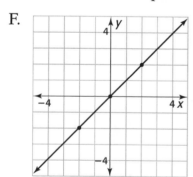

G.

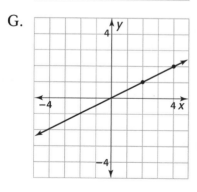

H.

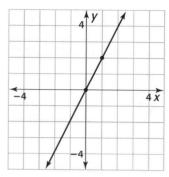

J.

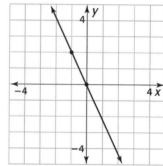

Short-Response Questions

11. On a coordinate plane, draw and label each line.

a. line *a* with a positive slope
b. line *b* with a negative slope
c. line *c* with a slope of 0
d. line *d* with undefined slope

12. In each case, graph both points and draw the line that they determine. Find the slope of the line.

a. (2, 1) and (4, 7)
b. (0, 0) and (−8, 2)
c. (5, −6) and (3, −10)
d. (1, 4) and (4, 6)

13. a. Graph the line $x + y = 4$.
b. Find the slope of the line $x + y = 4$. Show your work.

14. Find the slope of the line graphed below. Show your work.

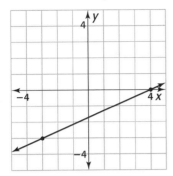

15. Find the slope of the line of ascent of the airplane shown.

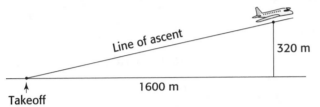

Line of ascent

320 m

1600 m

Takeoff

Extended-Response Questions

16. Little Hill has a slope of $\frac{1}{8}$. Big Hill has a slope of $\frac{2}{5}$. Draw a sketch of the two hills. Show the vertical rise of each for a horizontal distance of 40 feet.

17. a. Graph $\triangle ABC$ with vertices $A(-3, 5)$, $B(9, -7)$, and $C(-6, -1)$.
   b. Find the slope of each side of the triangle.

18. Find the value of the variable so that the line through the two given points has the given slope.

   a. $(6, a)$ and $(3, 4)$; slope $= 1$
   b. $(2, -3)$ and $(3, b)$; slope $= -1$

19. Draw a line with the given slope through the given point.

   a. $(0, 0)$; slope $= 2$
   b. $(1, 1)$; slope $= \frac{3}{4}$
   c. $(-2, 4)$; slope $= -3$
   d. $(-1, -3)$; slope $= -\frac{1}{2}$

20. Tell if each set of points lies on a line. Explain your answer.

   a. $P(4, 7)$, $Q(6, 8)$, $R(8, 9)$
   b. $S(6, 10)$, $T(8, 13)$, $U(12, 18)$

11.5 Exploring Nonlinear Functions

The graph of a function such as $y = 3x + 4$ is a straight line. For this reason, $y = 3x + 4$ is defined as a **linear function**. Knowing two or three ordered pairs is enough to draw the graph of a linear function correctly.

Not all functions are linear. The graph of a **nonlinear function** is not a straight line. More ordered pairs are needed to graph a nonlinear function.

A function with a variable raised to the second power is called a **quadratic function**.

Example $y = x^2$

A function in which a variable appears as an exponent is called an **exponential function**.

Example $y = 2^x$

 **Model Problems**

1. Graph the quadratic function $y = x^2$.

   Solution Make a table of values. Use positive and negative values for x to get an idea of how the graph should look.

x	x^2	y
-4	$(-4)^2$	16
-3	$(-3)^2$	9
-2	$(-2)^2$	4
-1	$(-1)^2$	1
0	0^2	0
1	1^2	1
2	2^2	4
3	3^2	9
4	4^2	16

   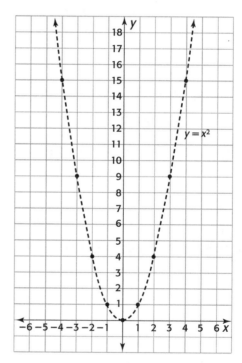

Since the square of a number is always nonnegative, the ordered pairs for the function above will only be in the first and second quadrants. Moreover, the points should be connected by a smooth curve, not line segments. To see why line segments do not accurately represent this graph, try some non-integral values for x such as:

$(0.5)^2 = 0.25$, $(-1.5)^2 = 2.25$, and so on.

Because this function "grows by multiplying," the y-value gets very large quickly. For each 1-unit change in the x-value, the change in the y-value is not the same. The slope of a nonlinear function is not constant.

For example,

x	x^2	y
1	1^2	1
2	2^2	4
3	3^2	9

+1 from 1 to 2, +1 from 2 to 3; +3 from 1 to 4, +5 from 4 to 9

Also observe that the y-axis is a line of symmetry for the graph of $y = x^2$.

---

2. Caroline's science class plans to collect data on the growth of bread mold. Caroline has hypothesized that the mold will grow according to the exponential function $y = 2^x$, where y represents units of bread mold and x represents what day of the experiment it is.

a. Complete the table showing Caroline's hypothesis from Day 0 (the start of the experiment) to Day 5.

b. Graph the ordered pairs from the table.

c. Describe the pattern you see in the table and the appearance of the graph.

Solution a.

Day		Units of Mold	
x	2^x	y	(x, y)
0	2^0	1	(0, 1)
1	2^1	2	(1, 2)
2	2^2	4	(2, 4)
3	2^3	8	(3, 8)
4	2^4	16	(4, 16)
5	2^5	32	(5, 32)

c. In the table, the units of mold (y) double each day. The graph is nonlinear and appears to get steeper as you read from left to right.

b.

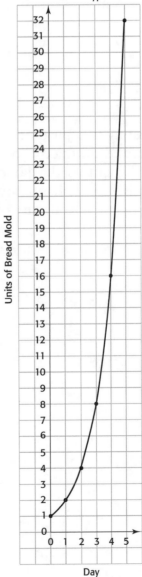

Mold Growth Hypothesis

Multiple-Choice Questions

1. The graph of which function is NOT a straight line?

   A. $y = 2x$

   B. $y = \frac{1}{2}x$

   C. $y = x + 2$

   D. $y = x^2$

2. To graph the function $y = x^2 + 1$, Roy made a table of values. The y-values when $x = -3, -2,$ and -1 in the table are (in order)

   F. 5, 2, 1
   G. 10, 5, 2
   H. −10, −5, −2
   J. −8, −3, 0

3. Billy put one penny in a jar on the first day of the year and doubled the number of pennies he put in the jar every day after that. How much money did he put in the jar on the 10th day of the year?

   A. $0.18
   B. $0.19
   C. $2.56
   D. $5.12

4. For the function $y = 3^x$, what is the value of y when $x = 4$?

   F. 12
   G. 27
   H. 34
   J. 81

5. Which could be the graph of $y = \frac{1}{2}x^2$?

   A.
   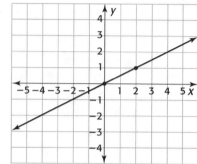

   B.
   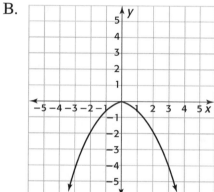

   C.
   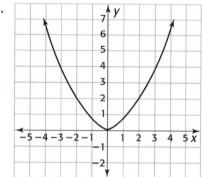

   D.
   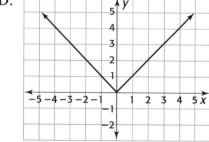

6. Which of the following is true about the graph of $y = 3x^2$?

   F. All the points of the graph are in Quadrants I and II.
   G. The y-axis is a line of symmetry for the graph.
   H. The graph is a smooth curve.
   J. All of the above are true.

7. What value of m will make the following statement true?

$$128 = 2^m$$

   A. 6   B. 7
   C. 8   D. 9

Short-Response Questions

8. Christopher's new fitness plan begins with one pull-up each day and then triples the number of pull-ups every week. If he sticks with the plan, how many pull-ups will Christopher be doing each day of the 5th week?

9. Draw the graphs of $y = 2x^2$ and $y = 2x$ on the same coordinate plane. Use integral values of x from $x = -3$, to $x = 3$. Name the ordered pair(s) that belong to both graphs.

Extended-Response Questions

10. a. Draw the graph of $y = -x^2$. Use integral values of x from $x = -3$ to $x = 3$.
   b. How is the graph of $y = -x^2$ related to the graph of $y = x^2$?

11. a. Draw the graphs of $y = x^2$ and $y = (x - 1)^2$ on the same coordinate plane. Use integral values of x from $x = -3$ to $x = 3$.
   b. Compare the two graphs you drew. How are the graphs related to each other?

12. a. On the same coordinate plane, draw the graphs of $y = x^2$ and $y = 2^x$. Use integral values of x from $x = 0$ to $x = 6$.
   b. As x increases, which function do you think increases more rapidly? Give reasons for your answer.

Chapter 11 Review

Multiple-Choice Questions

1. What is the domain of this relation?

x	0	1	4	7
y	0	2	8	14

   A. 0, 1, 4, 7
   B. 0, 2, 8, 14
   C. (0, 0), (1, 2), (4, 8), (7, 14)
   D. $y = 2x$

2. Which ordered pair is a solution of $2y - x = 8$?

   F. $(-5, 13)$
   G. $(12, -6)$
   H. $(6, 7)$
   J. $(0, -8)$

3. Which is the graph of $y < -2$?

A.

B.

C.

D.

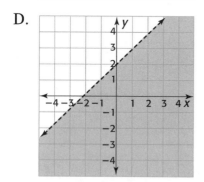

4. What must be the value of k if $(k, -4)$ lies on the line $y = 3x - 10$?

F. -2 G. 2

H. $-\dfrac{14}{3}$ J. $\dfrac{14}{3}$

5. The graph of $y = x^2 + 1$ is a

A. straight line B. point
C. curve D. half-plane

6. Which equation expresses the relationship between x and y shown in the table?

x	-1	0	1	3	4
y	2	4	6	10	12

F. $y = x + 3$ G. $y = 3x + 2$
H. $y = -2x + 6$ J. $y = 2x + 4$

7. Which ordered pair is a solution of $y > 3x - 2$?

A. $(4, 10)$ B. $(0, -5)$
C. $(-3, -8)$ D. $(7, 15)$

8. Which line has a positive slope?

F. 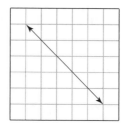 G.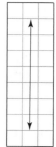

H.

J.

9. The slope of the line connecting the points $(1, 5)$ and $(8, -9)$ is

A. -7
B. -2
C. $\dfrac{4}{7}$
D. 2

10. The line connecting the points $(7, 3)$ and $(1, -1)$

F. slants up from left to right
G. slants down from left to right
H. is parallel to the x-axis
J. is parallel to the y-axis

Short-Response Questions

11. The table gives pairs of values for the variables c and n.

n	0	1	2	4
c	0	6	12	24

   a. Does c vary directly as n? Explain.
   b. Express the relationship between the variables using an equation.
   c. When $n = 8$, what is the value of c?

12. It will take 6 workers 10 days to remodel a restaurant.

   a. If only 4 workers are assigned to the job, how long will it take?
   b. If 10 workers are assigned to the job, how long will it take?

Extended-Response Questions

13. On the first day of summer there were two crickets in the garden. Every day after that the number of crickets doubled.

   a. Write an equation that expresses the relationship between the days and the number of crickets. Explain what each variable represents.
   b. Graph the equation from part a. Use only values that make sense.
   c. How many crickets are there on the 4th day? Label this ordered pair on your graph.

14. a. Write an equation that represents a line parallel to the x-axis.
   b. Graph the equation you wrote.
   c. What is the slope of the line you graphed?

15. a. Graph a straight line through points $(1, 1)$ and $(-2, -8)$.
   b. Write the coordinates of two other points on the line.
   c. Does the point $(3, -4)$ lie on the line that you graphed?
   d. Find the slope of the line you graphed. Show your work.

16. The graph shows a relation between two variables x and y. Is y a function of x? Explain why or why not.

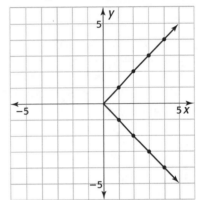

17. Draw the graph of $y \leq 2x - 5$.

18. Graph the point $P(-4, -2)$. Then draw a line through P whose slope is $\frac{4}{3}$. Explain the steps you use.

19. a. Make a table of values for $y = x^2 - 3$. Use integral values of x from -4 to 4.
   b. Draw the graph of $y = x^2 - 3$.
   c. Describe a line of symmetry for the graph.

20. a. Find the slope of the delivery ramp.
   b. Workers have complained the ramp is too steep. If a new ramp is to be built the same height but with a slope of $\frac{1}{6}$, how long will the ramp have to be?

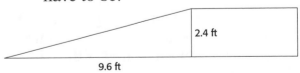

2.4 ft

9.6 ft

Multiple-Choice Questions

1. Which shows 0.00014 expressed in scientific notation?

   A. 14×10^{-5} B. 1.4×10^{-4}
   C. 0.14×10^{-3} D. 1.4×10^{-6}

2. How much greater is 2.176 than 0.8948?

   F. 6.772   G. 3.0708
   H. 1.2812   J. 0.6772

3. Which group of numbers is in order from least to greatest?

   A. $0.43, \frac{4}{5}, 16\%, 2.1, \frac{14}{8}$

   B. $16\%, \frac{4}{5}, \frac{14}{8}, 0.43, 2.1$

   C. $0.43, \frac{4}{5}, 2.1, \frac{14}{8}, 16\%$

   D. $16\%, 0.43, \frac{4}{5}, \frac{14}{8}, 2.1$

4. If $3^x = 81$, what is the value of x?

   F. 4
   G. 5
   H. 9
   J. 27

5. If a number is tripled and then increased by 4, the result is the square root of 100. What is the number?

   A. 32
   B. 6
   C. $4\frac{2}{3}$
   D. 2

6. One leg of a right triangle measures 3 feet and the hypotenuse measures 60 inches. What is the area of the triangle?

   F. 6 ft²   G. 7.5 ft²
   H. 24 ft²   J. 85.5 ft²

7. Two angles of a triangle measure 105° and 20°. Find the measure of the third angle and classify the triangle by its angles.

   A. 45°, acute   B. 55°, acute
   C. 55°, obtuse   D. 75°, obtuse

8. The area of circle A is 4π cm². The area of circle B is 100π cm². What is the difference between the length of the diameter of circle B and the length of the diameter of circle A?

   F. 4 cm   G. 8 cm
   H. 16 cm   J. 48 cm

9. One of the digits of the number 9,876,532 is picked at random. What is the probability that the digit is a prime number?

   A. $\frac{5}{7}$ B. $\frac{4}{7}$

   C. $\frac{3}{7}$ D. $\frac{2}{7}$

10. There are 6 runners in a race. How many different ways can the runners finish first, second, and third if all runners complete the race?

    F. 18   G. 30
    H. 120   J. 720

Short-Response Questions

11. Find the area of the shaded region. Show your work.

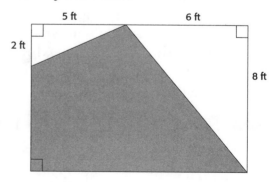

12. The bar graph shows the number of Supra and Jupiter cars sold in the town of Babylon from 1997–2000. Use the graph to answer the questions below.

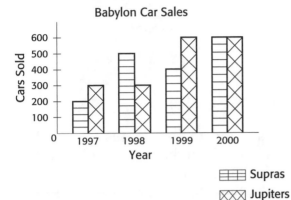

Babylon Car Sales

⊞ Supras
⊠ Jupiters

a. From 1997–2000, which car had the greater sales? How much greater?
b. What was the median for Jupiter sales?
c. To the nearest percent, what percent of total car sales took place in 1999?

13. Harry's Hardware is having a Holiday Hammer Sale. The sign shows the discounts for some hammer purchases. Assume the pattern continues. Whitney bought 24 hammers that would regularly cost $500. After she received her discount, what was the price per hammer?

Holiday Hammer Sale	
If you buy	You save a total of
3	$4
6	$8
9	$12
12	$16

Extended-Response Questions

14. Students in Mr. Gomez's science class received the following scores on a test:

70 43 58 82 93 81 76 54 68
74 85 87 77 92 71 88 83 66
90 80 65 79 83 87 78

a. Copy and complete the frequency table to group the data.

Interval	Tally	Frequency
40–49		
50–59		
60–69		
70–79		
80–89		
90–99		

b. Draw a histogram for the data.
c. What was the median score?
d. If a student is chosen at random, what is the probability the student scored at least 80 on the test? Write the answer as a decimal.

15. The Hurricanes have a 50% chance of winning any game against the Tornadoes.

a. Use a tree diagram or other method to show the possible outcomes of a 4-game series with the Tornadoes. Use W (win) and L (loss).
b. If the outcome of each game is independent, what is the probability that the Hurricanes will win exactly 2 games?
c. What is the probability the Hurricanes will lose at least 1 game?

16. a. Graph the figure with vertices $A(-6, 5)$, $B(-3, 5)$, $C(-1, 2)$, and $D(-7, 2)$.
b. Join the vertices in order and identify the type of figure.
c. Translate the figure 8 spaces right and 2 units up. Write the coordinates of the vertices A', B', C', D'.
d. Reflect the figure A' B' C' D' in the x-axis. Write the coordinates of the vertices A'', B'', C'', and D''.

17. The length of a box is represented by $2x + 1$, the width by $\frac{4}{3}x$, and the height by $\frac{1}{2}x + 9$.

   a. Find the volume of the box when $x = 6$ inches.

   b. If the volume of the box is increased by 50%, what will be the new volume?

   c. Suggest how the box can be redesigned to have the increased volume.

18. The cost of renting a car is $30 per day plus $0.25 per mile. Let x represent the number of miles a car was driven and let y represent the total cost, in dollars, for a day.

   a. Write an equation for the rental cost of the car in terms of the number of miles driven.

   b. Complete the table.

x (miles)	y (cost)
60	
120	
200	

   c. Use the ordered pairs from the table to graph the equation you wrote. Show only the part of the graph that makes sense for the problem.

   d. Use the graph to find the cost of driving 160 miles.

19. A 30-foot ladder leans against a building and makes an angle of 72° with the ground.

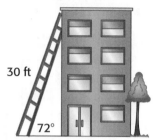

   a. Find to the nearest tenth of a foot the height at which the ladder touches the building.

   b. Find to the nearest tenth of a foot the distance between the foot of the ladder and the building.

20. The hour hand on an outdoor clock is 30 centimeters long. To the nearest tenth of a meter, how far does the tip of the hand travel in one week? Show your work.

Mathematics Reference Sheet

Formulas

 Rectangular Solid   Total Surface Area = $2(\ell w) + 2(hw) + 2(\ell h)$

 Right Circular Cylinder Total Surface Area = $2\pi rh + 2\pi r^2$ Volume = $\pi r^2 h$

 Pythagorean Theorem $c^2 = a^2 + b^2$

Trigonometric $\sin A = \dfrac{\text{opposite}}{\text{hypotenuse}}$ $\cos A = \dfrac{\text{adjacent}}{\text{hypotenuse}}$ $\tan A = \dfrac{\text{opposite}}{\text{adjacent}}$

Trigonometric Table			
Degrees	Sine	Cosine	Tangent
0	.0000	1.0000	.0000
5	.0872	.9962	.0875
10	.1736	.9848	.1763
15	.2588	.9659	.2679
20	.3420	.9397	.3640
25	.4226	.9063	.4663
30	.5000	.8660	.5774
35	.5736	.8192	.7002
40	.6428	.7660	.8391
45	.7071	.7071	1.0000
50	.7660	.6428	1.1918
55	.8192	.5736	1.4281
60	.8660	.5000	1.7321
65	.9063	.4226	2.1445
70	.9397	.3420	2.7475
75	.9659	.2588	3.7321
80	.9848	.1736	5.6713
85	.9962	.0872	11.4301
90	1.0000	.0000	....

PRACTICE TEST 1

Multiple-Choice Questions

1. What is the value of this expression?
$$16 - 4 \div (1 + 1)^2$$
   A. 3
   B. 6
   C. 14
   D. 15

2. What is the prime factorization of 120?
   F. $2 \times 4 \times 3 \times 5$
   G. $2^2 \times 3^2 \times 5$
   H. $2^3 \times 3 \times 5$
   J. $2^4 \times 5^2$

3. Which set of numbers is in order from least to greatest?
   A. $6\frac{2}{3}$, 6.6, 6.7, $6\frac{3}{7}$
   B. 6.6, $6\frac{2}{3}$, $6\frac{3}{7}$, 6.7
   C. $6\frac{3}{7}$, 6.6, $6\frac{2}{3}$, 6.7
   D. $6\frac{3}{7}$, $6\frac{2}{3}$, 6.6, 6.7

4. When full, a delivery truck weighs 2.4×10^4 pounds. How much would 10 of these trucks weigh?
   F. 2.4×10^5 lb
   G. 24×10^5 lb
   H. 2.4×10^6 lb
   J. 24×10^6 lb

5. Donna and Lucas volunteered to paint the fence around a local park. Donna painted $\frac{1}{4}$ of the fence and Lucas painted 40%. What percent of the fence remains to be painted?
   A. 15%
   B. 25%
   C. 35%
   D. 45%

6. A terrace is shaped like the trapezoid shown below. $\overline{AB}$ measures 50 feet, $\overline{DC}$ measures 40 feet, and the area of the terrace is 1,350 square feet. How long is $\overline{AD}$?

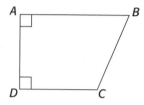

   F. 15 ft   G. 30 ft
   H. 45 ft   J. 60 ft

7. Chenique answered 5 out of 8 questions on the quiz correctly. What percent of the questions did she answer correctly?
   A. 5.8%
   B. 58%
   C. 62.5%
   D. 87.5%

8. Which number is the greatest?
   F. $(0.2)^5$
   G. $(0.3)^4$
   H. $(0.4)^3$
   J. $(0.5)^2$

9. A sales agent makes 40 calls a day. About 15% of the calls result in sales. How many days will it take for the sales agent to meet a quota of 120 sales?
   A. 3
   B. 8
   C. 20
   D. 34

10. $2.74 \div 0.8$ to the nearest hundredth is
   F. 34.25
   G. 3.43
   H. 0.34
   J. 0.03

11. What type of triangle is △ABC?

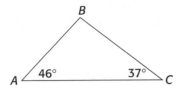

A. acute   B. obtuse
C. right   D. isosceles

12. Use your ruler to help you solve this problem. A playground is in the shape of a rectangle as shown in the figure below. The playground is to be enclosed by a fence. Based on the scale, how much fencing is needed to enclose the playground?

Scale: 0.5 centimeters = 2 meters

F. 14.5 meters   G. 58 meters
H. 72.5 meters   J. 116 meters

13. Mr. Lang was driving to Ithaca. He left home with a full tank and drove $\frac{1}{4}$ of the way there. Then he changed his mind and drove back home. When he got back, the gas tank was $\frac{1}{4}$ full. Which statement below is correct.

A. He used $\frac{1}{2}$ tank of gas.
B. If he had continued to Ithaca, he would have had $\frac{1}{4}$ tank left when he arrived there.

C. He did not have enough gas to make it to Ithaca.
D. The distance he drove was the same as the distance to Ithaca.

14. A delivery service charges $9.00 for a package up to 2 pounds and $1.50 for each additional one fourth of a pound. Thomas was charged $18 for a package delivery. How much did the package weigh?

F. 3 pounds   G. $3\frac{1}{2}$ pounds

H. 4 pounds   J. $4\frac{1}{2}$ pounds

15. A piece of land is shaped like a triangle with a semicircle on one side. A diagram of the land is shown below. What is the area of the piece of land? Use 3.14 for π.

A. 503.25 square feet
B. 653.25 square feet
C. 1,006.5 square feet
D. 1,306.5 square feet

Do problems 16–19 about the Hicksville Beautification Campaign.

16. New trees are being planted throughout Hicksville. The table shows the trees planted so far. If the pattern continues, how many trees will be planted in Week 8?

Tree Planting Schedule	
Week	Number of Trees
1	2
2	3
3	5
4	8
5	12

F. 16   G. 19
H. 23   J. 30

17. Trees cost $18.50 each. Customers get 1 tree free for every 7 trees purchased at full price. How much will it cost the Beautification Committee if it orders 192 trees?

A. $444.00
B. $507.43
C. $3,108.00
D. $3,422.50

18. The amounts of money contributed by some local businesses to the Beautification Campaign are shown in the table. Which of the following represents the mean number of dollars the businesses contributed?

Beautification Contributions	
Business	**Amount**
T & M Music	$150
Denim Duds	$150
The 89¢ Store	$300
Johnson Press	$200
Cybernetics	$0
Country Bank	$600
Sanchez & Sons	$350
Ross, Inc.	$250

F. $150
G. $225
H. $250
J. $285.71

19. A shipment contained 16 maple trees and 24 oak trees. If one of the trees is selected at random, what is the probability that a maple tree will be selected?

A. $\frac{2}{5}$

B. $\frac{2}{3}$

C. $\frac{3}{4}$

D. $\frac{3}{2}$

20. Doubleena the circus clown rides a bicycle that has one wheel with a radius twice as long as the radius of the other. How does the circumference of the larger wheel compare to the circumference of the smaller wheel?

F. twice the length
G. four times the length
H. six times the length
J. eight times the length

21. On a coordinate plane, point (m, n) is the same distance from the origin as point (p, n). Which of the following must be true?

A. $m = p$
B. $|m| = |p|$
C. $m = -p$
D. $m + p = 1$

22. Which inequality is graphed correctly on the number line shown?

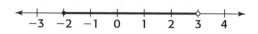

F. $-2 \geq x > 3$
G. $-2 < x \leq 3$
H. $-2 \leq x \leq 3$
J. $-2 \leq x < 3$

23. The combined mass of a class of 25 eighth-graders is about

    A. 120,000 milligrams
    B. 12,000 grams
    C. 1200 kilograms
    D. 120 kilograms

24. The two figures shown are similar. What is the perimeter of Figure N?

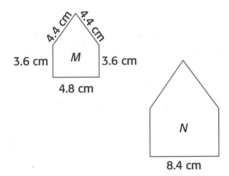

    F. 22.4 cm   G. 28.7 cm
    H. 30.1 cm   J. 36.4 cm

25. In the equation below, which value of x will make the statement true?

$$4(x - 1) - 2 = 14$$

    A. -3 B. 4
    C. 5   D. 6

26. Misty makes a beaded bracelet every 25 minutes. She sells the bracelets for $3.75 each. If she filled an order for bracelets worth $56.25, how many hours did she spend making bracelets?

    F. $6\frac{1}{4}$ hours   G. $8\frac{1}{3}$ hours

    H. $12\frac{1}{2}$ hours   J. 15 hours

27. Ted has a collection of 171 CDs. One day he made three stacks using all the CDs.

- The second stack had twice as many CDs as the first stack.
- The number of CDs in the third stack was the mean of the first and second stacks.

How many CDs were in the first stack?

    A. 23   B. 27
    C. 38   D. 41

Short-Response Questions

 You may use a calculator to solve problems on this part of the test.

28. Look at the number line below.

    a. Assume that the number line is scaled so that the distance between two whole numbers is divided into 3 equal parts. Give the fraction or mixed number that corresponds to each letter.

    X ____ Y ____ Z ____

    b. Assume that the number line is scaled so that the distance between two whole numbers is divided into 8 equal parts. Give the fraction or mixed number that corresponds to each letter.

    X ____ Y ____ Z ____

29. At the intersection shown, a car can go left (L), right (R), or straight ahead (S). Two cars arrive at the intersection.

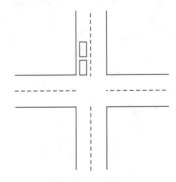

    a. Make a list of all the possible outcomes for the directions the two cars can take. Show your work.

    b. How many outcomes are there?

    c. If it is equally likely that a car will go in any one of the three directions and one car's direction does not influence the car behind it, what is the probability the two cars will go in the same direction?

30. The results of a survey of students' favorite fruits are shown.

Favorite Fruits	
Fruit	**Number**
Banana	20
Apple	15
Orange	14
Pear	8
Plum	3

a. What percent of the students surveyed picked apple or plum as their favorite?

b. If you were displaying this data in a circle graph, what would the central angle of the banana sector be?

c. Based on these results, if a larger survey of 900 students were conducted, how many students would you expect to name orange as their favorite fruit? Explain your answer.

31. A scale model of a new hotel was built for display. The scale used was 5 centimeters = 3 meters. The scale model required 13,500 square centimeters of plastic for the outside surfaces of the hotel.

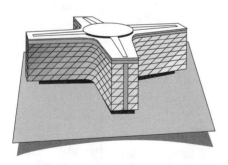

a. What will the area of the outside surfaces of the actual hotel be?

b. The volume of the scale model is 202,500 cubic centimeters. What will the volume of the actual hotel be?

32. Handy Mandy fixes major appliances such as refrigerators and air conditioners. She charges a fixed fee for a home visit plus an hourly rate for the repair work. Use the information given below to answer the questions.

Hours	3	7	10
Total Cost	$97	$205	$286

a. What is Mandy's fixed fee?
b. What is Mandy's hourly rate?

33. This year, Cobblestone Orchards harvested 680 pumpkins. They sold 90% of the harvest and they used the remaining pumpkins to make pumpkin pies. The baker uses about $\frac{1}{2}$ pumpkin for each pie. How many pies were made this year?

34. The table below shows the regular price and the sale price for several items at The Clothes Closet. The store uses the same percent discount for all items during this sale.

Item	Regular Price	Sale Price
Shirt	$24.00	$16.80
Sweater	$38.00	$26.60
Pants	$30.00	
Jacket	$65.00	
Belt	$13.00	

a. Find the sale price of the pants, jacket, and belt. Show your work.

b. The sales tax on clothing is $7\frac{1}{4}\%$. Determine the final cost of purchasing all five items above. Round your answer to the nearest cent.

35. A cable 60 meters long supports a cellular network tower. The cable is fastened to the ground at a point 36 meters from the base of the tower. Draw a diagram of the tower and cable. What is the height of the tower? Show your work.

36. The sun is about 93 million miles from Earth. If you drove at 60 miles per hour, about how many days would it take you to reach the sun? Show your work.

37. a. Graph points $N(-3, 0)$, $I(0, 4)$, $C(3, 0)$, and $E(0, -4)$. Connect the points in order.
  b. What type of quadrilateral is *NICE*? Give reasons for your answer.
  c. Find the area of *NICE*. Explain your method.

38. The variables a, b, c, and d each represent a different whole number. Fill in each blank with a, b, c, or d. Explain your reasoning and name properties you used.

$$a + b = c$$
$$b + d = b$$

$b + a = $ ____
$(a + b) + (a - a) = $ ____
$a + b + d = $ ____
$c - b = $ ____
$(a + b) - (b + a) = $ ____

39. Anthony needs $33 this week to buy a concert ticket. In addition to his $8 weekly allowance, his father agreed to pay him $5 an hour for cleaning out the garage. Let h represent the number of hours Anthony works.

  a. Write an inequality that can be solved to find the least number of hours Anthony must work to have enough money for the ticket.
  b. Solve your inequality and graph the solution on a number line.

 You may use a calculator to solve problems on this part of the test.

40. The ninth and tenth grades at Rockland High School are selling raffle tickets to raise money for a new playground in the community. The raffle winner gets a dinner for two at Umberto's Lobster House. The table below shows the number of tickets sold by each grade during a four-week period.

Playground Raffle Sales		
Week	9th grade	10th grade
1	190	230
2	140	170
3	180	160
4	250	200

  a. Construct a double-line graph to display the data in the table.
  b. Explain how you can use your graph to determine the week in which the difference between the ninth and tenth grade sales was the least?

41. Caitlin is competing in a 7-game bowling tournament. Her scores for the first 6 games are shown below

Caitlin's Tournament Results							
Game	1	2	3	4	5	6	7
Score	141	125	160	133	148	133	?

  a. Find the mean, median, mode, and range of Caitlin's scores.
  b. How would the mean change if Caitlin bowls 174 in game 7?
  c. Which measure(s) will change if Caitlin bowls 140 in game 7?
  d. Which measure(s) will **NOT** change if Caitlin bowls 160 in game 7?

42. a. Using the table, fill in the missing numbers for the four ordered pairs for the function $y + 4 = 2x$.

x	y
−3	
	−6
2	
	6

b. Graph the function $y + 4 = 2x$ using the ordered pairs from the table. Label the points.

c. Using any two points, find the slope of the line.

43. Look at Figures 1 to 3.

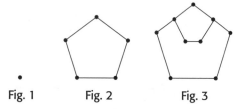

Fig. 1   Fig. 2   Fig. 3

a. Draw the next figure in the pattern.

b. Complete the table by writing the number of dots in each figure.

Figure	Number of Dots
1	
2	
3	
4	

c. Write a description of the pattern for the number of dots in each figure.

d. Predict the number of dots in the next three figures.

e. Will one of the figures in the pattern have 50 dots? Explain why or why not.

44. Triangle *ABC* is a right triangle. The tangent of $\angle B$ is $2\frac{2}{5}$.

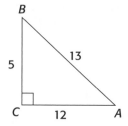

a. Find the sine, cosine, and tangent of $\angle A$.

b. What is the relationship between $\angle A$ and $\angle B$?

c. What is the relationship between the tangent of $\angle A$ and the tangent of $\angle B$?

45. The space station shuttle from Andromeda to Darth makes stops at Bluto and Candor. The travel distances are shown on the map below. The shuttle travels at an average speed of 50 light-years per hour. At each space station there is a 30-minute stopover to load and unload passengers and baggage. The shuttle departs from Andromeda at 8:30 A.M.

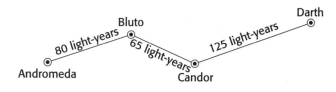

a. Complete the schedule of arrival and departure times for each space station. Show how you found the travel time needed for each part of the trip.

Space Station	Arrive	Depart
Andromeda	___	8:30 A.M.
Bluto		
Candor		
Darth		___

b. How long does a passenger spend getting from Andromeda to Darth?

Practice Test 2

Multiple-Choice Questions

1. Which of the numbers below is equivalent to 0.6?

   A. 60%
   B. 3.5
   C. $\sqrt{3.6}$
   D. $0.6\overline{1} - 1$

2. Vladimir read 27 pages in 3 hours. Each hour he read 4 pages more than the previous hour. How many pages did he read the third hour?

   F. 3
   G. 5
   H. 9
   J. 13

3. Junior had a 12-game bowling average of 140. He scored 179 in his next game. What was his new average?

   A. 141.6
   B. 143.0
   C. 154.9
   D. 159.5

4. Evaluate the expression.
   $15 \times 8 + 35 \div 5$

   F. 225
   G. 129
   H. 127
   J. 31

5. Express the quotient $\frac{0.007}{2,000}$ in scientific notation.

   A. 1.4×10^{-5}
   B. 3.5×10^{-6}
   C. 2.857×10^{5}
   D. 3.5×10^{6}

6. The figure shown consists of two squares and two isosceles right triangles. The length of the side of one square is s. The area of the figure is

   F. $\frac{3}{2}s^2$
   G. $2s^2$
   H. $3s^2$
   J. $4s^2 - s$

7. The circle graph shows the monthly budget for the Cruz family. The family's monthly income is $3,000. How much more do they spend on food than on recreation?

   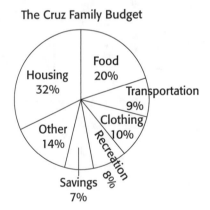

   The Cruz Family Budget

   A. $240
   B. $360
   C. $600
   D. $840

8. If a computer randomly chooses a letter in the word MULTIPLICATION, what is the probability that it will be a vowel?

   F. $\frac{5}{14}$ G. $\frac{3}{7}$

   H. $\frac{1}{2}$ J. $\frac{3}{4}$

9. How many cubic yards of dirt were removed from a hole that is 12 feet by 18 feet by 6 feet deep?

   A. 48 cubic yards
   B. 144 cubic yards
   C. 432 cubic yards
   D. 1,296 cubic yards

10. A store window is in the shape of a square with a semicircle on one side. What is the perimeter of the window? Use 3.14 for π.

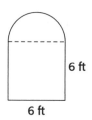

6 ft

6 ft

   F. 27.42 ft
   G. 45.42 ft
   H. 46.26 ft
   J. 55.68 ft

11. Which triangle is a 180° rotation of $\triangle RST$?

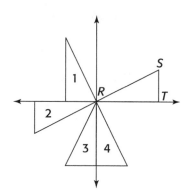

   A. $\triangle 1$
   B. $\triangle 2$
   C. $\triangle 3$
   D. $\triangle 4$

12. Suppose $a > 0$ and $b > 0$. If a increases and b remains the same, for which expression will c increase?

   F. $b \div a = c$
   G. $b - a = c$
   H. $b \times a = c$
   J. $a \times c = b$

Questions 13–16 are about Supreme Sushi.

13. A new restaurant, Supreme Sushi, opened for business on Monday. The owners recorded the number of customers each day. If the pattern continues, how many customers will the restaurant have had in all by closing time on Sunday?

Supreme Sushi	
Day	**Customers**
Monday	10
Tuesday	18
Wednesday	24
Thursday	32
Friday	38

   A. 218   B. 220
   C. 224   D. 228

14. A group of coworkers ate lunch at Supreme Sushi. The cost of the food ordered by each person is shown below.

Person	Orier	Natalie	Philip	Senaka
Food Cost	$14	$12	$15	$11
Person	Max	Teresa	André	Denise
Food Cost	$13	$12	$17	$14

Which of the following is true about the mean, median, and mode of the food costs?

F. mean = median
G. median > mean
H. median = mode
J. mode < mean

15. The entry room of Supreme Sushi is shown in the diagram. How much carpeting was used to cover the whole entry room?

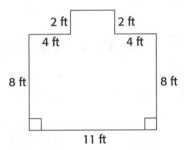

A. 92 square feet
B. 94 square feet
C. 102 square feet
D. 106 square feet

16. The name of the restaurant, Supreme Sushi, is above the front door. It is made up of individual tiles, one letter to a tile. If a fly lands on one of the tiles, what is the probability that it lands on an S?

F. $\frac{1}{12}$ G. $\frac{1}{6}$

H. $\frac{1}{4}$ J. $\frac{1}{3}$

17. 80% of a number is equal to 40% of

A. half the number
B. twice the number
C. one and one half times the number
D. four times the number

18. The scale of a blueprint is 3 inches = 5 feet. What is the actual length of a hallway that measures 10.5 inches on the blueprint?

F. 6.3 feet   G. 17.5 feet
H. 31.5 feet   J. 52.5 feet

19. The cost of a telephone call from Buffalo, New York to Mexico City is $2.00 for the first 3 minutes and $0.20 for each additional minute. Paulina called her sister in Mexico City from her home in Buffalo and was charged $6.40. How long was the call?

A. 22 minutes
B. 25 minutes
C. 28 minutes
D. 32 minutes

20. ABCD is a parallelogram. What number of degrees does x represent?

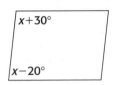

F. 65°   G. 75°
H. 85°   J. 105°

21. A farmer had exactly 100 tomatoes. He divided the tomatoes into 3 crates so that the following were true:

• There were no tomatoes left over.
• No crate had more than 45 tomatoes.
• The ratio of the tomatoes in two crates was 3 : 5.

Which of the following sets of numbers could be the contents of the crates?

A. 12, 33, 55   B. 12, 32, 38
C. 22, 34, 44   D. 24, 36, 40

22. In which triangle is there NOT enough information to find the measure of x?

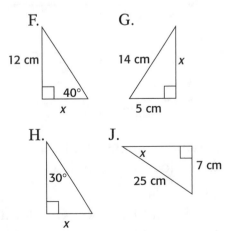

23. Lines ℓ and m are parallel. Which list includes angles that are all congruent to angle 3?

A. 1, 4, 8
B. 2, 7, 8
C. 2, 6, 7
D. 5, 6, 8

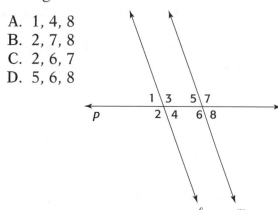

24. The value of an original *Arachnid Man* comic book rose by 50%. It is now worth about $600. What was the comic book valued at before this increase?

F. $200
G. $300
H. $400
J. $800

25. Which graph represents the solution to the inequality?

$$2x - 1 \le 5$$

A. ![number line -4 to 4]

B. ![number line -4 to 4]

C. ![number line -4 to 4]

D. ![number line -4 to 4]

26. In which quadrant of the coordinate plane does the graph of $P(|-2|, -3)$ appear?

F. Quadrant I   G. Quadrant II
H. Quadrant III   J. Quadrant IV

27. Determine the surface area of the cube shown.

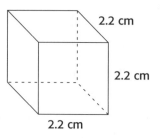

A. 10.648 cm²   B. 26.4 cm²
C. 29.04 cm²   D. 39.6 cm²

Short-Response Questions

 You may use a calculator to solve problems on this part of the test.

28. Sara, Tim, Janet, and Kit all have different subjects first period. English, math, Spanish, and biology are taught first period.

- Sara does not have Spanish.
- Tim has English first period.
- Janet and Sara are in the same fifth period biology class.

Which subject does each student have first period?

29. Tickets to the student production of *The Lion King* cost $4 in advance and $5 at the door. In all, 100 tickets were sold and $463 was collected. How many of each ticket were sold? Show your work.

30. Sean packed the following items for a trip.

Pants	Jackets	Shirts
Black	Blue	White
Blue	Tan	Red
Gray		Striped

a. Construct a tree diagram showing all the possible outfits consisting of one choice from each column. Assuming that everything matches, how many different outfits can be made?
b. Sean is thinking of packing either another pair of pants or another

jacket. Which should he pack in order to have the greatest number of possible outfits? Support your answer.

31. Rectangles *JKLM* and *PQRS* are congruent. Find the coordinates of points *L*, *M*, *P*, and *Q*.

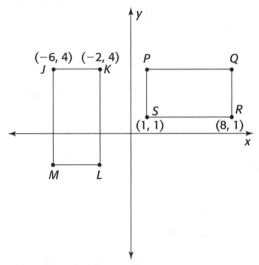

32. a. Find the missing values of the variables needed to complete the table for the equation $y + 2x = 5$.

x	y
−2	
	5
3	
	−5

b. Graph the points represented by the ordered pairs in the table and draw a line through the points.

33. In a park, a slide is built over a horizontal distance of 10 feet along the ground and makes an angle of 50° with the ground. Find the length of the slide to the nearest tenth of a foot. Show your work.

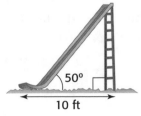

34. Jonas walks home by an empty rectangular field. He can either walk across the field diagonally, or he can walk the length and width. To the nearest foot, how much shorter is his walk if he takes the diagonal route? Show all work.

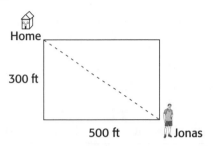

35. High school students voted for The Item I Can't Live Without. Some of the results are shown in the table. Complete the table, showing the number of votes and the percent for each item and the total number of students surveyed. Explain the reasoning used to find each missing piece of information.

The Item I Can't Live Without		
Item	**Votes**	**Percent**
TV		10%
Computer		25%
Cell Phone	1,092	
Stereo	260	10%
Other		
Total		100%

36. Find all the numbers that satisfy the following three conditions:

- I am a positive integer less than 100.
- Two more than my value is a multiple of 7.
- The sum of my digits is a prime number.

Show your work and explain the steps you used to solve the problem.

37. The first and fifth terms in a pattern are 3 and 35.

$$3, \underline{\quad}, \underline{\quad}, \underline{\quad}, 35$$

The rule for the pattern is "Add k to the previous number."

a. What is k? Explain how you found your answer.
b. What are the second, third, and fourth terms?
c. What is the tenth term in this pattern?

38. Use the histogram to answer the questions.

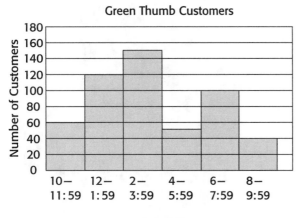

Green Thumb Customers

a. How many customers arrived from 6:00 P.M. until closing?
b. During which 4-hour interval did more than 50% of the day's customers arrive?
c. Estimate the probability that a customer planning to visit the Green Thumb will arrive between 4:00 and 5:59 P.M.

39. A gallery in a museum is 20 meters long, 15 meters wide, and 6 meters high.

a. If the walls and ceiling of the room are to be painted before an upcoming exhibit, what is the total area that must be painted? Show your work.
b. If each visitor needs at least 12 m³ of air, what is the maximum capacity of the room?

Extended-Response Questions

 You may use a calculator to solve problems on this part of the test.

40. Mr. N'Timm had 42 acres of land. He sold $\frac{2}{3}$ of the land to a hotel developer, and then sold $\frac{2}{3}$ of what remained to a neighbor who wanted more property.

a. How many acres of land does Mr. N'Timm have left?
b. What percent of his original property does the amount remaining represent?

41. Every Thursday night, Pronto Pizza gives out free pizza and soft drinks. Every 6th customer gets a free slice of pizza. Every 10th customer gets a free soft drink. Last Thursday, Pronto Pizza had 87 customers.

a. How many slices of pizza were given away?
b. How many soft drinks were given away?
c. Did any customer(s) get both items free? If so, which customers?

42. The route a runner takes can be modeled by connecting points on a coordinate plane. The route starts at $(-3, 4)$, goes to $(2, 4)$, then to $(2, -1)$, $(5, -1)$, $(5, -4)$, $(-2, -4)$, $(-2, 0)$, $(-3, 0)$, and back to $(-3, 4)$.

a. Graph the runner's route on a coordinate plane.
b. If it is 250 yards from the starting point to the first turn of the run, how many yards is the entire route?
c. Does the runner cover more or less than one mile? Explain.

43. a. Use a metric ruler and protractor to draw $\triangle DEF$, which is similar to, but not congruent to, $\triangle ABC$.

b. What is the ratio of the corresponding sides of $\triangle ABC$ and $\triangle DEF$ as you have drawn it?

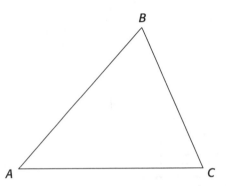

44. The formula $d = \frac{s + s^2}{20}$ gives the approximate stopping distance d in feet for a car traveling s miles per hour along a dry road.

a. Use the formula to complete the table.

Speed (s)	10	20	35	40	55
Stopping Distance (d)					

b. When the car's speed doubles from 20 mph to 40 mph, about how many times greater is the stopping distance?
c. If it takes about 100 feet for a car to stop, estimate the speed at which the car was traveling.

45. The five students whose names are shown are trying out for the school play.

a. If the director must cast the roles of a lawyer and a reporter, how many different choices are there using the five students?
b. If the director must choose two students to be extras in a courtroom crowd scene, how many different choices are there using the five students?
c. Do the situations above involve combinations or permutations? Explain your thinking.

PRACTICE TEST 3

Multiple-Choice Questions

1. Which would result in the largest number?

   A. 10^4
   B. $40^2 + 50^2$
   C. $10^3 + 20^3$
   D. $28 \times 31 + 37 \times 42$

2. The volume of this box is 540 cm^3. The length is triple the width. Find the length of the box.

   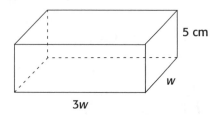

   5 cm

   w

   3w

   F. 6 cm   G. 12 cm
   H. 18 cm   J. 36 cm

3. Patrick wants to buy ice hockey skates that cost $195 including tax. He earns $45 a week at a part-time job. He spends 40% of his weekly earnings and saves the rest. What is the least number of weeks it will take him to save enough money to buy the skates?

   A. 5 weeks   B. 7 weeks
   C. 8 weeks   D. 11 weeks

4. Find the value of x.

   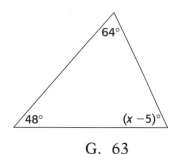

   64°

   48° $(x-5)°$

   F. 57   G. 63
   H. 68   J. 73

5. Which table satisfies the equation $y = 5 - x$?

   A.
x	y
1	6
0	5
−1	4

   B.
x	y
−2	3
0	5
2	−3

   C.
x	y
−3	8
1	4
4	1

   D.
x	y
−5	0
−1	6
3	2

6. A dripping faucet fills a 100-milliliter beaker in 30 seconds. At this rate, how long will it take to fill a 2-liter soda bottle?

   F. 5 minutes
   G. 10 minutes
   H. 20 minutes
   J. $66\frac{2}{3}$ minutes

7. A box contains 50 slips of paper numbered from 1 to 50. What is the probability of randomly drawing a card that is numbered with a multiple of 6?

A. $\frac{1}{50}$

B. $\frac{4}{25}$

C. $\frac{4}{21}$

D. $\frac{2}{5}$

8. Dante Motors sold 120 cars in October. How many red cars did it sell?

Car Colors for October

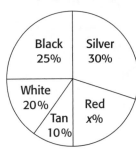

F. 12

G. 15

H. 18

J. 30

9. If these rational numbers were graphed on a number line, which of them would be closest to 0?

A. $-\frac{5}{16}$

B. 0.35

C. 32%

D. $-\frac{9}{32}$

10. Melissa has a 1:30 P.M. flight to Atlanta and she wants to get to the airport $\frac{3}{4}$ of an hour before the flight. She has some errands to do that will take 30 minutes, and then has to walk for 15 minutes to the airport van service to catch the 35-minute ride to the airport. What is the latest she should leave her apartment?

F. 11:25 A.M.　　G. 11:35 A.M.

H. 11:45 A.M.　　J. 11:55 A.M.

11. In centimeters, the lengths of the sides of a triangle are 8, 10, and 12. What is the perimeter of a similar triangle whose shortest side has a length of 20 centimeters?

A. 60 centimeters

B. 66 centimeters

C. 75 centimeters

D. 84 centimeters

12. A single card is drawn at random from a standard 52-card deck. For which of the following events is the probability of success greater than $\frac{1}{2}$?

F. drawing a heart or a club

G. drawing an ace or a queen

H. drawing a king or a diamond

J. drawing a queen or a red card

13. In how many different ways can Millie, William, Jill, and Phil take seats in a row that contains 6 chairs?

A. 720

B. 360

C. 120

D. 24

14. Simplify $4^2 + 9 \times 6 - 4$.

F. 50

G. 52

H. 66

J. 146

15. If $7 \ominus 8 = 31$, $4 \ominus 6 = 21$, and $5 \ominus 12 = 35$, what is $3 \ominus 2$?

A. 7

B. 11

C. 15

D. 16

16. What is the area of the trapezoid when $x = 5$ and $y = 4$?

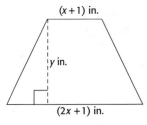

F. 18 sq in.　　G. 30 sq in.

H. 34 sq in.　　J. 108 sq in.

17. Three bananas cost as much as two apples. One apple costs the same as a plum and a banana. If Rita has enough money to buy four plums, how many bananas could she buy instead?

A. 2

B. 3

C. 6

D. 8

18. Which shows $y \leq 2$ graphed on a coordinate plane?

F.

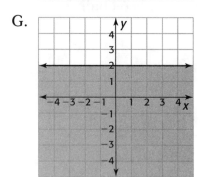

G.

H.

J.

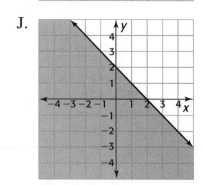

19. The ratio of girls to boys at Leewood High School is 8 to 7. If there are 1,005 students at Leewood, how many are boys?

A. 443
B. 465
C. 469
D. 536

20. What is the mean of the test scores shown in the frequency table below?

Social Studies Test Scores	
Score	Frequency
65	2
70	3
75	3
80	5
85	5
90	3
95	2
100	2

F. 82
G. 82.5
H. 83.5
J. 84

21. Figure Y is the image of Figure X after which transformation?

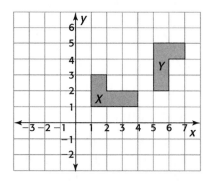

A. translation 3 units right and 2 units up
B. reflection in the line $y = 2$
C. 90° clockwise rotation about (5, 1)
D. 180° counterclockwise rotation about the origin

22. A sheet of plastic 6 in. by 10 in. is rolled along its shorter side to form a cylinder as shown. A second sheet of plastic of the same size is rolled along its longer side to form a second cylinder as shown. Which relationship is true?

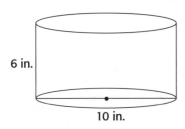

6 in.

10 in.

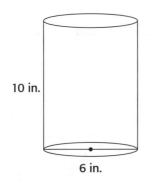

10 in.

6 in.

F. The short cylinder will have the greater volume.
G. The tall cylinder will have the greater volume.
H. Both cylinders will have the same volume, but the short cylinder will have the greater surface area.
J. Both cylinders will have the same volume, but the tall cylinder will have the greater surface area.

23. What is the tenth term in this pattern?

1, 3, 4, 7, 11, 18, . . .

A. 47
B. 56
C. 76
D. 123

24. A portable CD player marked $84 costs $78.12 after a discount. What is the discount?

F. 4%
G. 6%
H. 7%
J. 8%

25. Find the product of $6\frac{1}{4} \times 4.2$.

A. $24\frac{1}{20}$

B. $25\frac{1}{2}$

C. $26\frac{1}{8}$

D. $26\frac{1}{4}$

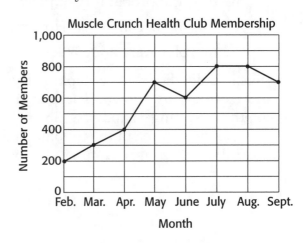

26. During which period did the membership at Muscle Crunch Health Club change twice as much as it did from February to March?

Muscle Crunch Health Club Membership

F. March–April
G. April–May
H. May–June
J. June–July

27. The cost of printing programs for the senior show is $35.00 for the first 100 programs and $0.10 for each additional program. Which equation could be used to find the total cost (*C*) of printing *x* programs if $x > 100$?

A. $C = 35 + 0.10x$

B. $C = 35 + 0.10(x - 100)$

C. $C = 35 + 0.10(100 + x)$

D. $C = 35 + 0.10(100 - x)$

Short-Response Questions

 You may use a calculator to solve problems on this part of the test.

28. A box contains 24 marbles that are either red or yellow.

a. The probability of picking a red marble is $\frac{7}{8}$. How many marbles of each color are in the box?

b. After more red marbles are added to the box, the probability of picking a red marble becomes $\frac{9}{10}$. How many red marbles were added?

29. Susan folded a square napkin in half horizontally as shown. The perimeter of the resulting shape was 36 inches. What was the area of the unfolded napkin? Explain your reasoning.

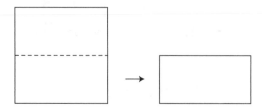

30. When water freezes, its volume increases by about 12%. The volume of a chunk of ice is determined to be 476 cubic centimeters. What was the volume of water that formed the ice?

31. The figure shows a scale drawing of a bedroom.

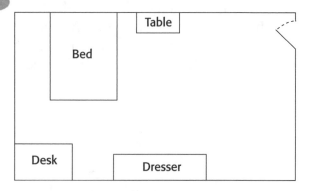

a. Use a customary ruler to measure the length and width of the drawing.

b. The actual length of the bedroom is 18 ft. What is the scale of the drawing?

c. Carpeting costs $24 per square yard including installation. Sales tax of 8% will be added. How much will wall-to-wall carpeting cost for the bedroom? Show all steps.

32. A circle has an area of 176.625 square inches. If the radius of the circle is doubled, what will be the area of the new circle? Show your work.

33. The area of Greenland is approximately 8.4×10^5 square miles. The area of Great Britain is approximately 84,200 square miles. Which island has the larger area? About how many times larger?

34. In the figure, $\overline{BC}$ represents the width of a river. $\overline{AB}$, the diagonal across the river, is 25m.

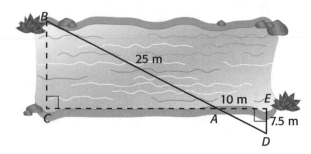

a. What is the geometric relationship between $\triangle ABC$ and $\triangle ADE$? Explain.

b. Find $\overline{BC}$, the width of the river. Show the steps you used.

35. During a storm, a utility pole broke and fell as shown. To the nearest tenth of a foot, what was the height of the pole before it broke? Show your work.

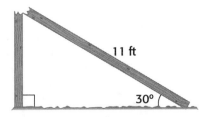

11 ft

30°

36. The number of students in each homeroom at the Oak Tree High School is shown below.

Number of Students in Each Homeroom
34 23 26 18 37 21 23 28 35 30
22 17 18 31 41 33 28 24 21 29

  a. Construct a stem-and-leaf plot to display the data.

  b. Find the range, mean, and median of the data.

  c. What percent of the classes have between 25 and 35 students?

37. Seven shipwreck survivors on an island estimate that they have enough food and water to last 6 days. Then an eighth survivor is spotted and pulled from the water. How long will the food and water last the larger group?

38. A survey was conducted one Saturday to determine where teenagers go at the mall. The following facts were recorded:

- 19 teens ate at the food court.
- 16 teens went to the video arcade.
- 21 teens saw a movie.
- 5 teens went to both the food court and the video arcade.
- 8 teens went to the food court and saw a movie.
- 7 teens went to the video arcade and saw a movie.
- Only 3 teens went to the food court, the video arcade, and saw a movie.
- Each teen surveyed had visited at least one of the three places.

  a. Complete the Venn diagram below to represent the results of the survey.

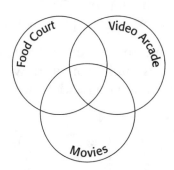

Teenagers at the Mall

Food Court Video Arcade Movies

  b. How many teens went only to the food court, only to the video arcade, and only to a movie?

  c. How many teens were surveyed?

39. Based on the results of the mall survey above, the mall manager wants to make some predictions.

  a. If 240 teenagers visit the mall on a Saturday, about how many can be expected to see a movie?

  b. How many of the 240 teenagers can be expected only to eat at the food court?

Extended-Response Questions

 You may use a calculator to solve problems on this part of the test.

40. Four people are waiting in line to be in the studio audience of a live TV show. Everyone who waits in line gets either a ticket for that day's show or a rain check to see the show on another day. An equal number of tickets and rain checks are given out randomly.

a. Let T represent "the person gets a ticket," and let R represent "the person gets a rain check." Construct a tree diagram to show all the possible outcomes for the four people.

b. How many outcomes are there?

c. What is the probability that at least two of the people get in to see the show? Explain your answer.

41. A manufacturer is considering three box models for a new cereal snack product. The models are shown below.

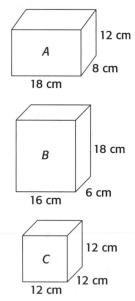

a. Compare the volumes of boxes A, B, and C.

b. Compare the surface areas of boxes A, B, and C.

c. Which box do you think the company should choose? Why?

42. The length of a spring is related to the weight that hangs from it. For a certain spring, the equation that relates the length ℓ in cm to the weight w in kg is $\ell = 10 + 2w$.

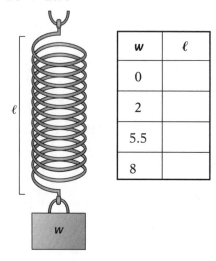

w	ℓ
0	
2	
5.5	
8	

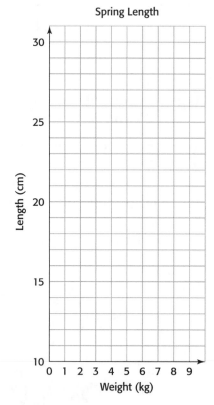

Spring Length

a. Complete the table of values for the equation $\ell = 10 + 2w$.

b. Use the ordered pairs (w, ℓ) to graph $\ell = 10 + 2w$ on the given axes.

c. What was the length of the spring before any weight was attached?

d. Explain how to use the graph to find how much weight was attached if the length of the spring is 18 cm.

43. Meg and her sister Lisa live 1.25 miles from their school. Lisa rides her bicycle and Meg walks to school along the same route.

- Meg left at 8:00 A.M. and walked at a steady rate of 3 miles per hour for 20 minutes, then stopped to talk to a friend for 5 minutes, then continued walking at the same rate until she reached the school.
- Lisa left at 8:15 A.M. and rode at a steady rate of 9 miles per hour until she reached the school.

a. Use the information given to complete the graph of both trips. Use your graph to answer b and c.

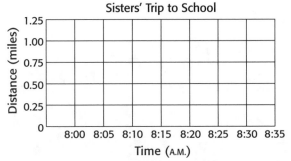

b. Who arrived at school earlier? About how much earlier?
c. When did the two sisters pass each other?

44. Ms. P'Oreo gave a math test with 60 questions. She announced that to penalize for guessing, one fourth of the wrong answers (*W*) would be deducted from the right answers (*R*) to calculate the final score (*S*).

a. Write a formula for the final score *S*.
b. What is the lowest possible score using this formula?
c. If Eileen answered every question and got 52 questions right, what was her final score?

d. If Nick got a score of 0, how many questions did he get right?

45. In the figure, line *m* is parallel to line *n*. Also, the measure of $\angle 4 = 3x - 20$ and the measure of $\angle 5 = x + 40$.

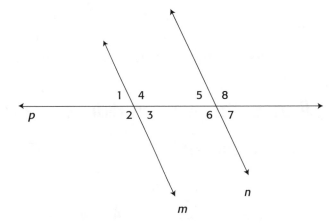

a. Find the value of *x*. Show your work.
b. Find the measure of each angle.

m∠1 ____   m∠5 ____
m∠2 ____   m∠6 ____
m∠3 ____   m∠7 ____
m∠4 ____   m∠8 ____

Using a Scientific Calculator

What calculator should you use?

For the mathematics in this review book, students should use at least a four function, square root key calculator. Scientific calculators are recommended. Graphing calculators are not required.

What calculator features are you likely to use?

Basic operations: [+] [−] [×] [÷]

Parentheses: [(] [)]

Parenthesis keys can be used to define the order in which your calculator performs operations. They are pressed in the same order in which they would be written.

    Example: To calculate $(3 + 5) \times 7$, press [(] 3 [+] 5 [)] [×] 7 [=].
    You will get [56.].

Square a Number: [x^2]

Use this key to raise a number to the second power.

    Example: 8 [x^2] [=] gives you [64.].

Raise a Number to a Power: [x^y]

To use this key, first enter the base (the number corresponding to x) and press [x^y]. Then enter the exponent (the number corresponding to y) and press [=].

    Example: To find 4^5, press 4 [x^y] 5 [=]. Your display should read
    [1024.].

Square Root: [$\sqrt{\ }$]

This key is used to find the square root of whatever follows.

    Example: To find the square root of 8, press [$\sqrt{\ }$] 8 [=] to get
    [2.828427125].

Trigonometric Functions: [SIN] [COS] [TAN]

These keys will give you the sine, cosine, and tangent, respectively, of the degree measure you enter.

    Example: To find the sine of 45°, press [SIN] 45 [=] to get
    [0.707106781].

Note: Many calculators operate in the opposite order. In this case, for the example above, you would press 45 [SIN] and the display would give you the value. Do not press [=].

[INV]   This key will find the degree measure of an angle when you know the value of the trigonometric ratio.

Example: To find the angle with a sine ratio of 0.5, enter [INV] [SIN] [.] 5 [=] to get [30.] .

Note: Your calculator may have keys labeled [SIN⁻¹], [COS⁻¹], and [TAN⁻¹] rather than [INV].

What should you keep in mind when using a calculator on a test?

You should be comfortable with the calculator you are using. You do not have time to figure out how to use the calculator during the test. Make sure you are familiar with the keypad and the features of the calculator.

To receive full credit for an answer, you must show clear, complete explanations and adequate work. When you use the calculator to solve a problem, make sure to write down the steps you performed on the calculator and the answers given.

Think before you use the calculator! The test questions will not tell you when to use your calculator. You must decide. Do not try to use the calculator for every question. Many questions involve reasoning that cannot be done by a calculator.

When you do use your calculator to solve a problem, estimate the answer first and check that the calculator answer is reasonable.

Index

A

Absolute value, 112, 118
Accuracy
  of graphs versus tables, 321
  of measurement, 220
Acute angles, 228
  cosine (cos) of, 273
  sine (sin) of, 273
  tangent (tan) of, 273
Acute triangles, 233
Addition
  associative property of, 45
  commutative property of, 45
  of decimals, 47
  distributive property of, 45
  estimation and, 47–48
  of fractions, 93–94
  identity property of, 45
  of integers, 118–119
  of mixed numbers, 93–94
  of whole numbers
  zero property in, 45
Additive inverse, 112, 121
Adjacent leg, 273
Algebraic expressions
  evaluating, 145–146
  writing, 142–143
Altitude, 255
  of triangle, 310
Amount of markup, 191
Angles, 228
  acute, 228, 273
  central, 251
  complementary, 229
  congruent, 230
  constructing and bisecting, 308
  corresponding, 242, 245
  exterior, 233
  interior, 233
  obtuse, 228
  remote interior, 233
  right, 228
  straight, 228
  supplementary, 229
  vertical, 230
Area, 255–57
  of circle, 255
  of parallelogram, 255
  of rectangle, 255
  of trapezoid, 255
  of triangle, 255
  surface, 260-261
Associative property
  of addition, 45
  of multiplication, 45
Average, 327

B

Bar graph, 320
Base, 35
  of cylinder, 260
  of parallelogram, 255
  of prism, 260
  of pyramid, 260
  of trapezoid, 255
  of triangle, 255
Box-and-whisker plot, 333–334

C

Capacity
commonsense comparisons of, 208
customary measures of, 199
metric measures of, 206
Celsius scale, 210
Center, 251
Centimeters, 206-208
cubic, 264
square, 255
Central angle, 251
Central Time, 215
Chart, place-value, 31
Chord, 251
Circles, 245, 251–252
area of, 255
center of, 251
circumference of, 251
diameter of, 251
radius of, 251
Circumference, 251
Combinations, 348–349
Commas
in algebraic expressions, 143
in place value, 32
Commission, 191
rate of, 191
Commutative property
of addition, 45
of multiplication, 45
Comparing and ordering integers, 115–116
Comparing and rounding fractions, 89–91
Complementary angles, 229
Composite number, 70, 71
Compound events, 342
Cone, volume of, 264
Congruent angles, 230
Congruent figures, 242–243
Construction, 308
Coordinate plane, 284
distance between two points on, 289–292
Coordinates, 285
in graphing points, 284–287
of a reflection image, 298
of a translation image, 294
Corresponding angles, 242, 245
Corresponding parts, 242
Corresponding sides, 242, 245
Cosine (cos) of acute angle, 273
Cost, 191
Counting numbers, 132
Counting Principle, 342
finding permutations using, 348
Cross products, 172
Cube, volume of, 264
Cubed, 35
Cubic centimeters, 264
Cubic feet, 264

Cubic inches, 264
Cubic meters, 264
Cumulative frequency, 325
Cumulative relative frequency, 353
Customary system
comparison to metric system, 208
computing with, 202–204
conversion table for, 199
of measure, 198–200
temperature in, 210
Cylinder
surface area of, 260
volume of, 264

D

Data, 319–320
accuracy of, 321
box-and-whisker plot for displaying, 333–334
displaying, with graphs, 319–321
finding the range, mean, median, and mode of, 327
stem-and-leaf plot for displaying, 333–334
Decagon, 237–238
Decimal points, invisible, 48
Decimals
addition of, 47
division of, 61–63
by power of 10, 63
multiplication of, 54
by power of ten, 55
repeating, 82–84
rounding, 42
subtraction of, 50
terminating, 82–84
writing, as fraction, 82
writing, as percent, 179
writing fraction as, 82
writing percent as, 179
Decrease, percent of, 188
Dependent events, 342
Diagonal, 237–238
Diagrams in problem solving, 12
Diameter, 251
Difference, 50
Direct variation, 367
Discount, 190
Distributive property
of addition, 45
of multiplication, 45
Dividend, 57
Divisibility, 73–75
Division
of decimals, 61–63
of fractions, 103–105
of integers, 126–127
involving zero, 60
of mixed numbers, 103–105
steps in, 57

strategy in, 58
of whole numbers, 57–60
Divisor, 57
Domain, 366

E

Eastern Time, 215
Elapsed time, 213
Equations, 147
   one-step, 151–152
   two-step, 154–156
   writing and solving, 139–168
Equilateral triangles, 233, 250
   perimeter of, 250
Equivalent fractions, 76–78
Estimation, 27-28
   addition and, 47–49
   multiplication and, 53–55
   subtraction and, 50-51
Evaluation
   of algebraic expressions, 145–146
   of formulas, 161
Events
   compound, 342
   dependent, 342
   independent, 342
   probability of, 337–339
Expanded notation, 32
Experimental probability, 352–353
Exponential functions, 384
Exponents, 35
   negative, 36-37
   positive, 36
   power, 35
Exterior angle, 233
Extremes, 172

F

Factor pair, 74
Factor tree, 71-72
Factors, 35-36, 53, 70, 71
   prime, 71
Fahrenheit scale, 210
Favorable outcomes, 337
Feet, 199
   cubic, 264
   square, 255
Formulas, 161–163
Fractions, 70–110
   addition of, 93–94
   comparing and ordering, 89–91
   division of, 103–105
   equivalent, 76–77
   improper, 79–80
   least common denominator of, 86-87
   like, 89
   multiplication of, 100–101
   subtraction of, 96–97

   unlike, 89
   writing, as decimal, 82
   writing, as percent, 179
   writing decimal as, 82
   writing percent as, 179
Frequency, 324–325
   cumulative, 325
   cumulative relative, 353
   relative, 352
Frequency table, 324–325
Functions, 365–392
   defined, 366
   exponential, 384
   linear, 383
   nonlinear, 383
   quadratic, 384

G

Geometry, 227–283
   angles in, 228, 233–235
   area in, 255–257
   circles in, 251–252
   congruent figures, 242–243
   lines and angles, 228–230
   perimeter in, 249–250
   polygons in, 237–239
   similar figures in, 245–246
   special quadrilaterals in, 238-239
   surface area in, 260–261
   triangles in, 233–235
   volume in, 264–265
Graphing
   inequalities, 158–159, 375–376
   linear equations, 371–372
   nonlinear functions, 383-385
   ordered pairs, 285
Graphs, 285
   accuracy of data in, 321
   bar, 320
   of equations, 371
   of functions, 383
   of inequalities, 375
   line, 320–321
Greater than (>), 39, 148
Greatest common factor (GCF), 73–75
Greatest possible error (GPE), 218–219
Guessing and checking, 7-8

H

Half-planes, 375
Hexagons, 237–238
Histograms, 324–325
Hypotenuse of right triangle, 268

I

Identity property
   of addition, 45
   of multiplication, 45

Image, 293
Improper fraction, 79–80
Inches, 199
   cubic, 264
   square, 255
Increase, percent of, 188
Independent events, 342
Inequalities, 147
   graphing, 158–159, 375–376
   symbols for, 39, 148
   writing, 147–148
Information, identifying, 4-5
Integers, 112, 132
   addition of, 118–119
   comparing and ordering, 115–116
   division of, 126–127
   multiplication of, 123–124
   negative/positive, 111
   signs of, 111
   subtraction of, 121–122
Interest
   calculating, 192
   simple, 192
Interest rate, 192
Interior angles, 233
   remote, 233
Inverse operations, 9–10, 151
Inverse variation, 367–368
Irrational numbers, 129, 132
Isosceles triangle, 233
   perimeter of, 250

L
Least common multiple (LCM), 86–87
Least common denominator (LCD), 86-87
Legs, 268
   adjacent, 273
   opposite, 273
Length
   commonsense comparisons of, 208
   customary measures of, 199
   metric measures of, 206
Less than (<), 39, 148
Like fractions
   comparing, 89-90
Linear equations, 371
   graphing, 371–372
Linear function, 383
Line graph, 320–321
Lines, 228-230
   parallel, 230
   perpendicular, 229, 310–311
   of reflection, 298
   slope of, 379–381
Line segment
   midpoint of, 310
   perpendicular bisector of, 310
Line symmetry, 299
Lists in problem solving, 15-16

Logic, 23
Logical reasoning, 23
Lowest terms, 76

M
Maps, 176
   time zone, 215
Markup, 191
Mean, 327
Means, 172
Measurement, 198–226
   accuracy of, 220
   customary units of, 198–204, 208, 210
   metric units of, 206–208, 210
   precision in, 218–221
   significant digits in, 220–221
   of temperature, 210–211
   of time, 213–216
Measures of central tendency, 327
Median, 327
Meters, 206-208
   cubic, 264
   square, 255
Metric system
   comparison to customary system, 208
   temperature in, 210
   units in, 206–208
Midpoint of line segment, 310
Mixed numbers, 79–80
   addition of, 93–94
   division of, 103–105
   multiplication of, 100–101
   rounding, 90
   subtraction of, 96–97
Mode, 327
Mountain Time, 215
Multiples, 86–87
Multiplication
   associative property of, 45
   commutative property of, 45
   of decimal by a power of 10, 55
   of decimal by a whole number or decimal, 54
   distributive property of, 45
   estimation and, 53–55
   of fractions, 100–101
   identity property of, 45
   of integers, 123–124
   of mixed numbers, 100–101
   symbols of, 123, 142
   of whole numbers, 53
   zero property in, 45
Multiplicative inverses, 103–105

N
Natural numbers, 132
Negative exponents, 36-37
Negative integers, 111
Negative powers, 35
Nonlinear functions, 383–385

Notation
  expanded, 32
  scientific, 35–37
Number line, 39, 112, 132
  adding integers on, 118
  comparing and ordering integers on, 115
  subtracting integers on, 121
Numbers
  composite, 70, 71
  counting, 132
  irrational, 129, 132
  mixed, 79–80
  natural, 132
  prime, 70, 71
  rational, 112, 132
  real, 132
  rounded, 42
  whole, 132
Numerical expressions, 139

O

Obtuse angle, 228
Obtuse triangle, 233
Octagon, 237–238
One-step equations, 151–152
Opposite leg, 273
Opposites, 112
Ordered pair, 284, 366, 371
  graphing, 285
Order of operations, 139, 145
Origin, 284
Outcomes, 337
  counting, 342
  favorable, 337

P

Pacific Time, 215
Parallel lines, 230
Parallelogram, 238, 239
  area of, 255
  perimeter of, 249
Parentheses
  as multiplication symbol, 123, 142
  order of operations and, 139-140
Partial products, 53
Patterns, recognizing, in problem solving, 18
Pentagon, 237–238
Percent, 179–181
  applications of, 190–192
  of decrease, 188
  of increase, 188
  shortcuts, 180
  solving problems, 183–185
  writing, as decimal, 179
  writing, as fraction, 179
  writing decimal as, 179
  writing fraction as, 179
Perfect square, 129
Perimeter of polygons, 249–250

Period, 32
Permutations, 348–349
Perpendicular bisector of line segment, 310
Perpendicular lines, 229
  constructing, 310–311
Pi (π), 82, 129, 251
Pictures in problem solving, 12
Place value, 31–33
Place-value chart, 32
Points
  distance between two, on coordinate plane, 289–292
  using coordinates to graph, 284–287
Point symmetry, 299
Polygons, 237–239
  perimeter of, 249–250
  regular, 237–238
  sum of angle measures in, 237
Population, 352
Positive exponents, 36
Positive integers, 111
Powers, 35
  negative, 35
  positive, 35
Powers of 10, 35
  division of decimal by, 63
  multiplication of decimal by, 55
Precision in measurement, 218–221
Prime factorization, 71
Prime factors, 71
Prime number, 70-71
Principal, 192
Prism
  surface area of, 260
  volume of, 264
Probability, 337–339
  experimental, 352–353
  theoretical, 353
Problem solving, 1–3
  defined, 1
  estimation in, 27
  guessing and checking in, 7
  identifying information in, 4
  lists in, 15
  logical reasoning in, 23
  percent in, 183–185
  pictures and diagrams in, 12
  recognizing patterns in, 18
  solving simpler related problem in, 20
  steps in, 2
  tables in, 15
  working backward in, 9–10
Products, 53
  partial, 53
Proportion, 172–174
  unit pricing and, 173–174
Protractor, 228

Pyramid
  surface area of, 260
  volume of, 264
Pythagorean Theorem, 268–270
Pythagorean triple, 268

Q

Quadrants, 285
Quadratic functions, 384
Quadrilaterals, 237–238
  special, 238–239
Quotient, 57, 126

R

Radical, 129
Radius, 251
Random sampling, 352
Range, 327, 366
Rates, 169
  of commission, 191
  of discount, 190
Rational numbers, 111-138
Ratios, 169–170
  trigonometric, 272
  writing, 169
Real numbers, 132
Reasonable answers, 27
Reasoning
  logical, 23
Reciprocals, 103
Rectangle, 238
  area of, 255
  perimeter of, 249, 250
Reflection, 298–301
  line of, 298
Regular pentagon, 250
  perimeter of, 250
Regular polygon, 237–238
  perimeter of, 250
Relations, 365–368
  defined, 366
Relative error, 220
Relative frequency, 352
  cumulative, 353
Remote interior angles, 233
Repeating decimal, 82–84
Rhombus, 238
Right angle, 228
Right triangle, 233
  hypotenuse of, 268
Pythagorean Theorem and, 268–270
  trigonometry of, 272–275
Rotational symmetry, 304
Rotations, 304–305
Rounded numbers, 42
Rounding, 42–43
  decimals, 42
  fractions, 89–91

  mixed numbers, 90
  whole numbers, 42
Rules of divisibility, 73–75

S

Sample space, 337
Scale drawing, 176–177
Scalene triangle, 233
Scientific notation, 35–37
Selling price, 190
Sets of numbers, 132–133
Sides, corresponding, 242, 245
Signed Numbers, 111-138
  addition of, 118–119
  comparing and ordering, 115–116
  division of, 126–127
  multiplication of, 123–124
  negative, 111
  positive, 111
  signs of, 111
  subtraction of, 121–122
  using, 111–113
Significant digits, 220–221
Similar figures, 245–246
Simple interest, 192
Simplest form, 76
Sine (sin) of acute angle, 273
Slope, 379–381
Solution, 148, 371
Square, 129, 238, 250
  area of, 255
  perfect, 129
  perimeter of, 250
Square root, 129
Standard form, 37
Statistics, 319
Stem-and-leaf plot, 333–334
Straight angle, 228
Subtraction
  of decimals, 50
  estimation and, 50-51
  of fractions, 96–97
  of integers, 121–122
  of mixed numbers, 96–97
  of whole numbers, 50
Supplementary angles, 229
Surface area, 260–261
Symbols
  = equal, 132
  > greater than, 39, 148
  ≥ greater than or equal to, 148
  inequality, 39, 148
  ~ is similar to, 245
  < less than, 39, 148
  ≤ less than or equal to, 148
  of multiplication, 123, 142
  $-\sqrt{}$ negative square root, 129
  ≠ not equal to, 148

∥ parallel to, 230
⊥ perpendicular to, 229
± plus or minus, 129
π pi, 82, 129, 251
√ radical, 129
Symmetry
   line, 299
   point, 299
   rotational, 304

T
Tables
   accuracy of data in, 321
   frequency, 324–325
   in problem solving, 15
Tangent (tan) of acute angle, 273
Temperature, 210–211
Terminating decimal, 82–84
Theoretical probability, 353
Time, 213–215
   elapsed, 213
Time zones, 215–216
Transformations, 293
   reflections as, 298–301
   rotations as, 304–305
   translations as, 293–296
Translations, 293–296
Transversal, 230
Trapezoid, 238, 239
   area of, 255
Triangles, 233–235, 237
   acute, 233
   altitude of, 255, 310
   area of, 255
   equilateral, 233
   isosceles, 233
   obtuse, 233
   right, 233
   scalene, 233
Trigonometric ratios, 272, 273
Trigonometry of right triangle, 272–275
Two-step equations, 154–156

U
Uncertainty. *See* Probability
Unit pricing, 173–174
Units of measure
   customary, 198-204, 208, 210
   metric, 206-208, 210

Unlike fractions
   comparing, 89-90

V
Value, 145
Variables, 142
Variation
   direct, 367
   inverse, 367–368
Venn diagram, 23–24
Vertex, 228
Vertical angles, 230
Volume, 264–265
   of cube, 264
   of prism, 264
   of pyramid, 264

W
Weight
   commonsense comparisons of, 208
   customary measures of, 199
   metric measures of, 206
Whole numbers, 132
   addition of, 47
   comparing and ordering, 39–40
   division of, 57–60
   division of decimals by, 61
   multiplication of, 53
   multiplication of decimals by, 54
   reading, when written as numerals, 32
   rounding, 42
   subtraction of, 50-51
Working backward, 9–11

X
x-axis, 284-285
x-coordinate, 285

Y
y-axis, 284-285
y-coordinate, 285

Z
Zero, 71, 111, 112, 220
   division involving, 60
Zero property
   of addition, 45
   of multiplication, 45